Linear Algebra

Tom M. Apostol

Linear Algebra

A First Course, with Applications to Differential Equations

A Wiley-Interscience Publication

JOHN WILEY & SONS, INC.

New York • Chichester • Weinheim • Brisbane • Singapore • Toronto

Library of Congress Cataloging in Publication Data:

Apostol, Tom M.
    Linear algebra : a first course, with applications to differential
equations / Tom M. Apostol.
        p.   cm.
    "A Wiley-Interscience publication."
    ISBN 0-471-17421-1 (cloth : acid-free paper)
    1. Algebra, Linear.   2. Differential equations.   I. Title.
QA184.A66   1997
512'.5—dc21   96-37131
                                                        CIP

Printed in the United States of America

10 9 8 7 6 5 4 3 2 1

To
Erica, Emily, and Caitlin Jane

CONTENTS

1. VECTOR ALGEBRA

2. APPLICATIONS OF VECTOR ALGEBRA TO ANALYTIC GEOMETRY

3. LINEAR SPACES

4. LINEAR TRANSFORMATIONS AND MATRICES

5. DETERMINANTS

6. EIGENVALUES AND EIGENVECTORS

7. EIGENVALUES OF OPERATORS ACTING ON EUCLIDEAN SPACES

8. APPLICATIONS TO LINEAR DIFFERENTIAL EQUATIONS

9. APPLICATIONS TO SYSTEMS OF DIFFERENTIAL EQUATIONS

10. THE METHOD OF SUCCESSIVE APPROXIMATIONS

PREFACE

For many years the author has been urged to develop a text on linear algebra based on material in the second edition of his two-volume *Calculus*, which presents calculus of functions of one or more variables, integrated with differential equations, infinite series, linear algebra, probability, and numerical analysis. To some extent this was done by others when the two *Calculus* volumes were translated into Italian and divided into three volumes,[*] the second of which contained the material on linear algebra. The present text is designed to be independent of the *Calculus* volumes.

To accommodate a variety of backgrounds and interests, this text begins with a review of prerequisites (Chapter 0). The review is divided into two parts: pre-calculus prerequisites, needed to understand the material in Chapters 1 through 7, and calculus prerequisites, needed for Chapters 8 through 10. Chapters 1 and 2 introduce vector algebra in n-space with applications to analytic geometry. These two chapters provide motivation and concrete examples to illustrate the more abstract treatment of linear algebra presented in Chapters 3 through 7.

Chapter 3 discusses linear spaces, subspaces, linear independence, bases and dimension, inner products, orthogonality, and the Gram-Schmidt process. Chapter 4 introduces linear transformations and matrices, with applications to systems of linear equations. Chapter 5 is devoted to determinants, which are introduced axiomatically through their properties. The treatment is somewhat simpler than that given in the author's *Calculus*. Chapter 6 treats eigenvalues and eigenvectors, and includes the triangularization theorem, which is used to deduce the Cayley-Hamilton theorem. There is also a brief section on the Jordan normal form. Chapter 7 continues the discussion of eigenvalues and eigenvectors in the setting of Euclidean spaces, with applications to quadratic forms and conic sections.

In Chapters 3 through 7, calculus concepts occur only occasionally in some illustrative examples, or in some of the exercises; these are clearly identified and can be omitted or postponed without disrupting the continuity of the text. This part of the text is suitable for a first course in linear algebra not requiring a calculus prerequisite. However, the level of presentation is more appropriate for readers who have acquired some degree of mathematical sophistication in a course such as elementary calculus or finite mathematics.

Chapters 8, 9, and 10 definitely require a calculus background. Chapter 8 applies linear algebra concepts to linear differential equations of order n, with special emphasis on

---

[*]*Calcolo*, Volume primo: Analisi 1; Volume Secondo: Geometria; Volume Terzo: Analisi 2. Published by Editore Boringhieri, 1977.

equations with constant coefficients. Chapter 9 uses matrix calculus to discuss systems of differential equations. This chapter focuses on the exponential matrix, whose properties are derived by an interplay between linear algebra and matrix calculus. Chapter 10 treats existence and uniqueness theorems for systems of differential equations, using Picard's method of successive approximations, which is also cast in the language of contraction operators.

Although most of the material in this book was extracted from the author's *Calculus*, some topics have been revised or rearranged, and some new material and new exercises have been added.

This textbook can be used by first- or second-year students in college, and it can also be of interest to more mature individuals, who may have studied mathematics many years ago without learning linear algebra, and who now wish to learn the basic concepts without undue emphasis on abstraction or formalization.

TOM M. APOSTOL

California Institute of Technology

Linear Algebra

0

REVIEW OF PREREQUISITES

Part 1 of this chapter summarizes some pre-calculus prerequisites for this book—facts about real numbers, rectangular coordinates, complex numbers, and mathematical induction. Part 2 does the same for calculus prerequisites. Chapters 1 and 2, which deal with vector algebra and its applications to analytic geometry, do not require calculus as a prerequisite. These two chapters provide motivation and concrete examples to illustrate the abstract treatment of linear algebra that begins with Chapter 3. In Chapters 3 through 7, calculus concepts occur only occasionally in some illustrative examples, or in some exercises; these are clearly identified and can be omitted or postponed without disrupting the continuity of the text.

Although calculus and linear algebra are independent subjects, some of the most striking applications of linear algebra involve calculus concepts—integrals, derivatives, and infinite series. Familiarity with one-variable calculus is essential to understand these applications, especially those referring to differential equations presented in the last three chapters. At the same time, the use of linear algebra places some aspects of differential equations in a natural setting and helps increase understanding.

Part 1. Pre-calculus Prerequisites

0.1   Real numbers as points on a line

Real numbers can be represented geometrically as points on a straight line. A point is selected to represent 0 and another, to the right of 0, to represent 1, as illustrated in Figure 0.1. This choice determines the scale, or unit of measure. If one adopts an appropriate set of axioms for Euclidean geometry, then each real number corresponds to exactly one point on this line and, conversely, each point on the line corresponds to one and only one real

FIGURE 0.1   Real numbers represented geometrically on a line.

number. For this reason, the line is usually called the *real line* or the *real axis*. We often speak of the *point x* rather than the point corresponding to the real number *x*. The set of all real numbers is denoted by **R**.

If $x < y$, point *x* lies to the left of *y* as shown in Figure 0.1. Each positive real number *x* lies at a distance *x* to the right of zero. A negative real number *x* is represented by a point located at a distance $|x|$ to the left of zero.

0.2 Pairs of real numbers as points in a plane

Points in a plane can be represented by *pairs* of real numbers. Two perpendicular reference lines in the plane are chosen, a horizontal *x* axis and a vertical *y* axis. Their point of intersection, denoted by 0, is called the *origin*. On the *x* axis a convenient point is chosen to the right of 0 to represent 1; its distance from 0 is called the *unit distance*. Vertical distances along the *y* axis are usually measured with the same unit distance. Each point in the plane is assigned a pair of numbers, called its *coordinates*, which tell us how to locate the point. Figure 0.2 illustrates some examples. The point with coordinates $(3, 2)$ lies three units to the right of the *y* axis and two units above the *x* axis. The number 3 is called the *x coordinate* or *abscissa* of the point, and 2 is its *y coordinate* or *ordinate*. Points to the left of the *y* axis have a negative abscissa; those below the *x* axis have a negative ordinate. The coordinates of a point, as just defined, are called its *Cartesian coordinates* in honor of René Descartes (1596–1650), one of the founders of analytic geometry.

When a pair of numbers is used to represent a point, we agree that the abscissa is written first, the ordinate second. For this reason, the pair (a, b) is referred to as an *ordered pair*: the first entry is *a*, the second is *b*. Two ordered pairs (a, b) and (c, d) represent the same point if and only if we have $a = c$ and $b = d$. Points (a, b) with both *a* and *b* positive are said to lie in the *first quadrant*; those with $a < 0$ and $b > 0$ are in the *second quadrant*; those with $a < 0$ and $b < 0$ are in the *third quadrant*; and those with $a > 0$ and $b < 0$ are in the *fourth quadrant*. Figure 0.2 shows one point in each quadrant.

The procedure for locating points in space is analogous. We take three mutually perpendicular lines in space intersecting at a point (the origin). These lines determine three

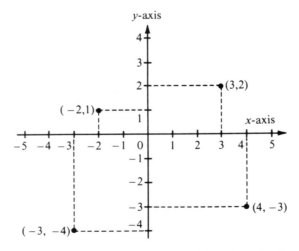

FIGURE 0.2   Points in the plane represented by pairs of real numbers.

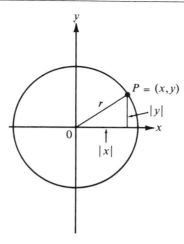

FIGURE 0.3   The circle represented by the Cartesian equation $x^2 + y^2 = r$.

mutually perpendicular planes, and each point in space can be completely described by specifying, with appropriate regard for signs, the distances from these planes. We shall discuss three-dimensional Cartesian coordinates in a later chapter; for the present we confine our attention to the two-dimensional case.

A geometric figure, such as a curve in the plane, is a collection of points satisfying one or more special conditions. By expressing these conditions in terms of the coordinates x and y we obtain one or more relations (equations or inequalitites) that characterize the figure in question. For example, consider a circle of radius r with its center at the origin, as shown in Figure 0.3.

Let (x, y) denote the coordinates of an arbitrary point P on this circle. The line segment OP is the hypotenuse of a right triangle whose legs have lengths $|x|$ and $|y|$ and, hence, by the theorem of Pythagoras, we have

$$x^2 + y^2 = r^2.$$

This equation, called a *Cartesian equation* of the circle, is satisfied by all points (x, y) on the circle and by no others, so the equation completely characterizes the circle. Points *inside* the circle satisfy the inequality $x^2 + y^2 < r^2$, while those *outside* satisfy $x^2 + y^2 > r^2$. This example illustrates how analytic geometry is used to reduce geometrical statements about points to algebraic relations about real numbers.

0.3   Polar coordinates

Points in a plane can also be located by using polar coordinates. This is done as follows. Let P be a point distinct from the origin. Suppose the line segment joining the origin to P has length $r > 0$ and makes an angle of θ radians with the positive x axis, as shown by the example in Figure 0.4. The two numbers r and θ are called *polar coordinates* of P. They are related to the rectangular coordinates x and y by the equations

(0.1) $$x = r \cos \theta, \qquad y = r \sin \theta.$$

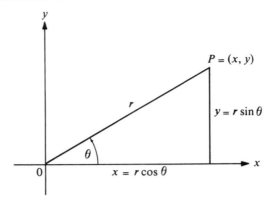

FIGURE 0.4   Polar coordinates.

The positive number r is called the *radial distance* of P, and θ is called *a polar angle*. We say *a* polar angle rather than *the* polar angle because if θ satisfies (0.1) so does $\theta + 2n\pi$ for any integer n. We agree to call all pairs of real numbers (r, θ) polar coordinates of P if they satisfy (0.1) with $r > 0$.

The radial distance r is uniquely determined by x and y: $r = \sqrt{x^2 + y^2}$, but the polar angle θ is determined only up to integer multiples of 2π.

When P is the origin, Eqs. (0.1) are satisfied with $r = 0$ and any θ. For this reason, we assign the radial distance $r = 0$ to the origin, and we agree that *any* real θ may be used as a polar angle.

Some curves are described more simply with polar coordinates rather than rectangular coordinates. For example, a circle of radius 2 with center at the origin has Cartesian equation $x^2 + y^2 = 4$. In polar coordinates the same circle is described by the simpler equation $r = 2$. The interior of the circle is described by the inequality $r < 2$, the exterior by $r > 2$.

0.4   Complex numbers

The quadratic equation $x^2 + 1 = 0$ has no solution in the real-number system because there is no real number whose square is negative. New types of numbers, called *complex numbers*, have been introduced to provide solutions to such equations.

As early as the 16th century, a symbol $\sqrt{-1}$ was introduced to provide solutions of the quadratic equation $x^2 + 1 = 0$. This symbol, later denoted by the letter i, was regarded as a fictitious or imaginary number, which could be manipulated algebraically like an ordinary real number, except that its square was -1. Thus, for example, the quadratic polynomial $x^2 + 1$ was factored by writing

$$x^2 + 1 = x^2 - i^2 = (x - i)(x + i),$$

and the solutions of the equation $x^2 + 1 = 0$ were exhibited as $x = \pm i$, without any concern regarding the meaning or validity of such formulas. Expressions such as $2 + 3i$ were called complex numbers, and they were used in a purely formal way for nearly 300 years before they were described in a manner that would be considered satisfactory by present-day standards.

Early in the 19th century, Carl Friedrich Gauss (1777–1855) and William Rowan Hamilton (1805–1865) independently and almost simultaneously proposed the idea of defining complex numbers as ordered pairs of real numbers (a, b) endowed with certain special properties. This idea is widely accepted today and is described in the next section.

0.5 Definition and algebraic properties of complex numbers

Complex numbers are defined as ordered pairs of real numbers, in the same way that we described the rectangular coordinates of points in the plane. The new feature is that we also define addition and multiplication so that we can perform algebraic operations on complex numbers.

DEFINITION.   *If a and b are real numbers, the pair* (a, b) *is called a complex number, provided that equality, addition, and multiplication of pairs is defined as follows:*
(a) *Equality:* $(a, b) = (c, d)$ *means* $a = c$ *and* $b = d$.
(b) *Sum:* $(a, b) + (c, d) = (a + c, b + d)$.
(c) *Product:* $(a, b)(c, d) = (ac - bd, ad + bc)$.

The definition of equality states that (a, b) is to be regarded as an *ordered pair*. Thus, the complex number $(2, 3)$ is distinct from the complex number $(3, 2)$. The numbers a and b are called *components* of the complex number. The first component, a, is also called the *real part* of the complex number; the second component, b, is called the *imaginary part*.

Note that the symbol $\sqrt{-1}$ does not appear anywhere in this definition. Presently we shall introduce i as a particular complex number that has all the algebraic properties ascribed to the fictitious symbol $\sqrt{-1}$ introduced by the early mathematicians. However, before we do this we discuss basic properties of the operations just defined.

THEOREM 0.1.   *Addition and multiplication of complex numbers satisfy the commutative, associative and distributive laws. That is, if x, y, and z are arbitrary complex numbers we have the following properties:*
Commutative laws: $x + y = y + x$, $xy = yx$.
Associative laws: $x + (y + z) = (x + y) + z$, $x(yz) = (xy)z$.
Distributive law: $x(y + z) = xy + xz$.

Proof.   All these laws are easily verified directly from the definition of sum and product. For example, to prove the associative law for multiplication, we express x, y, z in terms of their components, say $x = (x_1, x_2)$, $y = (y_1, y_2)$, $z = (z_1, z_2)$ and note that

$$x(yz) = (x_1, x_2)(y_1 z_1 - y_2 z_2, y_1 z_2 + y_2 z_1)$$
$$= \left(x_1(y_1 z_1 - y_2 z_2) - x_2(y_1 z_2 + y_2 z_1), x_1(y_1 z_2 + y_2 z_1) + x_2(y_1 z_1 - y_2 z_2)\right)$$
$$= \left((x_1 y_1 - x_2 y_2)z_1 - (x_1 y_2 + x_2 y_1)z_2, (x_1 y_2 + x_2 y_1)z_1 + (x_1 y_1 - x_2 y_2)z_2\right)$$
$$= (x_1 y_1 - x_2 y_2, x_1 y_2 + x_2 y_1)(z_1, z_2) = (xy)z.$$

The commutative and distributive laws may be similarly proved.

Further algebraic concepts, such as *zero, negative, reciprocal,* and *quotient,* analogous to those for real numbers, are defined as follows:

The complex number $(0, 0)$ is called the *zero* complex number. It is an identity element for addition because $(0, 0) + (a, b) = (a, b)$ for all complex numbers (a, b). Similarly, the complex number $(1, 0)$ is an identity for multiplication because

$$(a, b)(1, 0) = (a, b)$$

for all (a, b).

Since $(-a, -b) + (a, b) = (0, 0)$ we call the complex number $(-a, -b)$ the *negative* of (a, b) and we write $-(a, b)$ for $(-a, -b)$.

The *difference* $(a, b) - (c, d)$ of two complex numbers is defined to be the sum of (a, b) and the negative of (c, d).

Each nonzero complex number (a, b) has a *reciprocal* relative to the identity element $(1, 0)$, which we denote by $(a, b)^{-1}$. It is given by the ordered pair

$$(0.2) \qquad (a, b)^{-1} = \left(\frac{a}{a^2 + b^2}, \frac{-b}{a^2 + b^2} \right) \qquad \text{if } (a, b) \neq (0, 0),$$

and it has the property that $(a, b)(a, b)^{-1} = (1, 0)$. Note that $a^2 + b^2 \neq 0$ because $(a, b) \neq (0, 0)$.

The *quotient* $(a, b)/(c, d)$ of two complex numbers with $(c, d) \neq (0, 0)$ is defined to be the product $(a, b)(c, d)^{-1}$.

0.6   Complex numbers as an extension of real numbers

Let $\mathbf{C}$ denote the set of all complex numbers. Consider the subset $\mathbf{C}_0$ of $\mathbf{C}$ consisting of all complex numbers of the form $(a, 0)$, that is, all complex numbers with zero imaginary part. The sum or product of two members of $\mathbf{C}_0$ is again in $\mathbf{C}_0$. In fact we have

$$(a, 0) + (b, 0) = (a + b, 0) \qquad \text{and} \qquad (a, 0)(b, 0) = (ab, 0).$$

This shows that we can add or multiply two numbers in $\mathbf{C}_0$ by adding or multiplying the real parts alone. Or, in other words, with respect to addition and multiplication, the numbers in $\mathbf{C}_0$ act exactly as though they were real numbers. The same is true for subtraction and division because $-(a, 0) = (-a, 0)$, and $(b, 0)^{-1} = (b^{-1}, 0)$ if $b \neq 0$. For this reason, we make no distinction between the real number x and the complex number $(x, 0)$ whose real part is x. We agree to identify x and $(x, 0)$ and we write $x = (x, 0)$. In particular, we write $0 = (0, 0)$, $1 = (1, 0)$, $-1 = (-1, 0)$, and so on. Thus, we can regard the complex number system as an extension of the real number system.

This also makes sense geometrically. In a later section we will represent the complex number (x, y) by a point in the plane with Cartesian coordinates x and y; the subset $\mathbf{C}_0$ is represented geometrically by the points on the x axis.

0.7   The imaginary unit i

Complex numbers have some algebraic properties not possessed by real numbers. For example, the quadratic equation $x^2 + 1 = 0$, which has no solution among the real numbers, can now be solved with the use of complex numbers. In fact, the complex number $(0, 1)$ is a solution, because we have

$$(0, 1)^2 = (0, 1)(0, 1) = (0 \cdot 0 - 1 \cdot 1, 0 \cdot 1 + 1 \cdot 0) = (-1, 0) = -1.$$

DEFINITION.   *The complex number* $(0, 1)$ *is denoted by the symbol i and is called the imaginary unit.*

The imaginary unit has the property that its square is -1, $i^2 = -1$. Therefore the quadratic equation $x^2 + 1 = 0$ has the solution $x = i$. The reader can easily verify that $x = -i$ is another solution.

Now we can relate the ordered-pair idea with the notation used by the early mathematicians. First we note that the definition of multiplication gives us $(b, 0)(0, 1) = (0, b)$, and hence we have

$$(a, b) = (a, 0) + (0, b) = (a, 0) + (b, 0)(0, 1).$$

Therefore if we write $a = (a, 0)$, $b = (b, 0)$, and $i = (0, 1)$, we get $(a, b) = a + bi$. In other words, we have proved the following:

THEOREM 0.2.   *Every complex number* (a, b) *can be expressed in the form* $(a, b) = a + bi$.

This notation aids us in calculations involving addition and multiplication. For example, to multiply $a + bi$ by $c + di$, use the distributive and associative laws, and replace i^2 by -1. Thus,

$$(a + bi)(c + di) = ac - bd + (ad + bc)i,$$

which, of course, agrees with the definition of multiplication. Similarly, to compute the reciprocal of a nonzero complex number $a + bi$ we write

$$\frac{1}{a + bi} = \frac{a - bi}{(a + bi)(a - bi)} = \frac{a - bi}{a^2 + b^2} = \frac{a}{a^2 + b^2} - \frac{bi}{a^2 + b^2}.$$

This formula agrees with that given in (0.2).

With complex numbers we can solve not only the simple quadratic equation $x^2 + 1 = 0$, but also the more general quadratic equation $ax^2 + bx + c = 0$, where a, b, c are real and $a \neq 0$. By completing the square, we can write this quadratic equation in the form

$$\left(x + \frac{b}{2a} \right)^2 + \frac{4ac - b^2}{4a^2} = 0.$$

If $4ac - b^2 \leq 0$, the equation has the real roots $(-b \pm \sqrt{b^2 - 4ac})/(2a)$. If $4ac - b^2 > 0$, the left member is positive for every real x and the equation has no real roots. In this case, however, there are two complex roots, given by the formulas

(0.3) $$r_1 = -\frac{b}{2a} + i\frac{\sqrt{4ac - b^2}}{2a} \quad \text{and} \quad r_2 = -\frac{b}{2a} - i\frac{\sqrt{4ac - b^2}}{2a}.$$

In 1799, Gauss proved that every polynomial equation of the form

$$a_0 + a_1 x + a_2 x^2 + \cdots + a_n x^n = 0,$$

where $a_0, a_1, \ldots, a_n$ are arbitrary real numbers, with $a_n \neq 0$, has a solution among the complex numbers if $n \geq 1$. Moreover, even if the coefficients $a_0, a_1, \ldots, a_n$ are complex, a solution exists in the complex-number system. This fact is known as the *fundamental theorem of algebra*. It shows that there is no need to construct numbers more general than complex numbers to solve polynomial equations with complex coefficients.

0.8 Exercises

1. If the product of two complex numbers is zero, prove that at least one of the factors is zero.
2. Prove that $x = i$ and $x = -i$ are the *only* solutions of the quadratic equation $x^2 + 1 = 0$.
3. Instead of the definition of multiplication given in Section 0.5, suppose that the product of two complex numbers is defined by the simpler equation $(a, b)(c, d) = (ac, bd)$, which is analogous to that for addition.
   (a) Show that this new product is commutative and associative and also satisfies the distributive law.
   (b) Give two reasons why you think this simpler definition is not appropriate for multiplying complex numbers.

0.9 Geometric interpretation. Modulus and argument

Because a complex number (x, y) is an ordered pair of real numbers, it can be represented geometrically by a point in a plane, or by an arrow extending from the origin to the point (x, y), as shown in Figure 0.5. In this context, the xy plane is often referred to as the complex plane. The x axis is called the real axis; the y axis is the imaginary axis. It is customary to use the words *complex number* and *point* interchangeably. Thus, we refer to the point z rather than the point corresponding to the complex number z.

The operations of addition and subtraction of complex numbers have a simple geometric interpretation. If two complex numbers z_1 and z_2 are represented by arrows from the origin to z_1 and z_2, respectively, then the sum $z_1 + z_2$ is determined by the *parallelogram law*. The arrow from the origin to $z_1 + z_2$ is a diagonal of the parallelogram determined by 0, z_1, and z_2, as illustrated by the example in Figure 0.6. The other diagonal is related to the difference of z_1 and z_2. The arrow from z_1 to z_2 is parallel to and equal in length to the arrow from 0 to $z_2 - z_1$; the arrow in the opposite direction, from z_2 to z_1, is related in the same way to $z_1 - z_2$.

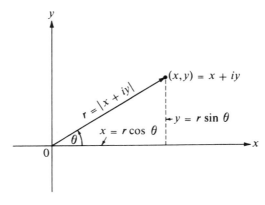

FIGURE 0.5   Geometric representation of the complex number $x + iy$.

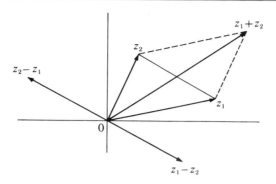

FIGURE 0.6   Addition and subtraction of complex numbers represented geometrically by the parallelogram law.

If $(x, y) \neq (0, 0)$ we can express x and y in polar coordinates,

$$x = r \cos \theta, \qquad y = r \sin \theta,$$

and we obtain

$$x + iy = r(\cos \theta + i \sin \theta).$$

(See Figure 0.5.) The positive number r, which represents the distance of (x, y) from the origin, is called the *modulus* or *absolute value* of $x + iy$ and is denoted by $|x + iy|$. Thus, we have

$$|x + iy| = \sqrt{x^2 + y^2}.$$

The polar angle θ is call an *argument* of $x + iy$. We say *an* argument rather than *the* argument because for a given point (x, y) the angle θ is determined only up to multiples of 2π. Sometimes it is desirable to assign a unique argument to a complex number. This may be done by restricting θ to lie in a half-open interval of length 2π. The intervals $[0, 2\pi)$ and $(-\pi, \pi]$ are commonly used for this purpose. We shall use the interval $(-\pi, \pi]$ and refer to the corresponding θ in this interval as the *principal argument* of $x + iy$; we denote this θ by $\arg(x + iy)$. Thus, if $x + iy \neq 0$ and $r = |x + iy|$, we define $\arg(x + iy)$ to be the unique real θ satisfying the conditions

$$x = r \cos \theta, \qquad y = r \sin \theta, \qquad -\pi < \theta \leq \pi.$$

For the zero complex number we assign the modulus 0 and agree that any real θ may be used as argument.

Since the absolute value of a complex number z is simply the length of a line segment, it is not surprising to learn that it has the usual properties of absolute values of real numbers. For example,

$$|z| > 0 \text{ if } z \neq 0, \qquad \text{and} \qquad |z_1 - z_2| = |z_2 - z_1|.$$

Geometrically, the absolute value $|z_1 - z_2|$ represents the distance between the points z_1 and z_2 in the complex plane.

0.10 Complex conjugates

If $z = x + iy$, the *complex conjugate* of z is the complex number $\bar{z} = x - iy$. Geometrically, $\bar{z}$ represents the reflection of z through the real axis; it has the same real part, but the imaginary part has opposite sign. The definition of conjugate implies that $|\bar{z}| = |z|$ and that

$$\overline{z_1 + z_2} = \bar{z_1} + \bar{z_2}, \qquad \overline{z_1 z_2} = \bar{z_1}\,\bar{z_2}, \qquad \overline{z_1/z_2} = \bar{z_1}/\bar{z_2}, \qquad z\bar{z} = |z|^2.$$

Using these properties we find that $|z_1 z_2|^2 = z_1 z_2 \overline{z_1 z_2} = z_1 \bar{z_1} z_2 \bar{z_2} = |z_1|^2 |z_2|^2$ and hence

(0.4) $$|z_1 z_2| = |z_1||z_2|.$$

Similarly, we find $|z_1/z_2| = |z_1|/|z_2|$ if $z_2 \neq 0$. The triangle inequality

(0.5) $$|z_1 + z_2| \leq |z_1| + |z_2|$$

is also valid. To prove this we write

$$|z_1 + z_2|^2 = (z_1 + z_2)(\bar{z_1} + \bar{z_2}) = |z_1|^2 + |z_2|^2 + z_1 \bar{z_2} + \bar{z_1} z_2.$$

Now observe that a complex number plus its conjugate is twice its real part; and since the real part of a complex number does not exceed its modulus, we have

$$z_1 \bar{z_2} + \bar{z_1} z_2 \leq 2|z_1 \bar{z_2}| = 2|z_1||z_2|.$$

Therefore

$$|z_1 + z_2|^2 \leq |z_1|^2 + |z_2|^2 + 2|z_1||z_2| = (|z_1| + |z_2|)^2,$$

from which we get the triangle inequality in (0.5).

If a quadratic equation with real coefficients has no real roots, its complex roots, given by (0.3), are conjugates. Conversely, if r_1 and r_2 are complex conjugates, say $r_1 = \alpha + i\beta$ and $r_2 = \alpha - i\beta$, where α and β are real, then r_1 and r_2 are roots of a quadratic equation with real coefficients. In fact,

$$r_1 + r_2 = 2\alpha \qquad \text{and} \qquad r_1 r_2 = \alpha^2 + \beta^2,$$

so

$$(x - r_1)(x - r_2) = x^2 - (r_1 + r_2)x + r_1 r_2$$

and the quadratic equation in question is

$$x^2 - 2\alpha x + \alpha^2 + \beta^2 = 0.$$

0.11 Exercises

1. Express each of the following complex numbers in the form $a + bi$.
   (a) $(1 + i)^2$.
   (b) $1/i$.
   (c) $1/(1 + i)$.
   (d) $(2 + 3i)(3 - 4i)$.
   (e) $(1 + i)/(1 - 2i)$.
   (f) $i^5 + i^{16}$.
   (g) $1 + i + i^2 + i^3$.
   (h) $\frac{1}{2}(1 + i)(1 + i^{-8})$.

2. Compute the absolute value of each of the following complex numbers.
   (a) $1 + i$.
   (b) $3 + 4i$.
   (c) $(1 + i)/(1 - i)$.
   (d) $1 + i + i^2$.
   (e) $i^7 + i^{10}$.
   (f) $2(1 - i) + 3(2 + i)$.

3. Compute the modulus and principal argument of each of the following complex numbers.
   (a) $2i$.
   (b) $-3i$.
   (c) -1.
   (d) 1.
   (e) $-3 + \sqrt{3}i$.
   (f) $(1 + i)/\sqrt{2}$.
   (g) $(-1 + i)^3$.
   (h) $(-1 - i)^3$.
   (i) $1/(1+i)$.
   (j) $1/(1 + i)^2$.

4. In each case, determine all real numbers x and y that satisfy the given relation.
   (a) $x + iy = x - iy$.
   (b) $x + iy = |x + iy|$.
   (c) $|x + iy| = |x - iy|$.
   (d) $(x + iy)^2 = (x - iy)^2$.
   (e) $(x + iy)/(x - iy) = x - iy$.
   (f) $\sum_{k=0}^{100} i^k = x + iy$.

5. Make a sketch showing the set of all z in the complex plane that satisfy each of the following conditions.
   (a) $|z| < 1$.
   (b) $z + \bar{z} = 1$.
   (c) $z - \bar{z} = i$.
   (d) $|z - 1| = |z + 1|$.
   (e) $|z - i| = |z + i|$.
   (f) $z + \bar{z} = |z|^2$.

6. Let f be a polynomial with real coefficients.
   (a) Show that $\overline{f(z)} = f(\bar{z})$.
   (b) Use part (a) to deduce that the nonreal zeros of f (if any exist) must occur in pairs of conjugate complex numbers.

7. Make a sketch showing the set of all complex z that satisfy each of the following conditions.
   (a) $|2z + 3| < 1$.
   (b) $|z + 1| < |z - 1|$.
   (c) $|z - i| \le |z + i|$.
   (d) $|z| \le |2z + 1|$.

8. Let $w = (az + b)/(cz + d)$, where a, b, c, d are real. Prove that

$$w - \bar{w} = (ad - bc)(z - \bar{z})/|cz + d|^2.$$

If $ad - bc > 0$, prove that the imaginary parts of z and w have the same sign.

0.12   Mathematical induction

There is no largest positive integer because when we add 1 to an integer k, we obtain $k + 1$, which is larger than k. Even though there are infinitely many positive integers, if we start with the number 1 we can reach any positive integer whatever in a finite number of steps, passing successively from k to $k + 1$ at each step. This is the basis for a type of reasoning that mathematicians call *proof by induction*.

Method of proof by induction.   Let $A(n)$ be an assertion (which may be true or false) involving an integer n. We conclude that $A(n)$ is true for every $n \ge 1$ if the following two statements hold:
   (a) $A(1)$ is true.
   (b) For every integer $k \ge 1$, $A(k)$ implies $A(k + 1)$.

The idea of induction can be illustrated in many nonmathematical ways. For example, imagine a column of toy soldiers, numbered consecutively, that extends to the right without end, as suggested by the dots in Figure 0.7. Suppose they are so arranged that if any one of them falls backward, say the one labeled k, it will knock over the one behind it, labeled

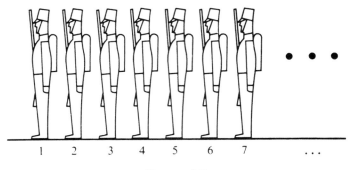

FIGURE 0.7

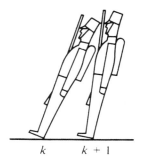

FIGURE 0.8

$k + 1$. (See Figure 0.8.) Then anyone can visualize what would happen if soldier number 1 were toppled backward. It is also clear that if a later soldier were knocked over first, say the one labeled n_1, then all soldiers behind *him* would fall. This illustrates a modified form of the principle of induction which states that $A(n)$ is true for all $n \geq n_1$, if $A(n_1)$ is true and, for every integer $k \geq n_1$, $A(k)$ implies $A(k + 1)$.

We illustrate the method of proof by induction with the following example.

EXAMPLE.   Let $A(n)$ be the assertion that $(2n)!/(n!)^2 > 2^{n+2}$. This inequality is false for $n = 1, 2$ and 3, but it is true for $n = 4$. We can use induction to prove it is true for all $n \geq 4$.

We will show that $A(k)$ implies $A(k + 1)$, not only for $k \geq 4$, but for all $k \geq 1$. Inequality $A(k)$ asserts that

(0.6)
$$\frac{(2k)!}{(k!)^2} > 2^{k+2},$$

while $A(k + 1)$ asserts that

(0.7)
$$\frac{(2k + 2)!}{((k + 1)!)^2} > 2^{k+3}.$$

The left member of inequality (0.7) can be written as follows:

$$\frac{(2k + 2)!}{((k + 1)!)^2} = \frac{(2k + 2)(2k + 1)(2k)!}{(k + 1)^2(k!)^2} = \frac{2k + 1}{k + 1} \cdot \frac{2(2k)!}{(k!)^2}.$$

Since $2k + 1 > k + 1$ the fraction $(2k + 2)!/((k + 1)!)^2$ is greater than $2(2k)!/(k!)^2$, which is twice the left member of (0.6). Therefore (0.6) implies (0.7) for all $k \geq 1$. Since $A(4)$ is true, it follows by induction that inequality $A(n)$ is true for all $n \geq 4$.

In the foregoing example we found that $A(k)$ implies $A(k + 1)$ for all $k \geq 1$, even though $A(n)$ is false for $n = 1, 2,$ and 3. But because $A(4)$ is true, the inequality holds for $n = 5$, $n = 6$, and all larger n.

Proofs by induction occur frequently in this text.

0.13 Exercises

1. Prove the following formulas for all integers $n \geq 1$ by induction:

   (a) $\displaystyle\sum_{k=1}^{n} k = \frac{n(n + 1)}{2}$.   (b) $\displaystyle\sum_{k=1}^{n} k^3 = \left(\sum_{k=1}^{n} k\right)^2$.

2. A real number x has the property that $x^2 = x + 1$. Prove by induction that there is a function $f(n)$ such that $x^{n+1} = f(n + 1)x + f(n)$ for all integers $n \geq 1$.

3. Let $A(n)$ denote the assertion: $\displaystyle\sum_{r=1}^{n} r = \frac{(2n + 1)^2}{8}$.

   (a) Prove that $A(k)$ implies $A(k + 1)$ for all integers $k \geq 1$.
   (b) Criticize the statement: "By induction it follows that $A(n)$ is true for all $n \geq 1$."
   (c) Amend $A(n)$ by replacing the equality by an inequality that is true for all integers $n \geq 1$.

0.14 Necessary and sufficient conditions

Much of this book discusses theorems and their proofs, and it is essential that the reader understand the meaning of some of the terminology that occurs in these discussions.

Every theorem involves one or more implications of the form "*A implies B*," where A is the hypothesis, and B is the conclusion. The statement "*A implies B*" means that "if A is true then B is true." This is sometimes described by saying that "A is a sufficient condition for B," or that "B is a necessary condition for A." For example, one of the simplest theorems of plane geometry states that *every equilateral triangle is isosceles*. This can be expressed as an implication: *if a triangle is equilateral, then it is isosceles*. This theorem has the form "*A implies B*," where A is the statement "a triangle is equilateral" and B is the statement "a triangle is isosceles."

If a theorem states that "*A implies B*," the converse of the theorem is the statement "*B implies A*," which may or may not be true. For example, the statement "*every equilateral triangle is isosceles*" is true, but the converse statement, "*every isosceles triangle is equilateral*," is false because there are some isosceles triangles that are not equilateral.

If a theorem, "*A implies B*," and its converse, "*B implies A*," are both true, this is described by saying that "A is necessary and sufficient for B," or more briefly, "*A if and only if B*." In this case we also say that A and B are logically equivalent.

To show that two statements A and B are logically equivalent, we must prove that "*A implies B*" and that "*B implies A*." To show that three statements A, B and C are logically equivalent, it suffices to prove that *A implies B*, that *B implies C*, and that *C implies A*. It then follows that *B implies A*, that *C implies B*, and that *A implies C*.

A *direct proof* of a statement *B* is a logically correct argument establishing the truth of *B*. A *proof by contradiction* assumes the negation of *B* (not *B*) and derives the truth of some statement *A* and of its negation (not *A*). This contradiction shows that the initial assumption (not *B*) cannot hold, thus establishing the truth of *B*. For example, let *B* denote the statement "*there is no largest positive integer.*" To prove this by contradiction, assume that *B* is false. Then there is a largest positive integer, which we can call *n*. By adding 1 to *n* we get *n* + 1, which is larger than *n*, contradicting the fact that *n* was the largest positive integer. This contradiction shows that *B* is true.

We conclude this section with the following somewhat surprising theorem related to the example just mentioned. The reader may find it instructive to prove this theorem by contradiction.

THEOREM.   *If there is a largest positive integer n, then n* = 1.

Part 2.   Calculus Prerequisites

The last three chapters of this text contain applications of linear algebra to linear differential equations and systems of differential equations. The second part of the present chapter reviews those calculus topics that are required as prerequisites for these applications. We begin with topics from differential calculus.

0.15   The concept of derivative

For a real-valued function *f* defined on an interval on the real axis, the derivative of *f* is a new function obtained from *f* by a process called *differentiation*, which can be described as follows:

Form the difference between two function values $f(x + h)$ and $f(x)$, and divide this difference by *h* to obtain the quotient

$$\frac{f(x + h) - f(x)}{h}, \qquad \text{where } h \neq 0.$$

This is called a *difference quotient*. Geometrically, it represents the slope of the chord joining the points $(x, f(x))$ and $(x + h, f(x + h))$ on the graph of the function. It is also called the *average rate of change* of the function in the interval from *x* to *x* + *h*.

Now let *h* approach zero. That is, let *h* take values that are smaller and smaller, and see what happens to the difference quotient during this process. If the difference quotient approaches a definite value, we call this value the *derivative of f* at *x* and we denote the limiting value by the special symbol $f'(x)$, read "*f* prime of *x*." The process is expressed in mathematical shorthand as follows:

$$f'(x) = \lim_{h \to 0} \frac{f(x + h) - f(x)}{h}.$$

The symbol $\lim_{h \to 0}$ is read "the limit as h approaches zero" and is meant to convey the idea that the derivative $f'(x)$ is the limiting value of the difference quotient as h approaches zero. When the quotient has a limit we say the derivative $f'(x)$ *exists*, and the function is called *differentiable* at x. Alternate notations for the derivative are

$$f'(x) = Df(x) = \frac{d}{dx}f(x).$$

Geometrically, the derivative $f'(x)$ represents the slope of the graph of f at the point $(x, f(x))$. The line with this slope through the point $(x, f(x))$ is called the *tangent line* to the graph at that point. The number $f'(x)$ is also called the *instantaneous rate of change* of the function at x.

The derivative is a new function obtained from f by the process of differentiation. Differential calculus is the study of derivatives and their properties.

0.16   Basic properties of derivatives

Calculation of derivatives is simplified by using the following basic properties that are immediate consequences of the definition. Proofs can be found in any calculus text.

(1) *The derivative of a constant times a function is that constant times the derivative of the function.* In symbols, for any constant c we have

$$(cf)' = cf'.$$

(2) *The derivative of the sum of two functions is the sum of their derivatives:*

$$(f + g)' = f' + g'.$$

Properties (1) and (2) can be combined into one property called *linearity*.

(3) For any constants a and b we have $(af + bg)' = af' + bg'$.

The next two properties are used to calculate derivatives of products and of quotients:

(4) *The product rule.* If $f \cdot g$ denotes the product of f and g, we have

$$(f \cdot g)' = f \cdot g' + f' \cdot g.$$

(5) *The quotient rule.* For the quotient f divided by g, we have

$$\left(\frac{f}{g}\right)' = \frac{g \cdot f' - f \cdot g'}{g^2}, \qquad \text{at points where } g(x) \neq 0.$$

The next property is used to calculate the derivative of the composition $f \circ g$ of two functions, where

$$(f \circ g)(x) = f[g(x)].$$

(6) *The chain rule.* If h denotes the composition $f \circ g$, then we have

$$h' = (f' \circ g) \cdot g'.$$

This can also be written as follows: If $h(x) = f[g(x)]$, then

$$h'(x) = f'[g(x)]g'(x).$$

(7) *The zero derivative theorem.* If the derivative of a function is zero at each point of an open interval, then the function is constant on that interval.

This implies that two functions with equal derivatives on an open interval differ only by a constant on that interval.

0.17 Derivatives of some elementary functions

The following brief table lists derivatives of some of the elementary functions that will be encountered in the last three chapters. In this table, n and k can be any nonzero real constants.

TABLE 0.1

Function $f(x)$	Derivative $f'(x)$
constant	zero
x^n	nx^{n-1}
$\sin kx$	$k \cos kx$
$\cos kx$	$-k \sin kx$
e^{kx}	ke^{kx}

Functions that can be obtained from the examples in the table by addition, multiplication, division, and composition can be differentiated using the rules discussed in Section 0.16.

0.18 Velocity and acceleration

Derivatives have applications to the study of motion. If a particle moves on a line so that its displacement at time t from some fixed position is $f(t)$, then the derivative $f'(t)$ is called the *velocity* of the particle at time t. The velocity is another function of t, say $v(t)$, and *its* derivative $v'(t)$ is called the *acceleration*. Acceleration is the derivative of a derivative, or the *second derivative* of displacement, and is denoted by the symbol $f''(t)$ (read: f double prime of t). For example, if the displacement of a particle at time t is

$$f(t) = \sin kt,$$

then the velocity of the particle is the first derivative of displacement,

$$v(t) = f'(t) = k \cos kt,$$

and its acceleration $v'(t)$ is the second derivative of displacement,

$$f''(t) = -k^2 \sin kt.$$

In other words, the displacement function $f(t) = \sin kt$ satisfies the equation

$$f''(t) = -k^2 f(t).$$

This is called the differential equation of simple harmonic motion. Simple harmonic motion is any motion in which acceleration is proportional to displacement but in the opposite direction (as indicated by the minus sign in the differential equation). The same differential equation is satisfied by the function $f(t) = \cos kt$ and, more generally, by any linear combination of the form

$$f(t) = a \cos kt + b \sin kt,$$

where a and b are constants. In fact, it is known that any function that satisfies the differential equation of simple harmonic motion must necessarily be a linear combination of $\sin kt$ and $\cos kt$.

0.19   The area problem and the history of integral calculus

We turn next to a brief review of topics from integral calculus. The history of the subject goes back to the ancient Greeks. A famous problem from anitiquity that challenged the world's best minds for nearly 2000 years was a problem on area called *quadrature*:

Given a plane region with curved boundaries, find a square having the same area.

The early Greeks attacked this problem with the method of exhaustion, which uses inscribed polygons with an increasing number of edges to approximate the region in question. Archimedes (287–212 B.C.E.) used this method to calculate the area of a circular disk and some other special figures. One of these special figures is the parabolic segment, a plane region lying below the parabola with Cartesian equation $y = x^2$ and above the interval $[0, t]$, where $t > 0$. This region lies inside a right triangle of base t and altitude t^2, so its area is less than $\frac{1}{2}t^3$, which is the area of the right triangle. Archimedes showed that the area of the parabolic segment is exactly equal to $\frac{1}{3}t^3$.

Further progress on the quadrature problem was delayed nearly eighteen centuries until widespread use of well-chosen algebraic symbols revived interest in the ancient method of exhaustion. Many fragmentary results were discovered in the 16th and 17th centuries by such pioneers as Cavalieri, Toricelli, Roberval, Fermat, Pascal, and Wallis. By the 17th century, the method of exhaustion was gradually transformed into what is today known as *integral calculus*, a systematic method for calculating areas and volumes of curved figures.

Isaac Newton (1642–1727) and Gottfried Wilhelm Leibniz (1646–1716) are generally regarded as the discoverers of integral calculus. Their great contribution was to unify earlier work and to relate the process of integration with the process of differentiation.

Just as the derivative is the basic concept of differential calculus, the basic concept of integral calculus is the *integral*. In the language of integral calculus, Archimedes' result for the area of a parabolic segment can be stated as follows:

The integral of x^2 from 0 to t is $\frac{1}{3}t^3$.

It is written symbolically as follows:

$$\int_0^t x^2 \, dx = \frac{1}{3}t^3.$$

The symbol $\int$ (an elongated S from the Latin word for sum) is called an *integral sign*, and was introduced by Leibniz in 1675. The notation was invented to suggest that the area of the parabolic segment is obtained by adding areas of many thin rectangles of height x^2 and width dx. The symbol dx is meant to suggest a small change in x. The process that produces the number $\frac{1}{3}t^3$ is called *integration*. The numbers 0 and t attached to the integral sign are called *limits of integration*. The symbol $\int_0^t x^2\,dx$ must be regarded as a single entity. Its definition treats it as such, just as the dictionary describes the word "lapidate" without reference to "lap," "id," or "ate."

0.20 Integration as a process for producing new functions

Although the integral was originally used to calculate areas, it evolved into a more general concept that can be used to compute not only area, but also a host of other quantities such as arc length, volume, work, average values, probability, and others. For the purpose of this review, we will not give a formal definition of the integral, but will simply note that integration is a process that produces new functions from given functions. If we start with an elementary function $f(x)$, say one of those given in the table in Section 0.17, the integral $\int_a^t f(x)\,dx$ is a new function of the upper limit t that we denote by $A(t)$:

$$A(t) = \int_a^t f(x)\,dx.$$

The function $f(x)$ under the integral sign is called the *integrand*. If the integrand is non-negative on the interval $[a, t]$ we can regard $A(t)$ as the *area function*, whose value at t is the area of the region below the graph of $y = f(x)$ and above the interval $[a, t]$.

The following brief table gives the integrals of some of the elementary functions that will be encountered in the last three chapters. The formulas are valid not only for intervals in which the integrand is nonegative but for any interval $[0, t]$ in which the integrand is continuous.

TABLE 0.2

Integrand $f(x)$	Integral $A(t) = \int_0^t f(x)\,dx$
constant c	ct
x^n	$t^{n+1}/(n+1)$ if $n \neq -1$
$\sin kx$	$(1 - \cos kt)/k$ if $k \neq 0$
$\cos kx$	$(\sin kt)/k$ if $k \neq 0$
e^{kx}	$(e^{kt} - 1)/k$ if $k \neq 0$
$(1 + x)^{-1}$	$\log(1 + t)$ if $t > -1$

0.21 Basic properties of the integral

Calculation of integrals is simplified by using the following basic properties.

(1) $\int_u^t f(x)\,dx = \int_a^t f(x)\,dx - \int_a^u f(x)\,dx$.

(2) $\int_u^t cf(x)\,dx = c \int_u^t f(x)\,dx$ *for any constant c.*

(3) $\int_u^t \{f(x) + g(x)\}\,dx = \int_u^t f(x)\,dx + \int_u^t g(x)\,dx$.

Properties (2) and (3) can be combined into one property called *linearity*. The integral of a linear combination of functions is the same linear combination of the integrals:

(4) $\int_u^t \{af(x) + bg(x)\}\,dx = a \int_u^t f(x)\,dx + b \int_u^t g(x)\,dx$.

(5) *Comparison property.* If $a < b$ and $f(x) \leq g(x)$ on the interval $[a, b]$, then

$$\int_a^b f(x)\,dx \leq \int_a^b g(x)\,dx.$$

This implies $|\int_a^b f(x)\,dx| \leq \int_a^b |f(x)|\,dx$.

The next two theorems relate integration and differentiation.

(6) FIRST FUNDAMENTAL THEOREM OF CALCULUS.

If $A(t) = \int_a^t f(x)\,dx$, *where a is a constant, then the derivative $A'(t)$ exists at each point of continuity of f and is equal to $f(t)$.*

In other words, as a function of the upper limit t, the integral of a continuous integrand is differentiable, and the derivative of the integral, $A'(t)$, is equal to the integrand evaluated at the upper limit.

(7) SECOND FUNDAMENTAL THEOREM OF CALCULUS.

If a continuous integrand $f(x)$ is the derivative of a function $P(x)$, then the integral $\int_a^t f(x)\,dx$ is equal to the difference $P(t) - P(a)$.

In other words, for integrands that are derivatives, calculation of integrals reduces to subtraction:

$$\int_a^t P'(x)\,dx = P(t) - P(a).$$

This remarkable theorem reduces the process of integration to that of finding a function whose derivative is the integrand. It can also be written in the following form:

$$P(t) = P(a) + \int_a^t P'(x)\,dx,$$

which tells us that a function is completely determined by its value at a particular point, $P(a)$, and a knowledge of its derivative throughout an interval. In particular, if the derivative $P'(x) = 0$ on the interval joining a and t, then $P(t) = P(a)$. In other words, a function with a zero derivative on an interval is constant on that interval. This is the *zero derivative theorem* mentioned earlier in Section 0.16.

Applying the second fundamental theorem to the derivative of the product $f \cdot g$ we find

$$\int_a^t (f \cdot g)'(x)\,dx = f(t)g(t) - f(a)g(a).$$

Because of the product rule for derivatives, $(f \cdot g)' = f \cdot g' + f' \cdot g$, the last equation can be written in the following form, which is known as the formula for integration by parts:

(8) *Integration by parts.* $\int_a^t f(x)g'(x)\,dx = f(t)g(t) - f(a)g(a) - \int_a^t g(x)f'(x)\,dx$.

0.22 The exponential function

Chapters 8 and 9 rely heavily on a knowledge of the exponential function e^x, often referred to as the most important of the real-valued elementary functions. There are several different ways to define e^x for real x. One method is to first define the natural logarithm by the integral

$$\log y = \int_1^y \frac{1}{x}\, dx \qquad \text{for } y > 0,$$

and then define the exponential by inversion of the logarithm, so that $y = e^x$ means that $x = \log y$. Another method defines e^x by the power series

$$e^x = \sum_{n=0}^{\infty} \frac{x^n}{n!} = 1 + x + \frac{x^2}{2!} + \frac{x^3}{3!} + \cdots + \frac{x^n}{n!} + \cdots,$$

which is known to converge for all real x.

All methods used to define the exponential are equivalent, in the sense that they produce a function with the same basic properties. This section summarizes these properties. The next section shows how the exponential can be extended so that e^z is meaningful for complex values of z; and in Chapter 9 the definition is extended even further to exponential matrices.

Basic Properties of the Real Exponential Function
1. *$e^0 = 1$.*
2. *Law of exponents: $e^x e^y = e^{x+y}$ for all real x and y.*
3. *Differential equation: If $f(x) = e^x$ then $f'(x) = e^x$.*
In other words, the exponential function is equal to its own derivative. The most general function satisfying the differential equation $f'(x) = f(x)$ is $f(x) = ce^x$, where c is constant.
4. *Positivity: $e^x > 0$ for every real x.*
5. *Relation to the natural logarithm. If $y > 0$, then $y = e^x$ if and only if $x = \log y$.*
6. *Power series representation: For every real x we have*

$$e^x = \sum_{n=0}^{\infty} \frac{x^n}{n!}.$$

0.23   Complex exponentials

The exponential can be extended so that e^x becomes meaningful when x is replaced by an arbitrary complex number z.

DEFINITION.   *If $z = x + iy$, we define e^z to be the complex number given by the equation*

$$e^z = e^x(\cos y + i \sin y).$$

Note that $e^z = e^x$ when $y = 0$; hence this exponential agrees with the usual exponential when z is real. The definition also implies the law of exponents.

THEOREM 0.3.   *If a and b are arbitrary complex numbers we have*

(0.8) $$e^a e^b = e^{a+b}.$$

Proof. Write $a = x + iy$, and $b = u + iv$, so that

$$e^a = e^x(\cos y + i \sin y), \quad \text{and} \quad e^b = e^u(\cos v + i \sin v).$$

The product of these two complex numbers is equal to

$$e^a e^b = e^x e^u \left[(\cos y \cos v - \sin y \sin v) + i(\cos y \sin v + \sin y \cos v) \right].$$

Using the addition formulas for $\cos(y + v)$ and $\sin(y + v)$ and the law of exponents for real exponents we find that the foregoing equation becomes

$$e^a e^b = e^{x+u} \left[\cos(y + v) + i \sin(y + v) \right],$$

which implies (0.8).

The complex exponential has properties that do not exist in the real case. For example, if x is real, then $e^x = 1$ if and only if $x = 0$. But in the complex case, if $e^z = 1$ for some $z = x + iy$, then

$$e^x \cos y = 1 \quad \text{and} \quad e^x \sin y = 0.$$

Because e^x is never zero, the second relation holds if and only if $\sin y = 0$, which means $y = 2n\pi$ for some integer n. In other words,

$$e^z = 1 \quad \text{if and only if } z = 2n\pi i$$

for some integer n.

Real exponentials are always positive. But complex exponentials can take negative values. For example, $e^{\pi i} = \cos \pi + i \sin \pi = -1$.

Because the sine and cosine are periodic with period 2π, complex exponentials are periodic with period $2\pi i$. That is, $e^{z+2\pi i} = e^z$. In fact, by the law of exponents we have $e^{z+2\pi i} = e^z e^{2\pi i} = e^z$ because $e^{2\pi i} = 1$.

0.24 Polar form of complex numbers

In Section 0.9 it was shown that every nonzero complex number $z = x + iy$ can be expressed in terms of polar coordinates by the equation

$$z = x + iy = r(\cos \theta + i \sin \theta),$$

where $r = |z|$ and $\theta = \arg z$. With complex exponentials this can now be expressed more simply as

$$z = re^{i\theta},$$

which is called the *polar form* of z. Because of periodicity of the complex exponential, this equation still holds if the angle θ is replaced by $\arg z$ plus any integer multiple of 2π.

The polar form is especially useful when complex numbers are multiplied. For example, if we multiply two nonzero complex numbers z and w expressed in polar form, say

$$z = re^{i\theta}, \quad \text{and} \quad w = Re^{i\phi},$$

their product is given by

$$zw = (re^{i\theta})(Re^{i\phi}) = rRe^{i(\theta+\phi)}.$$

In other words, to multiply two complex numbers we multiply their moduli and add their angles.

Complex exponentials can be used to derive trigonometric identities. For example, suppose that z has modulus 1, so that $z = e^{i\theta} = \cos\theta + i\sin\theta$. Then for any positive integer n we have

$$z^n = e^{in\theta} = \cos n\theta + i\sin n\theta.$$

This can be written in the form

$$(\cos\theta + i\sin\theta)^n = \cos n\theta + i\sin n\theta,$$

which is called *de Moivre's formula*, in honor of the mathematician Abraham de Moivre (1667–1754). It can be used to present a simple derivation of multiple angle formulas for the sine and cosine. For example, when $n = 3$, de Moivre's formula states that

$$(\cos\theta + i\sin\theta)^3 = \cos 3\theta + i\sin 3\theta.$$

Expand the left member by the binomial formula $(a + b)^3 = a^3 + 3a^2b + 3ab^2 + b^3$, with $a = \cos\theta$ and $b = i\sin\theta$, then separate real and imaginary parts to obtain

$$(\cos\theta + i\sin\theta)^3 = \cos^3\theta - 3\cos\theta\sin^2\theta + i(3\cos^2\theta\sin\theta - \sin^3\theta).$$

Now equate real and imaginary parts with those obtained from de Moivre's formula to obtain

$$\cos 3\theta = \cos^3\theta - 3\cos\theta\sin^2\theta \quad \text{and} \quad \sin 3\theta = 3\cos^2\theta\sin\theta - \sin^3\theta.$$

Use the Pythagorean identity $\sin^2\theta + \cos^2\theta = 1$ to replace each factor $\sin^2\theta$ by $1 - \cos^2\theta$, and rewrite these as follows:

$$\cos 3\theta = 4\cos^3\theta - 3\cos\theta, \quad \text{and} \quad \sin 3\theta = \sin\theta(4\cos^2\theta - 1).$$

This shows that $\cos 3\theta$ is a cubic polynomial in $\cos\theta$, and that $\sin 3\theta$ is $\sin\theta$ times a quadratic polynomial in $\cos\theta$. In a similar way we can show that, for any integer $n > 1$, $\cos n\theta$ is a polynomial in $\cos\theta$ of degree n, and $\sin n\theta$ is $\sin\theta$ times a polynomial in $\cos\theta$ of degree $n - 1$. The actual polynomials can be obtained by using the binomial theorem to expand $(\cos\theta + i\sin\theta)^n$, replacing each factor $\sin^2\theta$ by $1 - \cos^2\theta$.

0.25 Power series and series of functions

An infinite series of the form $c_0 + c_1x + c_2x^2 + \cdots$, written more briefly as $\sum_{k=0}^{\infty} c_k x^k$, is called a *power series* in x. The multipliers c_k are called its coefficients. If both x and all the coefficients c_k are real numbers, there is an open interval $(-r, r)$ associated with the series, called the *interval of convergence*, such that the series converges if $|x| < r$, and diverges if $|x| > r$. The number r can be zero, in which case the series converges only for $x = 0$, or it can be infinite, in which case the series converges for all real x.

Inside the interval of convergence, the sum of the series defines a function of x,

$$f(x) = \sum_{k=0}^{\infty} c_k x^k = \lim_{n \to \infty} \sum_{k=0}^{n} c_k x^k$$

that shares many properties of polynomials. For example, the derivative $f'(x)$ exists and can be obtained by differentiating the series term by term:

$$f'(x) = \sum_{k=1}^{\infty} k c_k x^{k-1}.$$

If $[a, b]$ is a subinterval of the interval of convergence the integral $\int_a^b f(x)\, dx$ can be obtained by integrating the series term by term:

$$\int_a^b f(x)\, dx = \sum_{k=0}^{\infty} c_k \int_a^b x^k \, dx.$$

Important examples of power series encountered in the later chapters are the *geometric series*

$$\frac{1}{1-x} = \sum_{k=0}^{\infty} x^k \qquad \text{(convergent for } |x| < 1),$$

and the *exponential series*

$$e^x = \sum_{n=0}^{\infty} \frac{x^n}{n!} \qquad \text{(convergent for all real } x).$$

In Chapter 10 we will also encounter more general series of the form $\sum_{k=0}^{\infty} u_k(x)$ that have a property called *uniform convergence on an interval*. A sufficient condition for uniform convergence on an interval I is that the series can be dominated by a convergent series of positive constants; that is, $|u_k(x)| \le M_k$ for all x in I, where the M_k are positive constants such that $\sum_{k=0}^{\infty} M_k$ converges. The following two important properties of uniformly convergent series will be used in Chapter 10.

(a) If each function $u_k(x)$ is continuous at a point x_0 in I, then the sum function

$$f(x) = \sum_{k=0}^{\infty} u_k(x) = \lim_{n \to \infty} \sum_{k=0}^{n} u_k(x)$$

is also continuous at x_0, and the limit of $f(x)$ as $x \to x_0$ can be obtained by passing to the limit term by term:

$$\lim_{x \to x_0} \sum_{k=0}^{\infty} u_k(x) = \sum_{k=0}^{\infty} \lim_{x \to x_0} u_k(x) = \sum_{k=0}^{\infty} u_k(x_0).$$

(b) If each function $u_k(x)$ is continuous on an interval $[a, b]$, the integral of the sum function can be obtained by integrating the series term by term:

$$\int_a^b f(x)\,dx = \sum_{k=0}^{\infty} \int_a^b u_k(x)\,dx.$$

0.26 Exercises

1. Let $f(x) = \frac{1}{3}x^3 - 2x^2 + 3x + 1$. Find all points on the graph of f at which the tangent line is horizontal.
2. Let $f(x) = x^2 + ax + b$, where a and b are constants. Find all values of a and b such that the line $y = 2x$ is tangent to the graph of f at the point $(2, 4)$.
3. Find values of the constants a, b, c for which the graphs of the two polynomials $f(x) = x^2 + ax + b$ and $g(x) = x^3 - c$ will intersect at $(1, 2)$ and have the same tangent line there.
4. The height of a projectile above the ground at time t seconds after being fired is $f(t) = 144t - 16t^2$ feet, where $0 \le t \le 9$.
   (a) Draw a graph of f.
   (b) Calculate the velocity of the projectile as a function of time.
   (c) Show that the velocity is equal to zero when the projectile reaches its maximum height.
   (d) Show that the projectile moves with constant acceleration.
5. In this exercise, $f(x) = (\log x)/x$ for $x > 0$.
   (a) Make a sketch of the graph of f over the interval $0 < x < e^2$. Describe those x for which $f'(x) > 0$ and those for which $f''(x) > 0$.
   (b) Determine a constant c such that $\int_1^t f(x)\,dx = c(\log t)^2$ for all $t > 0$.
   (c) A line $y = mx$ is tangent to the graph of f at $(a, f(a))$. Determine (in terms of e) the number m and the point of contact $(a, f(a))$.
   (d) Determine (in terms of e) a real number $b > 1$ such that the graph of f bisects the area of the triangle with vertices at $(0, 0)$, $(b, 0)$, and $(b, f(b))$.
6. Let $f(x) = 3 + \int_0^x e^t / (1 + t)\,dt$. (Do not attempt to evaluate this integral.) Find a polynomial $p(x)$ of degree 3 such that $p(0) = f(0)$, $p'(0) = f'(0)$, $p''(0) = f''(0)$, and $p'''(0) = f'''(0)$.
7. A function f satisfies the functional equation

$$f(x + y) = f(x) + f(y) + xy$$

   for all real x and y.
   (a) Show that $f(0) = 0$.
   (b) If $h \ne 0$, show that

$$\frac{f(x + h) - f(x)}{h} = \frac{f(h) - f(0)}{h} + x,$$

   and use this to deduce that if $f'(0)$ exists, then $f'(x)$ exists for all x and that $f'(x) = f'(0) + x$.
8. Let $f(x) = \int_0^x g(t) \cos(x - t)\,dt$, where g is a given function, differentiable everywhere.
   (a) Prove that $f'(x) = g(x) - \int_0^x g(t) \sin(x - t)\,dt$.
   [*Hint:* $\cos(a - b) = \cos a \cos b + \sin a \sin b$.]
   (b) Find constants a and b such that $f''(x) + f(x) = ag'(x) + bg(x)$.
9. Use de Moivre's formula, as suggested in Section 0.24, to derive the following identities:
   (a) $\cos 4x = 8 \cos^4 x - 8 \cos^2 x + 1$.
   (b) $\sin 4x = \sin x(8 \cos^3 x - 4 \cos x)$.
   (c) Show that each of (a) and (b) can be obtained from the other by differentiation.

1

VECTOR ALGEBRA

1.1 Historical introduction

The idea of using numbers to locate points on a line was known to the ancient Greeks. In the 17th century, René Descartes extended this idea, using pairs of numbers to locate points in the plane, and triples of numbers to locate points in space. In doing so, he founded analytic geometry, where properties of geometric figures are expressed in terms of algebraic relations involving coordinates.

Throughout their history, analytic geometry and calculus have been intimately related; a discovery in one field led to an improvement in the other. The problem of drawing tangents to curves resulted in the development of differential calculus. The problem of calculating areas of plane regions with curved boundaries led to the development of integral calculus.

Along with these accomplishments came other parallel developments in mechanics and mathematical physics. In 1788 Joseph-Louis Lagrange (1736–1813) published his masterpiece, *Méchanique analytique* (Analytical Mechanics), which showed the great flexibility and tremendous power attained by using analytical methods in the study of mechanics. In the 19th century, the Irish mathematician William Rowan Hamilton introduced his *Theory of Quaternions*, a new method and a new point of view, which contributed much to the understanding of both algebra and physics. The best features of quaternion analysis and Cartesian geometry were later united, largely through the efforts of J. W. Gibbs (1839–1903) and O. Heaviside (1850–1925), and a new subject called *vector algebra* sprang into being. It was soon realized that vectors are the ideal tools for the exposition and simplification of many important ideas in geometry and physics. The elements of vector algebra are discussed in this chapter. Applications to analytic geometry are given in Chapter 2.

There are three different ways to introduce vector algebra: *geometrically*, *analytically*, and *axiomatically*. In the geometric approach, vectors are represented by directed line segments, or arrows. Algebraic operations on vectors, such as addition, subtraction, and multiplication by real numbers, are defined and studied by geometric methods.

In the analytic approach, vectors and vector operations are described entirely in terms of *numbers*, called *components*. Properties of the vector operations are then deduced from corresponding properties of numbers. The analytic description of vectors arises naturally from the geometric description as soon as a coordinate system is introduced.

In the axiomatic approach, no attempt is made to describe the nature of a vector or of the algebraic operations on vectors. Instead, vectors and vector operations are thought of as *undefined concepts*, of which we know nothing except that they satisfy a certain set of axioms. Such an algebraic system, with appropriate axioms, is called a *linear space* or a *linear vector space*. Examples of linear spaces occur in all branches of mathematics, and we will study many of them in Chapter 3. The algebra of directed line segments and the algebra of vectors defined by components are merely two examples of linear spaces.

The study of vector algebra from the axiomatic point of view is perhaps the most mathematically satisfactory approach because it furnishes a description of vectors that is free of coordinate systems and free of any particular geometric representation. This study, carried out in Chapter 3 and subsequent chapters, provides an introduction to *linear algebra*, the principal subject matter of this text.

In this chapter we base our treatment on the analytic approach, but we also use directed line segments to interpret many of the results geometrically. When possible, we give proofs by coordinate-free methods. Thus, this chapter serves to provide familiarity with important concrete examples of linear spaces, and it also motivates the more abstract approach presented in Chapter 3.

1.2 The vector space of *n*-tuples of real numbers

In the 17th century, Descartes used pairs of numbers (a_1, a_2) to locate points in a plane, and triples of numbers (a_1, a_2, a_3) to locate points in space. The 19th century mathematicians A. Cayley (1821–1895) and H. G. Grassmann (1809–1877) realized that there is no need to stop with three numbers. One can just as well consider a *quadruple* of numbers (a_1, a_2, a_3, a_4) or, more generally, an *n-tuple* of real numbers

$$(a_1, a_2, \ldots, a_n)$$

for any integer $n \geq 1$. Such an *n*-tuple is called an *n-dimensional point* or an *n-dimensional vector*, the individual numbers $a_1, a_2, \ldots, a_n$ being referred to as *coordinates* of the point or *components* of the vector. The set of all *n*-dimensional vectors is called *n*-space or the vector space of *n*-tuples, and is denoted by the symbol $\mathbf{R}^n$. Here $\mathbf{R}$ is used to remind us that the components are real numbers.

Higher-dimensional spaces are studied not only for the sake of generalization but because they provide a natural setting for treating problems that cannot be described by at most three coordinates. For example, in physics some events are characterized not just by three coordinates in space, but also by a time coordinate. Thus, every such event corresponds to a point in $\mathbf{R}^4$. One often encounters systems with several degrees of freedom, or problems that involve a large number of simultaneous equations. These are more easily analyzed by introducing vectors in a suitable *n*-space and replacing all these equations by a single vector equation. Moreover, by studying *n*-space for a general *n* we are able to deal in one stroke with many properties common to 1-space, 2-space, 3-space, etc., that is, properties independent of the dimensionality of the space. This is in keeping with the spirit of modern mathematics, which favors development of comprehensive methods for attacking problems on a wide front.

Unfortunately, the geometric pictures that are a great help in motivating and illustrating vector concepts when $n = 1, 2$, and 3 are not available when $n > 3$; therefore, the study of vector algebra in higher-dimensional spaces must proceed entirely by analytic means.

In this chapter we shall usually denote vectors by capital letters $A, B, C, \ldots$, and components by the corresponding small letters $a, b, c, \ldots$. Thus, we write

$$A = (a_1, a_2, \ldots, a_n).$$

To convert $\mathbf{R}^n$ into an algebraic system, we introduce *equality* of vectors and two vector operations called *addition* and *multiplication by scalars*. The word *scalar* is used here as a synonym for *real number*. It was introduced by Hamilton from the Latin for *stairs* or *scale*.

DEFINITION. *Two vectors A and B in $\mathbf{R}^n$ are called equal whenever they agree in their respective components. That is, if $A = (a_1, a_2, \ldots, a_n)$ and $B = (b_1, b_2, \ldots, b_n)$, the vector equation $A = B$ means exactly the same as the n scalar equations*

$$a_1 = b_1, a_2 = b_2, \ldots, a_n = b_n.$$

The sum $A + B$ is defined to be the vector obtained by adding corresponding components:

$$A + B = (a_1 + b_1, a_2 + b_2, \ldots, a_n + b_n).$$

If c is a scalar, we define cA or Ac to be the vector obtained by multiplying each component of A by c:

$$cA = (ca_1, ca_2, \ldots, ca_n).$$

The following properties of these operations follow quickly from the definition.

THEOREM 1.1. *Vector addition is commutative,*

$$A + B = B + A,$$

and associative,

$$A + (B + C) = (A + B) + C.$$

Multiplication by scalars is associative,

$$c(dA) = (cd)A$$

and satisfies the two distributive laws

$$c(A + B) = cA + cB, \qquad and \qquad (c + d)A = cA + dA.$$

The vector with all its components equal to 0 is called the *zero vector* and is denoted by O. It has the property that $A + O = A$ for every vector A; in other words, O is an identity element for vector addition. The vector $(-1)A$ is also denoted by $-A$ and is called the *negative* of A. The sum of A and the negative of B is called the *difference* of A and B, written as $A - B$. Subtraction is the inverse of addition because $(A + B) - B = A$. Note that $0A = O$ and that $1A = A$ for every A.

The reader may have noticed the similarity between vectors in 2-space and complex numbers. Both are defined as ordered pairs of real numbers and both are added in exactly the same way. Thus, as far as addition is concerned, complex numbers and two-dimensional vectors are algebraically indistinguishable. They differ only when we introduce multiplication.

Multiplication of complex numbers generalizes many properties possessed by the real numbers. For example, if the product of two complex numbers is zero then at least one of the factors must be zero. This property is useful for solving polynomial equations by factoring. However, it is known that multiplication cannot be introduced in $\mathbf{R}^n$ for $n \geq 3$ to preserve this property together with the commutative, associative, and distributive laws. But there are other products that can be introduced in $\mathbf{R}^n$ that do not possess all these properties and yet serve a useful purpose. For example, in Section 1.5 we discuss the *dot product* of two vectors in $\mathbf{R}^n$. The result of this multiplication is a scalar, not a vector. We will learn that the dot product of two vectors can be zero without either factor being zero. Another product, called the *cross product*, is discussed in Section 2.10. This multiplication is applicable only in the case $n = 3$. The result is always a vector, but the cross product is not commutative. Also, the cross product of two vectors can be zero without either factor being zero.

1.3 Geometric interpretation for $n \leq 3$

Although the foregoing definitions are completely divorced from geometry, vectors and vector operations have an interesting geometric interpretation for spaces of dimension three or less. We shall draw pictures in 2-space to illustrate these concepts and ask the reader to provide the corresponding visualizations in 3-space and in 1-space.

A pair of points A and B is called a *geometric vector* if one of the points, say A, is called the *initial point* and the other, B, the *terminal point*, or *tip*. We visualize a geometric vector as an arrow from A to B, as shown in Figure 1.1, and denote it by the symbol $\overrightarrow{AB}$.

Geometric vectors are especially convenient for representing certain physical quantities such as force, displacement, velocity, and acceleration, which possess both magnitude and direction. The length of the arrow is a measure of the magnitude, and the arrowhead indicates the required direction. Hamilton first introduced the word *vector* (from the Latin for *carrier*) to denote a mathematical object that has both magnitude and direction.

Suppose we introduce a coordinate system with origin O. Figure 1.2 shows two geometric vectors $\overrightarrow{AB}$ and $\overrightarrow{CD}$ with $B - A = D - C$. In terms of components this means that we have

$$b_1 - a_1 = d_1 - c_1 \quad \text{and} \quad b_2 - a_2 = d_2 - c_2.$$

By comparison of the congruent triangles in Figure 1.2, we see that the two arrows representing $\overrightarrow{AB}$ and $\overrightarrow{CD}$ have equal lengths, are parallel, and point in the same direction. We call such geometric vectors *equivalent*. That is, we say $\overrightarrow{AB}$ is equivalent to $\overrightarrow{CD}$ whenever

$$(1.1) \qquad\qquad B - A = D - C.$$

This means that, if you slide a vector parallel to itself anywhere in space you get an

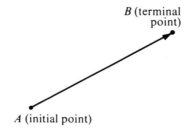

FIGURE 1.1   The geometric vector $\overrightarrow{AB}$ from A to B.

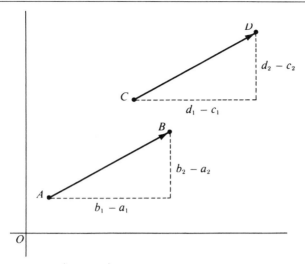

FIGURE 1.2 $\vec{AB}$ and $\vec{CD}$ are equivalent because $B - A = D - C$.

equivalent vector. Note that the four points A, B, C, D are vertices of a parallelogram. (See Figure 1.3.)

Equation (1.1) can also be written in the form $A + D = B + C$, which tells us that *opposite vertices of the parallelogram have the same sum*. In particular, if one of the vertices, say A, is the origin O, as in Figure 1.4, the geometric vector from O to the opposite vertex D corresponds to the vector sum $D = B + C$. This is described by saying that addition of geometric vectors follows the *parallelogram law*. The importance of vectors in physics stems from the remarkable fact that many physical quantities (such as force, velocity, and acceleration) combine by the parallelogram law.

For simplicity in notation, we shall use the same symbol to denote a point in $\mathbf{R}^n$ (when $n \leq 3$) and the geometric vector from the origin to this point. Thus, we write A instead of $\vec{OA}$, B instead of $\vec{OB}$, and so on. Sometimes, we also write A in place of any geometric vector equivalent to $\vec{OA}$. For example, Figure 1.5 illustrates the geometric meaning of

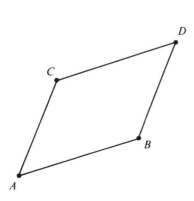

FIGURE 1.3   Opposite vertices of a parallelogram have the same sum: $A + D = B + C$.

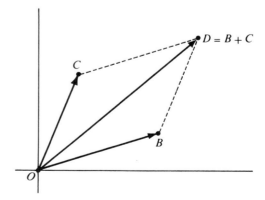

FIGURE 1.4   Vector addition interpreted geometrically by the parallelogram law.

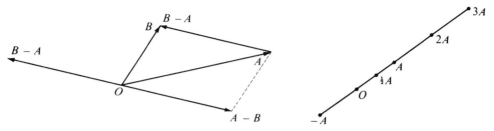

FIGURE 1.5   Geometric meaning of subtraction   FIGURE 1.6   Multiplication of
of vectors.   vectors by scalars.

vector subtraction. Two geometric vectors are labeled as $B - A$, but these geometric vectors
are equivalent. They have the same length and the same direction.

Figure 1.6 illustrates the geometric meaning of multiplication by scalars. If $B = cA$, the
geometric vector B has length $|c|$ times the length of A; it points in the same direction as A
if c is positive, and in the opposite direction if c is negative.

The geometric interpretation of vectors in n-space for $n \leq 3$ suggests a way to define
parallelism in a general n-space.

DEFINITION.   *Two vectors A and B in n-space are said to have the same direction if*
$B = cA$ *for some positive scalar c, and the opposite direction if* $B = cA$ *for some negative c.*
They are called parallel if $B = cA$ *for some nonzero c.*

Note that this definition makes every vector have the same direction as itself, a property
that we surely want. This definition also ascribes the following properties to the zero vector:
The zero vector is the only vector having the same direction as its negative and, therefore,
the only vector having the opposite direction to itself. The zero vector is the only vector
parallel to the zero vector.

1.4   Exercises

1. Let $A = (1, 3, 6)$, $B = (4, -3, 3)$ and $C = (2, 1, 5)$ be three vectors in $\mathbf{R}^3$. Determine the
   components of each of the following vectors:
   (a) $A + B$.   (c) $A + B - C$.   (e) $2A + B - 3C$.
   (b) $A - B$.   (d) $7A - 2B - 3C$.
2. Make a sketch showing the geometric vectors from the origin to points $A = (2, 1)$ and $B = (1, 3)$.
   On the same sketch, show the geometric vector from the origin to point $C = A + tB$ for each of
   the following values of t: $t = \frac{1}{3}$; $t = \frac{1}{2}$; $t = 1$; $t = 2$; $t = -1$; $t = -2$.
3. Solve Exercise 2 if $C = tA + B$.
4. Let $A = (2, 1)$, $B = (1, 3)$, and $C = xA + yB$, where x and y denote scalars.
   (a) Draw the geometric vector from the origin to C for each of the following pairs of values of x
       and y: $x = y = \frac{1}{2}$; $x = \frac{1}{4}$, $y = \frac{3}{4}$; $x = \frac{1}{3}$, $y = \frac{2}{3}$; $x = 2$, $y = -1$; $x = 3$, $y = -2$; $x = -\frac{1}{2}$,
       $y = \frac{3}{2}$; $x = -1$, $y = 2$.
   (b) Based on the results of part (a), make a sketch for what you think is the set of points C
       obtained as x and y run through all real numbers such that $x + y = 1$.
   (c) Make a sketch for what you think is the set of all points C obtained as x and y range
       independently over the intervals $0 \leq x \leq 1$, $0 \leq y \leq 1$.
   (d) Solve part (c) if x ranges over the interval $0 \leq x \leq 1$ and y ranges over all real numbers.
   (e) Solve part (c) if both x and y range over all real numbers.

5. Let $A = (2, 1)$ and $B = (1, 3)$. Show that every vector $C = (c_1, c_2)$ in $\mathbf{R}^2$ can be expressed in the form $C = xA + yB$. Express x and y in terms of c_1 and c_2.

6. Let $A = (1, 1, 1)$, $B = (0, 1, 1)$, $C = (1, 1, 0)$ be three vectors in $\mathbf{R}^3$ and let $D = xA + yB + zC$, where x, y, z are scalars.
   (a) Determine the components of D in terms of x, y, z.
   (b) If $D = O$, prove that $x = y = z = 0$.
   (c) Find x, y, z so that $D = (1, 2, 3)$.

7. Let $A = (1, 1, 1)$, $B = (0, 1, 1)$, $C = (2, 1, 1)$ be three vectors in $\mathbf{R}^3$ and let $D = xA + yB + zC$, where x, y, z are scalars.
   (a) Determine the components of D in terms of x, y, z.
   (b) Find x, y, z, not all zero, such that $D = O$.
   (c) Prove that no choice of x, y, z makes $D = (1, 2, 3)$.

8. Given vectors $A = (1, 1, 1, 0)$, $B = (0, 1, 1, 1)$, $C = (1, 1, 0, 0)$ in $\mathbf{R}^4$, let $D = xA + yB + zC$, where x, y, z are scalars
   (a) Determine the components of D in terms of x, y, z.
   (b) If $D = O$, prove that $x = y = z = 0$.
   (c) Find x, y, z so that $D = (1, 5, 3, 4)$.
   (d) Prove that no choice of x, y, z makes $D = (1, 2, 3, 4)$.

9. In $\mathbf{R}^n$ prove that two vectors parallel to the same vector are parallel to each other.

10. Given four nonzero vectors A, B, C, D in $\mathbf{R}^n$ such that $C = A + B$ and A is parallel to D. Prove that C is parallel to D if and only if B is parallel to D.

11. (a) For vectors in $\mathbf{R}^n$ prove the properties of addition and multiplication by scalars in Theorem 1.1.
    (b) Draw geometric vectors in the plane to illustrate the geometric meaning of the two distributive laws $c(A + B) = cA + cB$ and $(c + d)A = cA + dA$.

12. If a quadrilateral $OABC$ in $\mathbf{R}^2$ is a parallelogram with A and C as opposite vertices, prove that $A + \frac{1}{2}(C - A) = \frac{1}{2}B$. What geometrical theorem about parallelograms does this imply?

1.5 The dot product

Now we introduce another multiplication called the *dot product* or *scalar product* of two vectors in *n*-space.

DEFINITION.   *If* $A = (a_1, a_2, \ldots, a_n)$ *and* $B = (b_1, b_2, \ldots, b_n)$ *are two vectors in* $\mathbf{R}^n$, *their dot product is denoted by* $A \cdot B$ *and is defined by the equation*

$$A \cdot B = \sum_{k=1}^{n} a_k b_k.$$

Thus, to compute $A \cdot B$ we multiply corresponding components of A and B and then add all the products. This multiplication has the following algebraic properties:

THEOREM 1.2.   *For all vectors A, B, C in n-space and all scalars c we have:*
(a) $A \cdot B = B \cdot A$ *(commutative law),*
(b) $A \cdot (B + C) = A \cdot B + A \cdot C$ *(distributive law),*
(c) $c(A \cdot B) = (cA) \cdot B = A \cdot (cB)$ *(homogeneity),*
(d) $A \cdot A > 0$ *if* $A \neq O$ *(positivity),*
(e) $A \cdot A = 0$ *if* $A = O$.

Proof.   The first three properties are easy consequences of the definition and are left as exercises. To prove the last two, we use the relation $A \cdot A = \sum_{k=1}^{n} a_k^2$. Since each term is nonnegative, the sum is nonnegative. Moreover, the sum is zero if and only if each term in the sum is zero, and this can happen only if $A = O$.

The dot product has an interesting geometric interpretation that will be described in Section 1.9. Before we discuss this, however, we mention an important inequality concerning dot products that is fundamental in vector algebra.

THEOREM 1.3.   THE CAUCHY-SCHWARZ INEQUALITY.   *If A and B are vectors in n-space, we have*

$$(1.2) \qquad\qquad (A \cdot B)^2 \le (A \cdot A)(B \cdot B).$$

Moreover, the equality sign holds if and only if one of the vectors is a scalar multiple of the other.

Proof.   We shall give a proof of (1.2) based on the five properties listed in Theorem 1.1. This shows that the Cauchy-Schwarz inequality is a consequence of these five properties and does not depend on the particular definition that was used to deduce these properties.

First we note that (1.2) holds trivially if either A or B is the zero vector. Therefore, we may assume that both A and B are nonzero. Let C be the vector

$$C = xA - yB, \qquad \text{where} \qquad x = B \cdot B \qquad \text{and} \qquad y = A \cdot B.$$

Properties (d) and (e) of Theorem 1.2 imply that $C \cdot C \ge 0$. When we rewrite this inequality in terms of x and y it will yield (1.2). To express $C \cdot C$ in terms of x and y we use properties (a), (b) and (c) to obtain

$$C \cdot C = (xA - yB) \cdot (xA - yB) = x^2(A \cdot A) - 2xy(A \cdot B) + y^2(B \cdot B).$$

Using the definitions of x and y and the inequality $C \cdot C \ge 0$, we get

$$(B \cdot B)^2(A \cdot A) - 2(A \cdot B)^2(B \cdot B) + (A \cdot B)^2(B \cdot B) \ge 0.$$

Property (d) implies $B \cdot B > 0$ because $B \ne O$, so we can divide by $(B \cdot B)$ in the last inequality to obtain

$$(B \cdot B)(A \cdot A) - (A \cdot B)^2 \ge 0,$$

which is (1.2). This proof also shows that the equality sign holds in (1.2) if and only if $C = O$. But $C = O$ if and only if $xA = yB$. This equation holds, in turn, if and only if one of the vectors is a scalar multiple of the other.

Expressed in terms of components, the Cauchy-Schwarz inequality takes the following form:

$$\left(\sum_{k=1}^{n} a_k b_k \right)^2 \le \left(\sum_{k=1}^{n} a_k^2 \right) \left(\sum_{k=1}^{n} b_k^2 \right).$$

The Cauchy-Schwarz inequality has important applications to the properties of the *length* or *norm* of a vector, a concept to which we turn next.

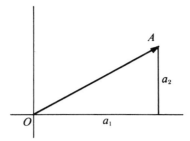

FIGURE 1.7   In $\mathbf{R}^2$ the length of A is $\sqrt{a_1^2 + a_2^2}$.

1.6   Length or norm of a vector

Figure 1.7 shows the geometric vector from the origin to a point $A = (a_1, a_2)$ in the plane. From the theorem of Pythagoras we find that the length of A is given by the formula

$$\text{length of } A = \sqrt{a_1^2 + a_2^2}.$$

A corresponding picture in 3-space is shown in Figure 1.8. Applying the theorem of Pythagoras twice, we find that the length of a geometric vector A in 3-space is given by

$$\text{length of } A = \sqrt{a_1^2 + a_2^2 + a_3^2}.$$

Note that in both cases the length of A is given by $(A \cdot A)^{1/2}$, the square root of the dot product of A with itself. This formula suggest a way to introduce the concept of length in n-space.

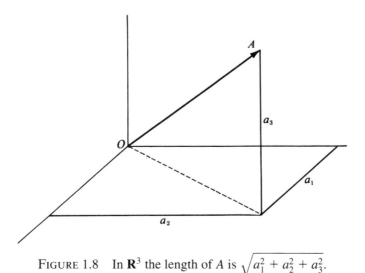

FIGURE 1.8   In $\mathbf{R}^3$ the length of A is $\sqrt{a_1^2 + a_2^2 + a_3^2}$.

DEFINITION. *If A is a vector in n-space, its length or norm is denoted by $\|A\|$ and is defined by the equation*

$$\|A\| = (A \cdot A)^{1/2}.$$

The fundamental properties of the dot product lead to corresponding properties of norms.

THEOREM 1.4. *For every vector A in n-space and every scalar c we have the following properties:*

(a) $\|A\| > 0$ *if* $A \neq O$ *(positivity),*
(b) $\|A\| = 0$ *if* $A = O$,
(c) $\|cA\| = |c| \|A\|$ *(homogeneity).*

Proof. Properties (a) and (b) follow at once from properties (d) and (e) of Theorem 1.1. To prove (c) we use the homogeneity property of dot products to obtain

$$\|cA\| = (cA \cdot cA)^{1/2} = (c^2 A \cdot A)^{1/2} = (c^2)^{1/2}(A \cdot A)^{1/2} = |c| \|A\|.$$

The Cauchy-Schwarz inequality can also be expressed in terms of norms. It states that

$$(1.3) \qquad\qquad (A \cdot B)^2 \leq \|A\|^2 \|B\|^2.$$

Taking the positive square root of each member, we can also write it in the equivalent form

$$(1.4) \qquad\qquad |A \cdot B| \leq \|A\| \|B\|.$$

Now we shall use the Cauchy-Schwarz inequality to deduce the triangle inequality.

THEOREM 1.5. TRIANGLE INEQUALITY. *For any two vectors A and B in n-space we have*

$$\|A + B\| \leq \|A\| + \|B\|.$$

Moreover, the equality sign holds if and only if $A = O$, or $B = O$, or $B = cA$ for some $c > 0$.

Note. The triangle inequality is illustrated geometrically in Figure 1.9. It states that the length of one side of a triangle does not exceed the sum of the lengths of the other two sides.

Proof. To avoid square roots, we write the triangle inequality in the equivalent form

$$(1.5) \qquad\qquad \|A + B\|^2 \leq \left(\|A\| + \|B\|\right)^2.$$

The left member is

$$\|A + B\|^2 = (A + B) \cdot (A + B) = A \cdot A + 2A \cdot B + B \cdot B = \|A\|^2 + 2A \cdot B + \|B\|^2,$$

whereas the right member is

$$\left(\|A\| + \|B\|\right)^2 = \|A\|^2 + 2\|A\| \|B\| + \|B\|^2.$$

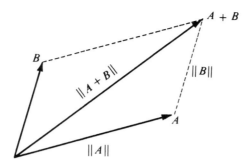

FIGURE 1.9 Geometric meaning of the triangle inequality $\|A + B\| \leq \|A\| + \|B\|$.

Comparing these two formulas, we see that (1.5) holds if and only if we have

(1.6) $$A \cdot B \leq \|A\|\|B\|.$$

But $A \cdot B \leq |A \cdot B|$ so (1.6) follows from the Cauchy-Schwarz inequality as expressed in (1.4). This proves that the triangle inequality is a consequence of the Cauchy-Schwarz inequality.

The converse statement is also true. That is, if the triangle inequality holds, then (1.6) also holds for A and for $-A$, from which we obtain (1.3).

If equality holds in (1.5), then $A \cdot B = \|A\|\|B\|$, so $B = cA$ for some scalar c. Hence $A \cdot B = c\|A\|^2$ and $\|A\|\|B\| = |c|\|A\|^2$. If $A \neq O$, this implies $c = |c|$ so $c \geq 0$. If $B \neq O$ then $B = cA$ with $c > 0$.

1.7 Orthogonality of vectors

In the course of the proof of the triangle inequality (Theorem 1.5) we obtained the formula

(1.7) $$\|A + B\|^2 = \|A\|^2 + \|B\|^2 + 2A \cdot B$$

which is valid for any two vectors A and B in n-space. Figure 1.10 shows two perpendicular geometric vectors in the plane. They determine a right triangle whose legs have lengths $\|A\|$ and $\|B\|$ and whose hypotenuse has length $\|A + B\|$.

Because A and B are perpendicular, the theorem of Pythagoras tells us that

$$\|A + B\|^2 = \|A\|^2 + \|B\|^2.$$

Comparing this with (1.7) we see that $A \cdot B = 0$. In other words, for vectors in $\mathbf{R}^2$, the dot product of two perpendicular vectors is zero. This property motivates the definition of perpendicularity of vectors in n-space.

DEFINITION. *Two vectors A and B in $\mathbf{R}^n$ are called perpendicular or orthogonal if* $A \cdot B = 0.$

Equation (1.7) shows that two vectors A and B in $\mathbf{R}^n$ are orthogonal if and only if $\|A + B\|^2 = \|A\|^2 + \|B\|^2$. This is called the Pythagorean identity in $\mathbf{R}^n$.

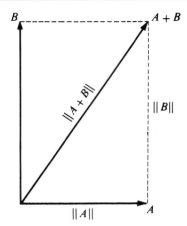

FIGURE 1.10   Two perpendicular vectors satisfy the Pythagorean identity
$\|A + B\|^2 = \|A\|^2 + \|B\|^2$.

1.8   Exercises

1. Let $A = (1, 2, 3, 4)$, $B = (-1, 2, -3, 0)$, $C = (0, 1, 0, 1)$ be three vectors in $\mathbf{R}^4$. Compute each of the following dot products:
   - (a) $A \cdot B$
   - (b) $B \cdot C$
   - (c) $A \cdot C$
   - (d) $A \cdot (B + C)$
   - (e) $(A - B) \cdot C$.

2. Given three vectors $A = (2, 4, -7)$, $B = (2, 6, 3)$, $C = (3, 4, -5)$. In each of the following there is only one way to insert parentheses to obtain a meaningful expression. Insert parentheses and perform the indicated operations.
   - (a) $A \cdot BC$
   - (b) $A \cdot B + C$
   - (c) $A + B \cdot C$
   - (d) $AB \cdot C$
   - (e) $A/B \cdot C$.

3. Prove or disprove the following statement about vectors in $\mathbf{R}^n$: *If $A \cdot B = A \cdot C$ and $A \neq O$, then $B = C$.*

4. Prove or disprove the following statement about vectors in $\mathbf{R}^n$: *If $A \cdot B = 0$ for every B, then $A = O$.*

5. If $A = (2, 1, -1)$ and $B = (1, -1, 2)$, find a nonzero C in $\mathbf{R}^3$ such that $A \cdot C = B \cdot C = 0$.

6. If $A = (1, -2, 3)$ and $B = (3, 1, 2)$, find scalars x and y such that $xA + yB$ is a nonzero vector C with $C \cdot B = 0$.

7. If $A = (2, -1, 2)$ and $B = (1, 2, -2)$, find two vectors C and D in $\mathbf{R}^3$ satisfying all the following conditions: $A = C + D$, $B \cdot D = 0$, C is parallel to B.

8. If $A = (1, 2, 3, 4, 5)$ and $B = \left(1, \frac{1}{2}, \frac{1}{3}, \frac{1}{4}, \frac{1}{5}\right)$, find two vectors C and D in $\mathbf{R}^5$ satisfying all the following conditions: $B = C + 2D$, $D \cdot A = 0$, C is parallel to A.

9. Let $A = (2, -1, 5)$, $B = (-1, -2, 3)$, $C = (1, -1, 1)$ be three vectors in $\mathbf{R}^3$. Calculate the norm of each of the following vectors:
   - (a) $A + B$
   - (b) $A - B$
   - (c) $A + B - C$
   - (d) $A - B + C$.

10. In each case find a vector B in $\mathbf{R}^2$ such that $B \cdot A = 0$ and $\|A\| = \|B\|$ if:
    - (a) $A = (1, 1)$
    - (b) $A = (1, -1)$
    - (c) $A = (2, -3)$
    - (d) $A = (a, b)$.

11. Let $A = (1, -2, 3)$ and $B = (3, 1, 2)$ be two vectors in $\mathbf{R}^3$. In each case, find a vector C of length 1 parallel to:
    - (a) $A + B$
    - (b) $A - B$
    - (c) $A + 2B$
    - (d) $A - 2B$
    - (e) $2A - B$.

12. Let $A = (4, 1, -3)$, $B = (1, 2, 2)$, $C = (1, 2, -2)$, $D = (2, 1, 2)$, $E = (2, -2, -1)$ be vectors in $\mathbf{R}^3$. Determine all orthogonal pairs.

13. Find all vectors in $\mathbf{R}^2$ that are orthogonal to A and have the same length as A if:
    (a) $A = (1, 2)$ (c) $A = (2, -1)$
    (b) $A = (1, -2)$ (d) $A = (-2, 1)$.

14. If $A = (2, -1, 1)$ and $B = (3, -4, -4)$, find a point C in 3-space such that A, B, C are the vertices of a right triangle.

15. If $A = (1, -1, 2)$ and $B = (2, 1, -1)$, find a nonzero C in $\mathbf{R}^3$ orthogonal to A and B.

16. Let $A = (1, 2)$ and $B = (3, 4)$ be two vectors in $\mathbf{R}^2$. Find vectors P and Q in $\mathbf{R}^2$ such that $A = P + Q$, P is parallel to B, and Q is orthogonal to B.

17. Solve Exercise 16 if the vectors are in $\mathbf{R}^4$, with $A = (1, 2, 3, 4)$ and $B = (1, 1, 1, 1)$.

18. Given vectors $A = (2, -1, 1)$, $B = (1, 2, -1)$, $C = (1, 1, -2)$ in $\mathbf{R}^3$. Find every vector D of the form $xB + yC$ that is orthogonal to A and has length 1.

19. Prove that for any two vectors A and B in $\mathbf{R}^n$ we have the identity

$$\|A + B\|^2 - \|A - B\|^2 = 4A \cdot B.$$

This implies that $A \cdot B = 0$ if and only if $\|A + B\| = \|A - B\|$. Interpreted geometrically in $\mathbf{R}^2$, it states that the diagonals of a parallelogram are of equal length if and only if the parallelogram is a rectangle.

20. Prove that for any two vectors A and B in $\mathbf{R}^n$ we have

$$\|A + B\|^2 + \|A - B\|^2 = 2\|A\|^2 + 2\|B\|^2.$$

What geometric theorem about the sides and diagonals of a parallelogram does this identity imply?

21. The following theorem from geometry suggests a vector identity involving three vectors A, B, C. Guess the identity and prove that it holds for vectors in $\mathbf{R}^n$.

    THEOREM.   *The sum of the squares of the sides of any quadrilateral exceeds the sum of the squares of the diagonals by four times the square of the length of the line segment that connects the midpoints of the diagonals.*

22. A vector A in $\mathbf{R}^n$ has length 6. A vector B in $\mathbf{R}^n$ has the property that for every pair of scalars x and y the vectors $xA + yB$ and $4yA - 9xB$ are orthogonal. Compute the length of B and of $2A + 3B$.

23. Given two vectors $A = (1, 2, 3, 4, 5)$ and $B = (1, \frac{1}{2}, \frac{1}{3}, \frac{1}{4}, \frac{1}{5})$ in $\mathbf{R}^5$. Find two vectors C and D in $\mathbf{R}^5$ satisfying the following three conditions: C is parallel to A, D is orthogonal to A, and $C + D = B$.

24. Given two vectors A and B in $\mathbf{R}^n$ with $A \neq O$, prove that there exist vectors C and D in $\mathbf{R}^n$ satisfying the three conditions in Exercise 23, and express C and D in terms of A and B.

25. Prove or disprove each of the following statements concerning vectors in $\mathbf{R}^n$:
    (a) If A is orthogonal to B, then $\|A + xB\| \geq \|A\|$ for all real x.
    (b) If $\|A + xB\| \geq \|A\|$ for all real x, then A is orthogonal to B.

1.9 Projections. Angle between vectors in n-space

The dot product of two vectors in the plane has an interesting geometric interpretation. Figure 1.11(a) shows two nonzero geometric vectors A and B making an angle θ with each other. In this example we have $0 < \theta \leq \frac{1}{2}\pi$. Figure 1.11(b) shows the same vector A and two perpendicular vectors whose sum is A. One of these, tB, is a scalar multiple of B that we call the *projection of A along B*. In this example, t is positive because $0 < \theta \leq \frac{1}{2}\pi$.

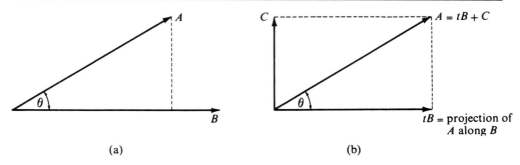

(a)   (b)

FIGURE 1.11   The vector tB is the projection of A along B.

We can use dot products to express t in terms of A and B. First we write $tB + C = A$ and then take the dot product of each member with B to obtain

$$tB \cdot B + C \cdot B = A \cdot B.$$

But $C \cdot B = 0$, because C was drawn perpendicular to B. Therefore $tB \cdot B = A \cdot B$, so we have

(1.8) $$t = \frac{A \cdot B}{B \cdot B} = \frac{A \cdot B}{\|B\|^2}.$$

On the other hand, the scalar t bears a simple relation to the angle θ. From Figure 1.11(b) we see that

$$\cos \theta = \frac{\|tB\|}{\|A\|} = \frac{t\|B\|}{\|A\|}.$$

Using (1.8) in this formula, we find that

(1.9) $$\cos \theta = \frac{A \cdot B}{\|A\|\|B\|}.$$

or

$$A \cdot B = \|A\|\|B\| \cos \theta.$$

In other words, the dot product of two nonzero vectors A and B in $\mathbf{R}^2$ is equal to the product of three numbers: the length of A, the length of B, and the cosine of the angle between A and B.

Equation (1.9) suggests a way to define the concept of angle in n-space. The Cauchy-Schwarz inequality, as expressed in (1.4), shows that the quotient on the right of (1.9) has absolute value ≤ 1 for any two nonzero vectors in $\mathbf{R}^n$. In other words, we have

$$-1 \leq \frac{A \cdot B}{\|A\|\|B\|} \leq 1.$$

Therefore, there is exactly one real θ in the interval $0 \leq \theta \leq \pi$ such that (1.9) holds. We define the angle between A and B to be this θ. The foregoing discussion is summarized in the following definition.

DEFINITION. *Let A and B be two vectors in $\mathbf{R}^n$, with $B \neq O$. The vector tB, where*

$$t = \frac{A \cdot B}{B \cdot B},$$

is called the projection of A along B. If both A and B are nonzero, the angle θ between A and B is defined by the equation

$$\theta = \arccos \frac{A \cdot B}{B \cdot B}.$$

Note. The definition of the arc cosine function restricts θ to the interval $0 \leq \theta \leq \pi$.

1.10 The unit coordinate vectors

In Chapter 0 we learned that every complex number (a, b) can be expressed in the form $a + bi$, where i denotes the complex number $(0, 1)$. Similarly, every vector (a, b) in $\mathbf{R}^2$ can be expressed in the form

$$(a, b) = a(1, 0) + b(0, 1).$$

The two vectors $(1, 0)$ and $(0, 1)$ that multiply the components a and b are called *unit coordinate vectors*. We now introduce the corresponding concept in $\mathbf{R}^n$.

DEFINITION. *The n vectors $I_1 = (1, 0, \ldots, 0)$, $I_2 = (0, 1, 0, \ldots, 0), \ldots, I_n = (0, 0, \ldots, 1)$ in $\mathbf{R}^n$ are called the unit coordinate vectors. It is understood that the kth component of I_k is 1 and all other components are 0.*

They are called *unit* vectors because each has length 1. Note that these vectors are mutually orthogonal; that is, the dot product of any two distinct unit coordinate vectors is zero:

$$I_k \cdot I_j = 0 \qquad \text{if } k \neq j.$$

THEOREM 1.6. *Every vector $X = (x_1, \ldots, x_n)$ in $\mathbf{R}^n$ can be expressed in the form*

$$X = x_1 I_1 + \cdots + x_n I_n = \sum_{k=1}^{n} x_k I_k.$$

Moreover, this representation is unique. That is, if

$$X = \sum_{k=1}^{n} x_k I_k \qquad \text{and} \qquad X = \sum_{k=1}^{n} y_k I_k$$

then $x_k = y_k$ for each $k = 1, 2, \ldots, n$.

Proof. The first statement follows immediately from the definition of addition and multiplication by scalars. The uniqueness property follows from the definition of vector equality.

A sum of the type $\sum_{k=1}^{n} c_k A_k$ is called a *linear combination* of the vectors $A_1, \ldots, A_n$. Theorem 1.6 tells us that every vector in $\mathbf{R}^n$ can be expressed as a linear combination of the unit coordinate vectors. We describe this by saying that the unit coordinate vectors *span* the space $\mathbf{R}^n$. We also say that they span $\mathbf{R}^n$ *uniquely* because each representation of a vector as a linear combination of $I_1, \ldots, I_n$ is unique. Some collections of vectors other than the unit coordinate vectors also span $\mathbf{R}^n$ uniquely, and in Section 1.12 we turn to the study of such collections.

In 2-space the unit coordinate vectors I_1 and I_2 are often denoted, respectively, by the symbols i and j in bold-face italic type. In 3-space the symbols i, j, and k are also used in place of I_1, I_2 and I_3. Sometimes a bar or arrow is placed over the symbol. The geometric meaning of Theorem 1.6 is illustrated in Figure 1.12 for $n = 3$.

When vectors are expressed as linear combinations of the unit coordinate vectors, algebraic manipulations can be performed by treating the sums $\sum_{k=1}^{n} x_k I_k$ according to the usual rules of algebra. The various components can be recognized at any stage in the calculation by collecting the coefficients of the unit coordinate vectors. For example, to add two vectors, say $A = (a_1, \ldots, a_n)$ and $B = (b_1, \ldots, b_n)$ we write

$$A = \sum_{k=1}^{n} a_k I_k, \qquad B = \sum_{k=1}^{n} b_k I_k,$$

and apply the linearity property of finite sums to obtain

$$A + B = \sum_{k=1}^{n} a_k I_k + \sum_{k=1}^{n} b_k I_k = \sum_{k=1}^{n} (a_k + b_k) I_k.$$

The coefficient of I_k on the right is the kth component of the sum $A + B$.

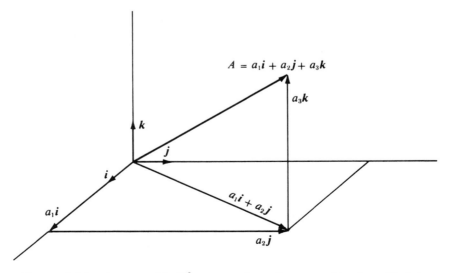

FIGURE 1.12   A vector A in $\mathbf{R}^3$ expressed as a linear combination of i, j, k.

1.11 Exercises

1. Determine the projection of A along B if $A = (1, 2, 3)$ and $B = (1, 2, 2)$.
2. Determine the projection of A along B if $A = (4, 3, 2, 1)$ and $B = (1, 1, 1, 1)$.
3. (a) Let $A = (6, 3, -2)$, and let a, b, c denote the angles between A and the unit coordinate vectors i, j, k, respectively. Compute $\cos a$, $\cos b$, and $\cos c$. These are called the *direction cosines* of A.
   (b) Find all vectors in $\mathbf{R}^3$ of length 1 parallel to A.
4. Prove that the angle between the two vectors $A = (1, 2, 1)$ and $B = (2, 1, -1)$ is twice that between $C = (1, 4, 1)$ and $D = (2, 5, 5)$.
5. Use vector methods to determine the cosines of the angles of the triangle in 3-space, whose vertices are at the points $(2, -1, 1)$, $(1, -3, -5)$, and $(3, -4, -4)$.
6. Three vectors A, B, C in $\mathbf{R}^3$ satisfy the following properties: $\|A\| = \|C\| = 5$, $\|B\| = 1$, $\|A - B + C\| = \|A + B + C\|$. If the angle between A and B is $\pi/8$, find the angle between B and C.
7. Given three nonzero vectors A, B, C in $\mathbf{R}^n$. Assume that the angle between A and C is equal to the angle between B and C. Prove that C is orthogonal to the vector $\|B\|A - \|A\|B$.
8. Let θ denote the angle between the following two vectors in $\mathbf{R}^n$: $A = (1, 1, \dots, 1)$ and $B = (1, 2, \dots, n)$. Find the limiting value of θ as $n \to \infty$.
9. Solve Exercise 8 if $A = (2, 4, 6, \dots, 2n)$ and $B = (1, 3, 5, \dots, 2n - 1)$.
10. Given vectors $A = (\cos\theta, -\sin\theta)$ and $B = (\sin\theta, \cos\theta)$ in $\mathbf{R}^2$.
    (a) Prove that A and B are orthogonal unit vectors. Make a sketch showing A and B when $\theta = \pi/6$.
    (b) Find all vectors (x, y) in $\mathbf{R}^2$ such that $(x, y) = xA + yB$. Be sure to consider all values of θ.
11. Use vector methods to prove that the diagonals of a rhombus are perpendicular.
12. By forming the dot product of the two vectors $(\cos a, \sin a)$ and $(\cos b, \sin b)$, deduce the trigonometric identity $\cos(a - b) = \cos a \cos b + \sin a \sin b$.
13. If θ is the angle between two nonzero vectors A and B in $\mathbf{R}^n$, prove that

$$\|A - B\|^2 = \|A\|^2 + \|B\|^2 - 2\|A\|\|B\| \cos\theta.$$

When interpreted geometrically in $\mathbf{R}^2$, this is the law of cosines of trigonometry.

14. Suppose that instead of defining the dot product of two vectors $A = (a_1, \dots, a_n)$ and $B = (b_1, \dots, b_n)$ by the formula $A \cdot B = \sum_{k=1}^{n} a_k b_k$, we used the following definition:

$$A \cdot B = \sum_{k=1}^{n} |a_k b_k|.$$

Which of the properties in Theorem 1.2 are valid with this definition? Is the Cauchy-Schwarz inequality still valid?

15. Suppose that in $\mathbf{R}^2$ we define the dot product of two vectors $A = (a_1, a_2)$ and $B = (b_1, b_2)$ by the formula

$$A \cdot B = 2a_1 b_1 + a_2 b_2 + a_1 b_2 + a_2 b_1.$$

Prove that all the properties of Theorem 1.2 are valid with this definition of dot product. Is the Cauchy-Schwarz inequality still valid?

16. Solve Exercise 15 if the dot product of two vectors $A = (a_1, a_2, a_3)$ and $B = (b_1, b_2, b_3)$ in $\mathbf{R}^3$ is defined by the formula $A \cdot B = 2a_1 b_1 + a_2 b_2 + a_3 b_3 + a_1 b_3 + a_3 b_1$.
17. Suppose that instead of defining the norm of a vector $A = (a_1, \dots, a_n)$ by the formula $\|A\| = (A \cdot A)^{1/2}$, we used the following definition:

$$\|A\| = \sum_{k=1}^{n} |a_k|.$$

(a) Prove that this definition of norm satisfies all the properties in Theorems 1.4 and 1.5.
(b) Use this definition in $\mathbf{R}^2$ and describe on a figure the set of all points (x, y) of norm 1.

(c) State properties of Theorems 1.4 and 1.5 that would hold if we used the definition

$$\|A\| = \left| \sum_{k=1}^{n} a_k \right|.$$

18. Suppose that the norm of a vector $A = (a_1, \ldots, a_n)$ was defined by the formula $\|A\| = \max_{1 \le k \le n} |a_k|$, where the symbol on the right means the maximum of the n numbers $|a_1|$, $|a_2|, \ldots, |a_n|$.
    (a) Which properties of Theorems 1.4 and 1.5 are valid with this definition?
    (b) Use this definition in $\mathbf{R}^2$ and describe on a figure the set of all (x, y) of norm 1.
19. If $A = (a_1, \ldots, a_n)$ is a vector in $\mathbf{R}^n$, define two new norms as follows:

$$\|A\|_1 = \sum_{k=1}^{n} |a_k| \quad \text{and} \quad \|A\|_2 = \max_{1 \le k \le n} |a_k|.$$

Prove that $\|A\|_2 \le \|A\| \le \|A\|_1$. Interpret this inequality geometrically in the plane.
20. If A and B are two points in n-space, the distance from A to B is denoted by $d(A, B)$ and is defined by the equation $d(A, B) = \|A - B\|$. Prove that distance has the following properties:
    (a) $d(A, B) = d(B, A)$.
    (b) $d(A, B) = 0$ if and only if $A = B$.
    (c) $d(A, B) \le d(A, C) + d(C, B)$.

1.12 The linear span of a finite set of vectors

Let $S = \{A_1, \ldots, A_k\}$ be a nonempty set consisting of k vectors in $\mathbf{R}^n$, where k, the number of vectors, may be less than, equal to, or greater than n, the dimension of the space. If a vector X in $\mathbf{R}^n$ can be expressed as a linear combination of $A_1, \ldots, A_k$, say

$$X = \sum_{i=1}^{k} c_i A_i,$$

then the set S is said to *span* the vector X.

DEFINITION. *The set of all vectors spanned by S is called the linear span of S and is denoted by $L(S)$.*

In other words, the linear span of S is simply the set of all possible linear combinations of vectors in S obtained by letting each coefficient c_i run through all possible scalars. Note that linear combinations of vectors in $L(S)$ are again in $L(S)$. If $L(S) = \mathbf{R}^n$ we say that S *spans the whole space* $\mathbf{R}^n$.

EXAMPLE 1. Let $S = \{A_1\}$. Then $L(S)$ consists of all scalar multiples of A_1.

EXAMPLE 2. A linear combination $\sum_{i=1}^{k} c_i A_i$ with each scalar $c_i = 0$ produces the zero vector O. Therefore *every* nonempty set spans O. This representation of O with all multipliers equal to 0 is called the *trivial representation* of the zero vector. There may be nontrivial linear combinations that represent O. For example, suppose one of the vectors in S is a scalar multiple of another, say $A_2 = 2A_1$. Then there are many nontrivial representations of O, for example,

$$2tA_1 - tA_2 + 0A_3 + \cdots + 0A_k = O,$$

for any scalar t.

We are especially interested in sets S that span vectors in exactly one way.

DEFINITION.   *Let $S = \{A_1, \ldots, A_k\}$ be a set of vectors in $\mathbf{R}^n$ that spans a vector X. We say that S spans X uniquely if*

$$(1.10) \quad X = \sum_{i=1}^{k} c_i A_i \quad and \quad X = \sum_{i=1}^{k} d_i A_i \quad implies\ c_i = d_i \quad for\ all\ i.$$

In the two sums appearing in (1.10) it is understood that the vectors $A_1, \ldots, A_k$ are written in the same order. It is also understood that the implication (1.10) is to hold for a fixed but arbitrary ordering of the vectors $A_1, \ldots, A_k$.

THEOREM 1.7.   *A set S spans every vector in L(S) uniquely if and only if S spans the zero vector uniquely.*

Proof.   If S spans every vector in $L(S)$ uniquely, then it certainly spans O uniquely. To prove the converse, assume S spans O uniquely, and choose any vector X in $L(S)$. Suppose S spans X in two ways, say,

$$X = \sum_{i=1}^{k} c_i A_i \quad and \quad X = \sum_{i=1}^{k} d_i A_i.$$

By subtraction we find that $O = \sum_{i=1}^{k}(c_i - d_i)A_i$. But since S spans O uniquely, each multiplier $c_i - d_i$ must be 0. Therefore $c_i = d_i$ for all i, so S spans X uniquely.

1.13   Linear independence

Theorem 1.7 demonstrates the importance of sets that span the zero vector uniquely. Such sets are given a special name.

DEFINITION.   *A set $S = \{A_1, \ldots, A_k\}$ of vectors in $\mathbf{R}^n$ that spans the zero vector uniquely is said to be a linearly independent set of vectors. Otherwise, S is called linearly dependent.*

In other words, *independence* means that S spans O with only the trivial representation; hence

$$\sum_{i=1}^{k} c_i A_i = O \quad implies\ all\ c_i = 0.$$

Dependence means that S spans O in some nontrivial way. That is, $\sum_{i=1}^{k} c_i A_i = O$ for some choice of scalars $c_1, \ldots, c_k$, not all 0.

Although dependence and independence are properties of sets of vectors, it is common practice to also apply these terms to the vectors themselves. For example, the vectors in a linearly independent set are often called *linearly independent vectors*. We also agree to call the empty set linearly independent.

The following examples may give further insight into the meaning of dependence and independence.

EXAMPLE 1. If a subset T of a set S is dependent, then S itself is dependent, because if T spans O nontrivially, then so does S. This is logically equivalent to the statement that every subset of an independent set is independent.

EXAMPLE 2. The n unit coordinate vectors in $\mathbf{R}^n$ span O uniquely, so they are linearly independent.

EXAMPLE 3. Any set containing the zero vector is dependent. To see why, multiply the zero vector by 1 and all other vectors in the set by 0 to get a nontrivial representation of O.

EXAMPLE 4. In $\mathbf{R}^2$ the set of vectors $S = \{i, j, i + j\}$ is linearly dependent because we have the following nontrivial representation of the zero vector:

$$O = i + j + (-1)(i + j).$$

In this example, we have three vectors in a space spanned by i and j. As the next theorem shows, *any* set of three vectors in a space spanned by two vectors must be dependent.

THEOREM 1.8. *Let $S = \{A_1, \ldots, A_k\}$ be a linearly independent set of k vectors in $\mathbf{R}^n$, and let $L(S)$ be the linear span of S. Then every set of $k + 1$ vectors in $L(S)$ is linearly dependent.*

Proof. The proof is by induction on k, the number of vectors in S. First suppose $k = 1$, so that S consists of one vector A_1, which is not zero since S is independent. Now take any two distinct vectors B_1 and B_2 in $L(S)$. Each is a scalar multiple of A_1, say $B_1 = c_1 A_1$ and $B_2 = c_2 A_2$, where not both c_1, c_2 are zero. Multiply B_1 by c_2 and B_2 by c_1 and subtract to get

$$c_2 B_1 - c_1 B_2 = O,$$

a nontrivial representation of O. Hence B_1 and B_2 are dependent, which proves the theorem when $k = 1$.

Now we assume the theorem is true for $k - 1$ and prove that it is also true for k. Take any set T of $k + 1$ vectors in $L(S)$, say $T = \{B_1, B_2, \ldots, B_{k+1}\}$. We will prove that T is linearly dependent. Each B_i is in $L(S)$ so we can write it as a linear combination of elements of S, say

(1.11)
$$B_i = \sum_{j=1}^{k} a_{ij} A_j$$

for each $i = 1, 2, \ldots, k + 1$. We examine the scalars a_{i1} that multiply A_1 and split the argument into two cases, according to whether all these scalars are zero or not.

CASE 1. $a_{i1} = 0$ *for every* $i = 1, 2, \ldots, k + 1$. In this case the sum in (1.11) does not involve A_1, so each B_i is in the linear span of the set $S' = \{A_2, \ldots, A_k\}$. But S' is linearly independent and consists of $k - 1$ vectors. By the induction hypothesis, the theorem is true for $k - 1$ so the k vectors $B_1, \ldots, B_k$ in $L(S')$ are dependent, hence T is dependent. This proves the theorem in Case 1.

CASE 2. *Not all the scalars* a_{i1} *are zero.* Let us assume that $a_{11} \neq 0$. (If necessary, we can renumber the B's to achieve this.) Take $i = 1$ in Eq. (1.11) and multiply both members by the scalar $c_i = a_{i1}/a_{11}$ to get

$$c_i B_1 = a_{i1} A_1 + \sum_{j=2}^{k} c_i a_{1j} A_j.$$

When we subtract Eq. (1.11) from this last equation, the terms involving A_1 cancel and we get

$$c_i B_1 - B_i = \sum_{j=2}^{k} (c_i a_{1j} - a_{ij}) A_j,$$

for $i = 2, \ldots, k + 1$. This equation expresses each of the k vectors $c_i B_1 - B_i$ as a linear combination of $k - 1$ linearly independent vectors $A_2, \ldots, A_k$. By the induction hypothesis, the k vectors $c_i B_1 - B_i$ must be dependent. Hence, for some choice of scalars $t_2, \ldots, t_{k+1}$, not all zero, we have

$$\sum_{i=2}^{k+1} t_i (c_i B_1 - B_i) = O,$$

from which we find

$$\left(\sum_{i=2}^{k+1} t_i c_i \right) B_1 - \sum_{i=2}^{k+1} t_i B_i = O.$$

But this is a nontrivial linear combination of $B_1, \ldots, B_{k+1}$ that represents the zero vector, so the vectors $B_1, \ldots, B_{k+1}$ must be dependent. This completes the proof.

Next we relate orthogonality to linear independence.

DEFINITION. *A set* $S = \{A_1, \ldots, A_k\}$ *of vectors in* $\mathbf{R}^n$ *is called an orthogonal set if* $A_i \cdot A_j = 0$ *whenever* $A_i \neq A_j$. *In other words, any two distinct vectors in an orthogonal set are perpendicular.*

THEOREM 1.9. *Every orthogonal set* $S = \{A_1, \ldots, A_k\}$ *of nonzero vectors in* $\mathbf{R}^n$ *is linearly independent. Moreover, if S spans a vector X, say*

(1.12)
$$X = \sum_{i=1}^{k} c_i A_i,$$

then the scalar multipliers $c_1, \ldots, c_n$ are given by the formulas

(1.13)
$$c_j = \frac{X \cdot A_j}{A_j \cdot A_j} \qquad for \ j = 1, 2, \ldots, k.$$

Note. If each A_j has norm 1 these formulas simplify to $c_j = X \cdot A_j$.

Proof. First we prove that S is linearly independent. Assume a linear combination of the A's is zero, say $\sum_{i=1}^{k} c_i A_i = O$. Take the dot product of each member with A_1 and use the fact that $A_1 \cdot A_i = 0$ for each $i \neq 1$ to obtain $c_1(A_1 \cdot A_1) = 0$. But $(A_1 \cdot A_1) \neq 0$ because A_1 is nonzero, so $c_1 = 0$. Now repeat the argument with A_1 replaced by A_j, and we find that each multiplier $c_j = 0$. Therefore S spans O uniquely, so S is linearly independent.

Now suppose that S spans a vector X as indicated in (1.12). Take the dot product of each member of (1.12) with A_j as above to obtain $c_j(A_j \cdot A_j) = X \cdot A_j$. This implies Eq. (1.13).

DEFINITION. *An orthogonal set of vectors, each of which has norm 1, is called an orthonormal set.*

The unit coordinate vectors $I_1, \ldots, I_n$ provide an example of an orthonormal set.

1.14 Bases

It is natural to study sets of vectors that span every vector in $\mathbf{R}^n$ uniquely. Such sets are called *bases*.

DEFINITION. *A finite set S of vectors in $\mathbf{R}^n$ is called a basis for $\mathbf{R}^n$ if S is linearly independent and spans $\mathbf{R}^n$. If, in addition, S is orthogonal, then S is called an orthogonal basis for $\mathbf{R}^n$.*

Thus, a basis spans every vector in $\mathbf{R}^n$ uniquely. The set of unit coordinate vectors is an example of a basis. This particular basis is also an orthogonal basis. Now we prove that every basis for $\mathbf{R}^n$ contains exactly n elements.

THEOREM 1.10. *Bases for $\mathbf{R}^n$ have the following properties:*
(a) *Every basis for $\mathbf{R}^n$ contains exactly n vectors.*
(b) *Any linearly independent set of vectors in $\mathbf{R}^n$ is a subset of some basis for $\mathbf{R}^n$.*
(c) *Any set of n linearly independent vectors in $\mathbf{R}^n$ is a basis for $\mathbf{R}^n$.*

Proof. We know that the unit coordinate vectors $I_1, \ldots, I_n$ constitute one basis for $\mathbf{R}^n$. To prove (a) we will show that any two bases contain the same number of vectors.

Let S and T be two bases for $\mathbf{R}^n$, where S has k vectors and T has r vectors. If $r > k$ then T contains at least $k + 1$ vectors in $L(S)$ so, by Theorem 1.8, T must be linearly dependent, contradicting the fact that T is a basis. Therefore $r \leq k$. The same argument with S and T interchanged shows that $k \leq r$. Therefore $k = r$, so part (a) is proved.

To prove (b), let $S = \{A_1, \ldots, A_k\}$ be any linearly independent set of vectors in $\mathbf{R}^n$. If $L(S) = \mathbf{R}^n$, then S is a basis. If not, there is some vector X in $\mathbf{R}^n$ that is not in $L(S)$. Adjoin this vector to S and let $S' = \{A_1, \ldots, A_k, X\}$. We will show that S' is an independent set.

If S' was dependent, there would be scalars $c_1, \ldots, c_{k+1}$, not all zero, such that

$$\sum_{i=1}^{k} c_i a_i + c_{k+1} X = O.$$

But $c_{k+1} \neq 0$ because $A_1, \ldots, A_k$ are independent. Hence we can solve for X and find that X is in the linear span of S, contradicting the fact that X is not in $L(S)$. Therefore S' is linearly independent but contains $k + 1$ vectors. If S' is a basis for $\mathbf{R}^n$, part (b) is proved because S is a subset of S'. If S' is not a basis for $\mathbf{R}^n$ we can argue with S' as we did with S and produce a new set S'' that contains $k + 2$ vectors and is linearly independent. If S'' is a basis, part (b) is proved. If not, we repeat the process. We must arrive at a basis in a finite number of steps, otherwise we would eventually obtain an independent set with $n + 1$ vectors, contracting Theorem 1.8. Therefore part (b) is proved.

Finally, we use (a) and (b) to prove (c). Let S be any linearly independent set consisting of n vectors. By part (b), S is a subset of some basis, say B. But by (a), the basis B has exactly n elements, so $S = B$.

1.15 Exercises

1. Let i and j denote the unit coordinate vectors in $\mathbf{R}^2$. In each case find scalars x and y such that the vector $x(i - j) + y(i + j)$ is equal to:
   (a) i (b) j (c) $3i - 5j$ (d) $7i + 5j$.
2. If $A = (1, 2)$, $B = (2, -4)$, and $C = (2, -3)$ are three vectors in $\mathbf{R}^2$, find scalars x and y such that $C = xA + yB$. How many such pairs are there?
3. If $A = (2, -1, 1)$, $B = (1, 2, -1)$, and $C = (2, -11, 7)$ are three vectors in $\mathbf{R}^3$, find scalars x and y such that $C = xA + yB$.
4. Prove that Exercise 3 has no solution if C is replaced by the vector $(2, 11, 7)$.
5. Let A and B be two nonzero vectors in $\mathbf{R}^n$.
   (a) If A and B are parallel, prove that A and B are linearly dependent.
   (b) If A and B are not parallel, prove that A and B are linearly independent.
6. If (a, b) and (c, d) are two vectors in $\mathbf{R}^2$, prove that they are linearly independent if and only if $ad - bc \neq 0$.
7. Find all real t for which the two vectors $(1 + t, 1 - t)$ and $(1 - t, 1 + t)$ in $\mathbf{R}^2$ are linearly independent.
8. Let i, j, k be the unit coordinate vectors in $\mathbf{R}^3$. Prove that the four vectors $i, j, k, i + j + k$ are linearly dependent, but that any three of them are linearly independent.
9. Let i and j be the unit coordinate vectors in $\mathbf{R}^2$ and let $S = \{i, i + j\}$.
   (a) Prove that S is linearly independent.
   (b) Prove that j is in the linear span of S.
   (c) Express $3i - 4j$ as a linear combination of i and $i + j$.
   (d) Prove that $L(S) = \mathbf{R}^2$.
10. Consider the three vectors $A = i$, $B = i + j$, and $C = i + j + 3k$ in $\mathbf{R}^3$.
   (a) Prove that the set $\{A, B, C\}$ is linearly independent.
   (b) Express each of j and k as a linear combination of A, B, C.
   (c) Express $2i - 3j + 5k$ as a linear combination of A, B, C.
   (d) Prove that $\{A, B, C\}$ is a basis for $\mathbf{R}^3$.
11. Let $A = (1, 2)$, $B = (2, -4)$, $C = (2, -3)$, and $D = (1, -2)$ be four vectors in $\mathbf{R}^2$. Display all nonempty subsets of $\{A, B, C, D\}$ that are linearly independent.
12. Let $A = (1, 1, 1, 0)$, $B = (0, 1, 1, 1)$ and $C = (1, 1, 0, 0)$ be three vectors in $\mathbf{R}^4$.
   (a) Determine whether A, B, C are linearly dependent or independent.
   (b) Exhibit a nonzero vector D such that A, B, C, D are dependent.

    (c) Exhibit a vector E such that A, B, C, E are independent.

    (d) Using E in part (c), express the vector $X = (1, 2, 3, 4)$ as a linear combination of A, B, C, E.

13. (a) Prove that the vectors $(\sqrt{3}, 1, 0), (1, \sqrt{3}, 1), (0, 1, \sqrt{3})$ in $\mathbf{R}^3$ are linearly independent.

    (b) Prove that the following three are dependent: $(\sqrt{2}, 1, 0), (1, \sqrt{2}, 1), (0, 1, \sqrt{2})$.

    (c) Find all real t for which the following three vectors are dependent: $(t, 1, 0), (1, t, 1), (0, 1, t)$.

14. For each of the following sets of vectors in $\mathbf{R}^4$ find a linearly independent subset containing as many vectors as possible.

    (a) $\{(1, 0, 1, 0), (1, 1, 1, 1), (0, 1, 0, 1), (2, 0, -1, 0)\}$.

    (b) $\{(1, 1, 1, 1), (1, -1, 1, 1), (1, -1, -1, 1), (1, -1, -1, -1)\}$.

    (c) $\{(1, 1, 1, 1), (0, 1, 1, 1), (0, 0, 1, 1), (0, 0, 0, 1)\}$.

15. Given three linearly independent vectors A, B, C in $\mathbf{R}^n$. Prove or disprove each of the following:

    (a) $A + B, B + C, A + C$ are linearly independent.

    (b) $A - B, B + C, A + C$ are linearly independent.

16. (a) Prove that a set S of three vectors in $\mathbf{R}^3$ is a basis for $\mathbf{R}^3$ if and only if its linear span $L(S)$ contains the three unit coordinate vectors i, j, k.

    (b) State and prove a generalization of part (a) for $\mathbf{R}^n$.

17. Find two different bases for $\mathbf{R}^3$ containing the two vectors $(0, 1, 1)$ and $(1, 1, 1)$.

18. Find two bases for $\mathbf{R}^4$ having only the two vectors $(0, 1, 1, 1)$ and $(1, 1, 1, 1)$ in common.

19. Consider the following three sets of vectors in $\mathbf{R}^3$:

$$S = \{(1, 1, 1), (0, 1, 2), (1, 0, -1)\}, \ T = \{(2, 1, 0), (2, 0, -2)\}, \ U = \{(1, 2, 3), (1, 3, 5)\}.$$

    (a) Prove that $L(T) \subseteq L(S)$.

    (b) Determine all inclusion relations that hold among the sets $L(S), L(T)$, and $L(U)$.

20. Let A and B denote two finite subsets of vectors in $\mathbf{R}^n$, and let $L(A)$ and $L(B)$ denote their linear spans. Prove each of the following statements:

    (a) If $A \subseteq B$, then $L(A) \subseteq L(B)$.

    (b) $L(A \cap B) \subseteq L(A) \cap L(B)$.

    (c) Give an example in which $L(A \cap B) \neq L(A) \cap L(B)$.

1.16 The vector space $\mathbf{C}^n$ of n-tuples of complex numbers

In Section 1.2 the vector space $\mathbf{R}^n$ was defined to be the collection of all n-tuples of real numbers. Equality, vector addition, and multiplication by scalars were defined in terms of components as follows: If $A = (a_1, \ldots, a_n)$ and $B = (b_1, \ldots, b_n)$, then

$$A = B \text{ means } a_i = b_i \qquad \text{for each } i = 1, 2, \ldots, n,$$

$$A + B = (a_1 + b_1, \ldots, a_n + b_n), \qquad cA = (ca_1, \ldots, ca_n).$$

If all the scalars a_i, b_i and c in these relations are allowed to be *complex* numbers, the new algebraic system so obtained is called *complex n-space* and is denoted by $\mathbf{C}^n$. Here $\mathbf{C}$ is used to remind us that the scalars are complex.

Because real and complex numbers share the same properties concerning addition, multiplication and division, those theorems in this chapter that involve only vector addition and multiplication by scalars are also valid for $\mathbf{C}^n$, provided all scalars are allowed to be complex.

This extension is not made simply for the sake of generalization. Complex vector spaces arise naturally in the theory of linear differential equations and in modern quantum mechanics, so their study is of considerable importance. Fortunately, many of the theorems about $\mathbf{R}^n$ carry over without change to $\mathbf{C}^n$. However, some minor changes have to be made in those theorems involving dot products. In proving that the dot product of a nonzero vector with itself is positive, we used the fact that a sum of squares of real numbers is positive. Since a sum of squares of complex numbers can be negative, we must modify the definition of dot product if we wish to retain the positivity property. In $\mathbf{C}^n$ we use the following definition of dot product.

DEFINITION.   *If $A = (a_1, \ldots, a_n)$ and $B = (b_1, \ldots, b_n)$ are two vectors in $\mathbf{C}^n$, we define the dot product $A \cdot B$ by the formula*

$$A \cdot B = \sum_{k=1}^{n} a_k \overline{b}_k$$

where $\overline{b}_k$ is the complex conjugate of b_k.

This definition agrees with that given earlier for $\mathbf{R}^n$ because $\overline{b}_k = b_k$ when b_k is real. The basic properties of the dot product, corresponding to those in Theorem 1.2, now take the following form:

THEOREM 1.11.   *For all vectors A, B, C in $\mathbf{C}^n$ and all complex scalars c, we have*
(a) $A \cdot B = \overline{B \cdot A}$,
(b) $A \cdot (B + C) = A \cdot B + A \cdot C$,
(c) $c(A \cdot B) = (cA) \cdot B = A \cdot (\overline{c}B)$,
(d) $A \cdot A > 0$ *if $A \neq O$,*
(e) $A \cdot A = 0$ *if $A = O$.*

All these properties are easy consequences of the definition, and their proofs are left as exercises. The reader should note that conjugation takes place in property (a) when the order of the factors is reversed. Also, conjugation of the scalar multiplier occurs in property (c) when the scalar c is moved from one side of the dot to the other.

The Cauchy-Schwarz inequality now takes the form

(1.14) $|A \cdot B|^2 \leq (A \cdot A)(B \cdot B)$.

The proof is similar to that given for Theorem 1.3. We consider the vector $C = xA - yB$, where $x = B \cdot B$ and $y = A \cdot B$, and compute $C \cdot C$. The inequality $C \cdot C \geq 0$ leads to (1.14). Details are left as an exercise for the reader.

Since the dot product of a vector with itself is nonnegative, we can introduce the norm of a vector in $\mathbf{C}^n$ by the usual formula

$$\|A\| = (A \cdot A)^{1/2}.$$

The fundamental properties of norms, as stated in Theorem 1.4, are also valid without change in $\mathbf{C}^n$. The triangle inequality, $\|A + B\| \leq \|A\| + \|B\|$, also holds in $\mathbf{C}^n$.

Orthogonality of vectors in $\mathbf{C}^n$ is defined by the relation $A \cdot B = 0$. As in the real case, two vectors A and B in $\mathbf{C}^n$ are orthogonal if and only if they satisfy the Pythagorean identity,

$$\|A + B\|^2 = \|A\|^2 + \|B\|^2.$$

The concepts of linear span, linear independence, linear dependence, and basis, are defined in $\mathbf{C}^n$ exactly as in $\mathbf{R}^n$. Theorems 1.7 through 1.10 (and their proofs) are all valid without change in $\mathbf{C}^n$.

1.17 Exercises

1. Let $A = (1, i)$, $B = (i, -i)$, and $C = (2i, 1)$ be three vectors in $\mathbf{C}^2$. Compute each of the following dot products:

   (a) $A \cdot B$ (e) $(iA) \cdot (iB)$ (i) $(A - C) \cdot B$

   (b) $B \cdot A$ (f) $B \cdot C$ (j) $(A - iB) \cdot (A + iB)$.

   (c) $(iA) \cdot B$ (g) $A \cdot C$

   (d) $A \cdot (iB)$ (h) $(B + C) \cdot A$

2. If $A = (2, 1, -i)$ and $B = (i, -1, 2i)$, find a nonzero vector C in $\mathbf{C}^3$ orthogonal to both A and B.
3. Prove that for any two vectors A and B in $\mathbf{C}^n$, we have the identity

$$\|A + B\|^2 = \|A\|^2 + \|B\|^2 + A \cdot B + \overline{A \cdot B}.$$

4. Prove that for any two vectors A and B in $\mathbf{C}^n$ we have the identity

$$\|A + B\|^2 - \|A - B\|^2 = 2(A \cdot B + \overline{A \cdot B}).$$

5. Prove that for any two vectors A and B in $\mathbf{C}^n$ we have the identity

$$\|A + B\|^2 + \|A - B\|^2 = 2\|A\|^2 + 2\|B\|^2.$$

6. (a) Prove that for any two vectors A and B in $\mathbf{C}^n$ the sum $A \cdot B + \overline{A \cdot B}$ is real.
   (b) If A and B are nonzero vectors in $\mathbf{C}^n$, prove that

$$-2 \leq \frac{A \cdot B + \overline{A \cdot B}}{\|A\|\|B\|} \leq 2.$$

7. We define the angle θ between two nonzero vectors A and B in $\mathbf{C}^n$ by the equation

$$\theta = \arccos \frac{\frac{1}{2}(A \cdot B + \overline{A \cdot B})}{\|A\|\|B\|}.$$

   The inequality in Exercise 6 shows that a unique angle θ exists in the closed interval $0 \leq \theta \leq \pi$ satisfying this equation. Prove that we have

$$\|A - B\|^2 = \|A\|^2 + \|B\|^2 - 2\|A\|\|B\| \cos \theta.$$

8. Use the definition in Exercise 7 to compute the angle between the following two vectors in $\mathbf{C}^3$: $A = (1, 0, i, i, i)$, and $B = (i, i, i, 0, i)$.
9. (a) Prove that the following vectors form a basis for $\mathbf{C}^3$: $A = (1, 0, 0)$, $B = (0, i, 0)$, $C = (1, 1, i)$.
   (b) Express the vector $(5, 2 - i, 2i)$ as a linear combination of A, B, C.
10. Prove that the basis of unit coordinate vectors $I_1, \ldots, I_n$ in $\mathbf{R}^n$ is also a basis for $\mathbf{C}^n$.

2

APPLICATIONS OF VECTOR ALGEBRA
TO ANALYTIC GEOMETRY

2.1 Introduction

This chapter discusses applications of vector algebra to the study of lines, planes, and conic sections. Properties of lines and planes that are familiar in 2-space and 3-space are extended to n-space.

The study of geometry as a deductive system, as expounded by Euclid around 300 B.C.E., begins with a set of axioms or postulates that describe properties of points and lines. The concepts *point* and *line* are taken as primitive notions and remain undefined. Other concepts are defined in terms of points and lines, and theorems are systematically deduced from the axioms. Euclid listed ten axioms from which he attempted to deduce all his theorems. It has since been shown that these axioms are not adequate for the theory. For example, in the proof of his very first theorem Euclid made a tacit assumption concerning the intersection of two circles that is not covered by his axioms.

Since then, other lists of axioms have been formulated that do give all of Euclid's theorems. The most famous list was formulated by the German mathematician David Hilbert (1862–1943) in his now classic *Grundlagen der Geometrie*, published in 1899. (An English translation exists; *The Foundations of Geometry*, Open Court Publishing Co., 1947.) This work, which went through seven German editions in Hilbert's lifetime, is said to have inaugurated the abstract mathematics of the twentieth century.

Hilbert starts his treatment of plane geometry with five undefined concepts: *point*, *line*, *on* (a relation holding between a point and a line), *between* (a relation between a point and a pair of points), and *congruent* (a relation between pairs of points). He then lists fifteen axioms from which he develops all of plane Euclidean geometry. His treatment of solid geometry is based on twenty-one axioms involving six undefined concepts.

The approach in analytic geometry is somewhat different. We define concepts such as point, line, on, between, etc., but we do so in terms of real numbers, which are left undefined. The resulting mathematical structure is called an *analytic model* of Euclidean geometry. In this model, properties of real numbers are used to deduce Hilbert's axioms as theorems. We shall not attempt to describe all of Hilbert's axioms. Instead, we merely indicate how the primitive concepts can be defined in terms of numbers, and then give a few proofs to illustrate the methods of analytic geometry.

2.2   Lines in *n*-space

This section uses real numbers to define the concepts of *point, line,* and *on.* The definitions are formulated to fit our intuitive ideas about two- and three-dimensional Euclidean geometry, but they are meaningful in *n*-space for any $n \geq 1$.

A *point* is simply a vector in $\mathbf{R}^n$, that is, an ordered *n*-tuple of real numbers. We shall use the words *point* and *vector* interchangeably. The vector space $\mathbf{R}^n$ is called an analytic model of *n*-dimensional Euclidean space, or simply *Euclidean n-space.* To define *line* we employ the algebraic operations of addition and multiplication by scalars that are available in $\mathbf{R}^n$.

DEFINITION.   *Let P be a given point and A a given nonzero vector. The set of all points of the form P + tA, where t runs through all real numbers, is called a line through P parallel to A. We denote this line by L(P; A) and write*

$$L(P; A) = \{P + tA : t \ real\} \qquad or, \ more \ briefly, \qquad L(P; A) = \{P + tA\}.$$

A point Q is said to be on the line L(P; A) if Q = P + tA for some scalar t.

In the notation $L(P; A)$ the point P, written first, is on the line since it corresponds to $t = 0$. The second point, A, is called a *direction vector* for the line. The line $L(O; A)$ through the origin O is the linear span of A; it consists of all scalar multiples of A. The line through P parallel to A is obtained by adding P to each vector in the linear span of A.

Figure 2.1 shows the geometric interpretation of this definition in $\mathbf{R}^3$. Each point $P + tA$ can be visualized as the tip of a geometric vector drawn from the origin. As t varies over all the real numbers, the corresponding point $P + tA$ traces out a line through P parallel to the vector A. Figure 2.1 shows points corresponding to a few values of t on both lines $L(P; A)$ and $L(O; A)$.

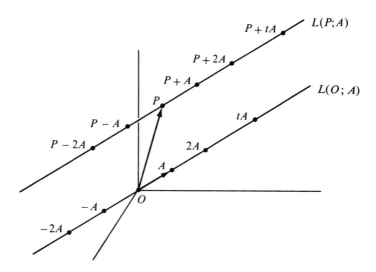

FIGURE 2.1   The line $L(P; A)$ through P parallel to A and its geometric relation to the line through O parallel to A.

2.3   Some simple properties of straight lines in **R**^{n}

First we show that the direction vector A that occurs in the definition of $L(P; A)$ can be replaced by any vector parallel to A. (We recall that two vectors A and B are called parallel if $A = cB$ for some nonzero scalar c.)

THEOREM 2.1.   *Two lines $L(P; A)$ and $L(P; B)$ through the same point P are equal if and only if the direction vectors A and B are parallel.*

Proof.   Assume first that $L(P; A) = L(P; B)$. Take a point on $L(P; A)$ distinct from P, for example $P + A$. This point is also on $L(P; B)$ so $P + A = P + cB$ for some scalar c. Hence we have $A = cB$, and $c \neq 0$ since $A \neq O$. Therefore A and B are parallel.

Now we prove the converse. Assume A and B are parallel, say $A = cB$ for some nonzero c. If Q is on $L(P; A)$ then we have $Q = P + tA = P + t(cB) = P + (tc)B$, so Q is on $L(P; B)$. Therefore $L(P; A) \subseteq L(P; B)$. Similarly, $L(P; B) \subseteq L(P; A)$, so $L(P; A) = L(P; B)$.

Next we show that the point P that occurs in the definition of $L(P; A)$ can be replaced by any other point Q on the same line.

THEOREM 2.2.   *Two lines $L(P; A)$ and $L(Q; A)$ with the same direction vector A are equal if and only if Q is on the line $L(P; A)$.*

Proof.   Assume $L(P; A) = L(Q; A)$. Since Q is on $L(Q; A)$, Q is also on $L(P; A)$. To prove the converse, assume that Q is on $L(P; A)$, say, $Q = P + cA$. We wish to prove that $L(P; A) = L(Q; A)$. If $X \in L(P; A)$ then $X = P + tA$ for some t. But $P = Q - cA$, so we have $X = Q - cA + tA = Q + (t - c)A$, and hence X is also on $L(Q; A)$. Therefore $L(P; A) \subseteq L(Q; A)$. Similarly, we find $L(Q; A) \subseteq L(P; A)$, so the two lines are identical.

One of Euclid's famous postulates is the parallel postulate, which is logically equivalent to the statement that "through a given point there exists one and only one line parallel to a given line." We shall deduce this property as an easy consequence of Theorem 2.1. First we need to define parallelism of lines.

DEFINITION.   *Two lines $L(P; A)$ and $L(Q; B)$ in n-space are called parallel if their direction vectors A and B are parallel.*

THEOREM 2.3.   *Given a line L and a point Q not on L, then there is one and only one line L' containing Q and parallel to L.*

Proof.   Suppose the given line has direction vector A. Consider the line $L' = L(Q; A)$. This line contains Q and is parallel to L. Theorem 2.1 tells us that this is the only line with these two properties.

Note.   For a long time mathematicians suspected that the parallel postulate could be deduced from the other Euclidean postulates, but all attempts to prove this resulted in failure. Then in the early 19th century the mathematicians Carl F. Gauss (1777–1855), J. Bolyai (1802–1860), and N. I. Lobatchevski (1791–1856) became convinced that the parallel postulate could not be derived from the others and proceeded to develop non-Euclidean geometries, that is to say, geometries in which the parallel postulate does not hold. The work of these men inspired other mathematicians and scientists to enlarge their points of view about "accepted truths" and to challenge other axioms that had been considered sacred for centuries.

It is also easy to deduce the following property of lines, which Euclid stated as an axiom.

THEOREM 2.4.   *Two distinct points determine a line. That is, if $P \neq Q$, there is one and only one line containing both P and Q. It can be described as the set $\{P + t(Q - P)\}$.*

Proof.   Let L be the line through P parallel to $Q - P$; that is, let

$$L = L(P; Q - P) = \{P + t(Q - P)\}.$$

This line contains both P and Q (take $t = 0$ to get P, and $t = 1$ to get Q). Now let L' be any line containing both P and Q. We shall prove that $L' = L$. Since L' contains P, we have $L' = L(P; A)$ for some nonzero A. But L' also contains Q so $P + cA = Q$ for some c. Hence we have $Q - P = cA$, where $c \neq 0$ because $Q \neq P$. Therefore $Q - P$ is parallel to A so, by Theorem 2.2, we have $L' = L(P; A) = L(P; Q - P) = L$.

EXAMPLE.   Theorem 2.4 gives us an easy way to test whether or not a point Q is on a given line $L(P; A)$. It tells us that Q is on $L(P; A)$ if and only if $Q - P$ is parallel to A. For example, consider the line $L(P; A)$ in 3-space, where $P = (1, 2, 3)$ and $A = (2, -1, 5)$. To test if the point $Q = (1, 1, 4)$ is on this line, we examine the difference $Q - P = (0, -1, 1)$. Since $Q - P$ is not a scalar multiple of A, the point $(1, 1, 4)$ is not on this line. On the other hand, if $Q = (5, 0, 13)$, we find that $Q - P = (4, -2, 10) = 2A$, so this Q is on the line.

Linear dependence of two vectors in $\mathbf{R}^n$ can be expressed in geometric language.

THEOREM 2.5.   *Two vectors A and B in $\mathbf{R}^n$ are linearly dependent if and only if they lie on the same line through the origin.*

Proof.   If either A or B is zero, the result holds trivially. If both are nonzero, then A and B are dependent if and only if $B = tA$ for some scalar t. But $B = tA$ if and only if B lies on the line through the origin parallel to A.

2.4   Lines and vector-valued functions in *n*-space

It is often helpful to regard a line as the track of a moving particle. The position of the particle at time t can be described by a vector-valued function X that associates to each real t a vector $X(t)$ given by the equation

(2.1) $X(t) = P + tA$.

This X is an example of a vector-valued function of a real variable. Its domain is the set of all real numbers, and its range is the line $L(P; A)$.

The scalar t in Eq. (2.1) is often called a *parameter*, and Eq. (2.1) is called a *vector parametric equation* or, simply, a *vector equation* of the line $L(P; A)$. If we think of t as representing *time*, we can regard $X(t)$ as the position vector that tells us where the particle is located at time t. The function point of view is important because it provides a natural method for describing not only lines, but more general space curves as well.

If a line passes through two distinct points P and Q, we can use $Q - P$ as the direction vector A in Eq. (2.1); and the vector equation of the line becomes

$$X(t) = P + t(Q - P) \qquad \text{or} \qquad X(t) = tQ + (1 - t)P.$$

Note that two points $X(a)$ and $X(b)$ on a given line $L(P; A)$ are equal if and only if $P + aA = P + bA$, or $(a - b)A = O$. Since A is nonzero, this last relation holds if and only if $a = b$. Thus, distinct values of the parameter t lead to distinct points on the line.

Now consider three distinct points on a given line, say $X(a)$, $X(b)$, and $X(c)$, where $a < b$. We say that $X(c)$ is *between* $X(a)$ and $X(b)$ if c is between a and b, that is, if $a < c < b$.

Congruence can be defined in terms of norms. A pair of points P, Q is called *congruent* to another pair P', Q' if $\|P - Q\| = \|P' - Q'\|$. The norm $\|P - Q\|$ is called the distance between P and Q.

This completes the definitions of the concepts of *point, line, on, between,* and *congruence* in our analytic model of Euclidean n-space. We turn next to some remarks concerning parametric equations for lines in 3-space and in 2-space.

2.5 Lines in 3-space and in 2-space

Let's express the vector equation (2.1) in terms of components when the line is in 3-space. Write $P = (p, q, r)$, $A = (a, b, c)$, and $X(t) = (x, y, z)$. Then (2.1) is equivalent to three scalar equations

$$(2.2) \qquad x = p + ta, \qquad y = q + tb, \qquad z = r + tc.$$

These are called *scalar parametric equations* or simply *parametric equations* for the line; they are useful in computations involving components. However, the vector equation is simpler and more natural for studying general properties of lines.

If all the vectors are in 2-space, only the first two parametric equations in (2.2) are needed. In this case, we can eliminate the parameter t from the two parametric equations. Multiply the first equation by b and the second by a, and subtract to obtain the relation

$$(2.3) \qquad b(x - p) - a(y - q) = 0,$$

which is called a *Cartesian equation* for the line. If $a \neq 0$, this can be written in the *point-slope form*:

$$y - q = \frac{b}{a}(x - p).$$

The point (p, q) is on the line; the number b/a is the slope of the line.

The Cartesian equation in (2.3) can also be written in terms of dot products. If we let

$$N = (b, -a), \qquad X = (x, y), \qquad \text{and} \qquad P = (p, q)$$

then Eq. (2.3) becomes $(X - P) \cdot N = 0$, or

$$X \cdot N = P \cdot N.$$

The vector N is perpendicular to the direction vector $A = (a, b)$ because $N \cdot A = ba - ab = 0$; vector N is called a *normal vector* to the line. The line consists of all points X satisfying the equation $X \cdot N = P \cdot N$.

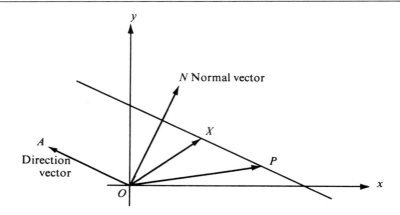

FIGURE 2.2   A line in the xy plane through P with normal vector N. Each point on the line satisfies $X \cdot N = P \cdot N$.

The geometric meaning of this relation is shown in Figure 2.2. The points P and X are on the line, and the normal vector N is orthogonal to $X - P$. The figure suggests that among all points X on the line, the smallest length $\|X\|$ occurs when X is the projection of P along N. We now give an algebraic proof of this fact.

THEOREM 2.6.   *Let L be the line in $\mathbf{R}^2$ consisting of all points X satisfying*

$$X \cdot N = P \cdot N,$$

where P is on the line and N is a nonzero vector normal to the line. Let

$$d = \frac{|P \cdot N|}{\|N\|}.$$

Then every X on L has length $\|X\| \geq d$. Moreover, $\|X\| = d$ if and only if X is the projection of P along N; that is, if and only if

$$X = tN, \qquad where \ t = \frac{P \cdot N}{N \cdot N}.$$

Proof.   If $X \in L$, we have $X \cdot N = P \cdot N$. By the Cauchy-Schwarz inequality we have

$$|P \cdot N| = |X \cdot N| \leq \|X\|\|N\|,$$

which implies $\|X\| \geq |P \cdot N|/\|N\| = d$. The equality sign holds if and only if $X = tN$ for some scalar t, in which case $P \cdot N = X \cdot N = tN \cdot N$, so $t = P \cdot N/(N \cdot N)$. This completes the proof.

In the same way we can prove that if Q is a given point in $\mathbf{R}^2$ not on the line L, then for all X on L the smallest value of $\|X - P\|$ is $|(P - Q) \cdot N|/(N \cdot N)$, and this occurs when

$X - Q$ is the projection of $P - Q$ along the normal vector N. The number

$$\frac{|(P - Q) \cdot N|}{\|N\|}$$

is called the *distance from the point Q to the line L*. The reader should illustrate these concepts with a diagram similar to that in Figure 2.2.

2.6 Exercises

1. A line L in $\mathbf{R}^2$ contains the two points $P = (-3, 1)$ and $Q = (1, 1)$. Determine which of the following points are on L.
   - (a) $(0, 0)$
   - (b) $(0, 1)$
   - (c) $(1, 2)$
   - (d) $(2, 1)$
   - (e) $(-2, 1)$.

2. Solve Exercise 1 if $P = (2, -1)$ and $Q = (-4, 2)$.

3. A line L in $\mathbf{R}^3$ contains the point $P = (-3, 1, 1)$ and is parallel to the vector $(1, -2, 3)$. Determine which of the following points are on L.
   - (a) $(0, 0, 0)$
   - (b) $(2, -1, 4)$
   - (c) $(-2, -1, 4)$
   - (d) $(-4, 3, -2)$
   - (e) $(2, -9, 16)$.

4. A line L in $\mathbf{R}^3$ contains the points $P = (-3, 1, 1)$ and $Q = (1, 2, 7)$. Determine which of the following points are on L.
   - (a) $(-7, 0, 5)$
   - (b) $(-7, 0, -5)$
   - (c) $(-11, 1, 11)$
   - (d) $(-11, -1, 11)$
   - (e) $(-1, \frac{3}{2}, 4)$
   - (f) $(-\frac{5}{3}, \frac{4}{3}, 3)$
   - (g) $(-1, \frac{3}{2}, -4)$.

5. In each case, determine if all three points P, Q, R lie on a line in $\mathbf{R}^3$.
   - (a) $P = (2, 1, 1)$, $Q = (4, 1, -1)$, $R = (3, -1, 1)$.
   - (b) $P = (2, 2, 3)$, $Q = (-2, 3, 1)$, $R = (-6, 4, 1)$.
   - (c) $P = (2, 1, 1)$, $Q = (-2, 3, 1)$, $R = (5, -1, 1)$.

6. Among the following eight points in $\mathbf{R}^3$, the three points A, B, C lie on a line. Determine all subsets of three or more points that lie on a line: $A = (2, 1, 1)$, $B = (6, -1, 1)$, $C = (-6, 5, 1)$, $D = (-2, 3, 1)$, $E = (1, 1, 1)$, $F = (-4, 4, 1)$, $G = (-13, 9, 1)$, $H = (14, -6, 1)$.

7. A line in $\mathbf{R}^3$ through the point $P = (1, 1, 1)$ is parallel to the vector $A = (1, 2, 3)$. Another line through $Q = (2, 1, 0)$ is parallel to the vector $B = (3, 8, 13)$. Prove that the two lines intersect and determine the point of intersection.

8. (a) Prove that two lines $L(P; A)$ and $L(Q; B)$ in $\mathbf{R}^n$ intersect if and only if $P - Q$ is in the linear span of A and B.
   (b) Determine whether or not the following two lines in $\mathbf{R}^3$ intersect:
   $$L = \{(1, 1, -1) + t(-2, 1, 3)\}, \quad L' = \{(3, -4, 1) + t(-1, 5, 2)\}.$$

9. Let $X(t) = P + tA$ be an arbitrary point on the line $L(P; A)$, where $P = (1, 2, 3)$ and $A = (1, -2, 2)$, and let $Q = (3, 3, 1)$.
   (a) Compute $\|Q - X(t)\|^2$, the square of the distance between Q and $X(t)$.
   (b) Prove that there is exactly one point $X(t_0)$ for which the distance $\|Q - X(t)\|$ is a minimum, and compute this minimum distance.
   (c) Prove that $Q - X(t_0)$ is orthogonal to A.

10. Let Q be a point not on the line $L(P; A)$ in $\mathbf{R}^n$.
    (a) Let $f(t) = \|Q - X(t)\|^2$, where $X(t) = P + tA$. Prove that $f(t)$ is a quadratic polynomial in t and that this polynomial takes on its minimum value at exactly one t, say at $t = t_0$.
    (b) Prove that $Q - X(t_0)$ is orthogonal to A.

11. Given two parallel lines $L(P; A)$ and $L(Q; A)$ in $\mathbf{R}^n$. Prove that either $L(P; A) = L(Q; A)$ or the intersection $L(P; A) \cap L(Q; A)$ is empty.

12. Given two lines $L(P; A)$ and $L(Q; A)$ in $\mathbf{R}^n$ that are not parallel. Prove that the intersection is either empty or consists of exactly one point.

2.7   Planes in Euclidean *n*-space

A line in *n*-space was defined to be a set of the form $\{P + tA\}$ obtained by adding to a given point *P* all vectors in the linear span of a nonzero vector *A*. A plane is defined in a similar fashion, except that we add to *P* all vectors in the linear span of two linearly independent vectors *A* and *B*. To make certain that $\mathbf{R}^n$ contains two linearly independent vectors, we will assume at the outset that $n \geq 2$. Most of our applications will be concerned with the case $n = 3$.

DEFINITION.   *A set M of points in* $\mathbf{R}^n$ *is called a plane if a point P and two linearly independent vectors A and B exist such that*

$$M = \{P + sA + tB : s \text{ and } t \text{ are real}\}.$$

We shall denote the set more briefly by writing $M = \{P + sA + tB\}$. Each point of *M* is said to lie on the plane. In particular, taking $s = t = 0$, we see that *P* is on this plane. The set of points $\{P + sA + tB\}$ is also called the plane through *P* spanned by *A* and *B*. When $P = O$, the plane is simply the linear span of *A* and *B*. Figure 2.3 shows a plane in 3-space through the origin spanned by *A* and *B* and also a plane through a nonzero point *P* spanned by the same two vectors.

Now we shall deduce some properties of planes analogous to the properties of lines given in Theorems 2.1 through 2.4. First we show that in the definition of the plane $\{P + sA + tB\}$ the pair of vectors *A* and *B* can be replaced by any other pair having the same linear span.

THEOREM 2.7.   *Two planes* $M = \{P + sA + tB\}$ *and* $M' = \{P + sC + tD\}$ *through the same point P are equal if and only if the linear span of A and B is equal to the linear span of C and D.*

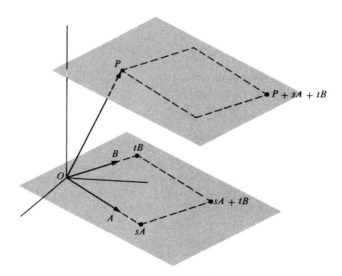

FIGURE 2.3   The plane through *P* spanned by *A* and *B* and its geometric relation to the plane through *O* spanned by *A* and *B*.

Proof. If the linear span of A and B is equal to that of C and D, then it is clear that $M = M'$. Conversely, assume that $M = M'$. Plane M contains both $P + A$ and $P + B$. Since both these points are also on M', each of A and B must be in the linear span of C and D. Similarly, each of C and D is in the linear span of A and B. Therefore the linear span of A and B is equal to that of C and D.

The next theorem shows that point P, which occurs in the definition of plane $\{P+sA+tB\}$, can be replaced by any other point Q in that plane.

THEOREM 2.8.   *Two planes $M = \{P + sA + tB\}$ and $M' = \{Q + sA + tB\}$ spanned by the same vectors A and B are equal if and only if Q is on M.*

Proof. If $M = M'$, then Q is certainly on M. To prove the converse, assume Q is on M, say, $Q = P + aA + bB$. Take any point X in M. Then $X = P + sA + tB$ for some scalars s and t. But $P = Q - aA - bB$, so $X = Q + (s - a)A + (t - b)B$. Therefore X is in M', so $M \subseteq M'$. Similarly, we find that $M' \subseteq M$, so the two planes are equal.

Euclid's parallel postulate (Theorem 2.3) has an analog for planes. Before we state this theorem we need to define parallelism of two planes. The definition is suggested by the geometric representation in Figure 2.2.

DEFINITION.   *Two planes $M = \{P + sA + tB\}$ and $M' = \{Q + sC + tD\}$ are said to be parallel if the linear span of A and B is equal to the linear span of C and D. We also say that a vector X is parallel to plane M if X is in the linear span of A and B.*

THEOREM 2.9.   *Given a plane M and a point Q not on M, there is one and only one plane M' that contains Q and is parallel to M.*

Proof. Let $M = \{P + sA + tB\}$ and consider the plane $M' = \{Q + sA + tB\}$. This plane contains Q and is spanned by the same vectors A and B that span M. Therefore M' is parallel to M. If M'' is another plane through Q parallel to M, then

$$M'' = \{Q + sC + tD\},$$

where the linear span of C and D is equal to that of A and B. By Theorem 2.7, we must have $M'' = M'$. Therefore M' is the only plane through Q that is parallel to M.

Theorem 2.4 tells us that two distinct points determine a line. The next theorem shows that three distinct points determine a plane, provided that the three points are not collinear.

THEOREM 2.10.   *If P, Q and R are three points in n-space not on the same line, then there is one and only one plane M containing these three points. It can be described as the set*

(2.4) $$M = \{P + s(Q - P) + t(R - P)\}.$$

Proof. We assume first that one of the points, say P, is the origin O. Then Q and R are not on the same line through the origin, so they are linearly independent. Therefore, they span a plane through the origin, say the plane

$$M' = \{sQ + tR\}.$$

This plane contains all three points O, Q, R.

Now we prove that M' is the only plane that contains all three points O, Q, R. Any other plane through the origin has the form

$$M'' = \{sA + tB\}$$

where A and B are linearly independent. If M'' contains Q and R, we have

(2.5) $Q = aA + bB, \qquad R = cA + dB,$

for some scalars a, b, c, d. Hence, every linear combination of Q and R is also a linear combination of A and B, so $M' \subseteq M''$.

To prove that $M'' \subseteq M'$, it suffices to prove that each of A and B is a linear combination of Q and R. Multiply the first equation in (2.5) by d, the second by b, and subtract to eliminate B. We find

$$(ad - bc)A = dQ - bR.$$

Now $ad - bc$ cannot be zero, otherwise Q and R would be dependent. Therefore we can divide by $ad - bc$ and express A as a linear combination of Q and R. Similarly, we can eliminate A from the two equations (2.5) and express B as a linear combination of Q and R, so we have $M'' \subseteq M'$. This proves the theorem when one of the points P, Q, R is the origin.

To prove the theorem in the general case, let M be the set in (2.4), and let $C = Q - P$, and $D = R - P$. First we show that C and D are linearly independent. If not we would have $D = tC$ for some scalar t, giving us $R - P = t(Q - P)$, or $R = P + t(Q - P)$, contradicting the fact that P, Q, R are not on the same line. Therefore M is a plane through P spanned by the linearly independent pair C and D. This plane contains all three points P, Q, and R. (Take $s = 1, t = 0$ to get Q, and $s = 0, t = 1$ to get R.) Now we must prove that this is the only plane containing P, Q, and R.

Let M' be any plane containing P, Q, and R. Since M' contains P, we have

$$M' = \{P + sA + tB\}$$

for some linearly independent pair A and B. Let $M_0 = \{sA + tB\}$ be the plane through the origin spanned by the same pair A and B. Clearly, M' contains a vector X if and only if M_0 contains $X - P$. Since M' contains Q and R, the plane M_0 contains $C = Q - P$ and $D = R - P$. But we have just shown that there is one and only one plane containing O, C, and D, because C and D are linearly independent. Therefore $M_0 = \{sC + tD\}$ so $M' = \{P + sC + tD\} = M$. This completes the proof.

In Theorem 2.5 we proved that two vectors in n-space are linearly dependent if and only if they lie on a line through the origin. The next theorem is the corresponding result for three vectors.

THEOREM 2.11.   *Three vectors A, B, C in n-space are linearly dependent if and only if they lie on the same plane through the origin.*

Proof. Assume A, B, C are dependent. Then we can express one of the vectors as a linear combination of the other two, say, $C = sA + tB$. If A and B are independent, they span a plane through the origin, and C is on this plane. If A and B are dependent, then A, B, C lie on a line through the origin, and hence they lie on any plane through the origin that contains all three points A, B, C.

To prove the converse, assume that A, B, C lie on the same plane through the origin, say, the plane M. If A and B are dependent, then A, B, and C are dependent, and there is nothing more to prove. If A and B are independent, they span a plane M' through the origin. By Theorem 2.10, there is one and only one plane through O containing A and B. Therefore $M' = M$. Since C is on this plane, we must have $C = sA + tB$, so A, B, C are dependent.

2.8   Planes and vector-valued functions

The correspondence that associates to each pair of real numbers s and t the vector $P + sA + tB$ on the plane $M = \{P + sA + tB\}$ is another example of a vector-valued function. In this case, the domain of the function is $\mathbf{R}^2$, the set of all pairs of real numbers (s, t), and its range is the plane M. If we denote the function by X and the function values by $X(s, t)$, then for each pair (s, t) we have

$$(2.6) \qquad\qquad X(s, t) = P + sA + tB.$$

We call X a vector-valued function of two real variables. The scalars s and t are called parameters, and Eq. (2.6) is called a *parametric* or *vector equation* of the plane. This is analogous to the representation of a line by a vector-valued function of one real variable. The presence of two parameters in Eq. (2.6) gives the plane a two-dimensional quality.

If we express each of the vectors in Equation (2.6) in terms of components, the vector equation is equivalent to n scalar equations. For example, in $\mathbf{R}^3$ suppose we write

$$P = (p_1, p_2, p_3), \qquad A = (a_1, a_2, a_3) \qquad B = (b_1, b_2, b_3) \qquad \text{and } X(s, t) = (x, y, z).$$

Then the vector equation in (2.6) can be replaced by three scalar equations:

$$x = p_1 + sa_1 + tb_1, \qquad y = p_2 + sa_2 + tb_2, \qquad z = p_3 + sa_3 + tb_3.$$

The parameters s and t can always be eliminated from these three equations to give one linear equation of the form $ax + by + cz = d$, called ·a *Cartesian equation of the plane*. We illustrate with an example.

EXAMPLE.   Let $M = \{P + sA + tB\}$, where $P = (1, 2, 3)$, $A = (1, 2, 1)$, and $B = (1, -4, -1)$. The corresponding vector equation is

$$X(s, t) = (1, 2, 3) + s(1, 2, 1) + t(1, -4, -1).$$

From this we obtain three scalar parametric equations

$$x = 1 + s + t, \qquad y = 2 + 2s - 4t, \qquad z = 3 + s - t.$$

To obtain a Cartesian equation we rewrite the first and third equations in the form $x - 1 = s + t$. $z - 3 = s - t$. Add these equations, then subtract them to find $2s = x + z - 4$, and

$2t = x - z + 2$. Substituting these values in the equation for y we are led to the Cartesian equation

$$x + y - 3z = -6.$$

We shall return to a further study of linear Cartesian equations in Section 2.17.

2.9 Exercises

1. Given a plane $M = \{P + sA + tB\}$ in $\mathbf{R}^3$, where $P = (1, 2, -3)$, $A = (3, 2, 1)$, $B = (1, 0, 4)$. Determine which of the following points are on M.
   (a) $(1, 2, 0)$ (c) $(6, 4, 6)$ (e) $(6, 6, -5)$.
   (b) $(1, 2, 1)$ (d) $(6, 6, 6)$
2. The three points $P = (1, 1, -1)$, $Q = (3, 3, 2)$, $R = (3, -1, -2)$ determine a plane M in $\mathbf{R}^3$. Determine which of the following points are on M.
   (a) $(2, 2, \frac{1}{2})$ (c) $(-3, 1, -3)$ (e) $(0, 0, 0)$.
   (b) $(4, 0, -\frac{1}{2})$ (d) $(3, 1, 3)$
3. Determine scalar parametric equations for each of the following planes in $\mathbf{R}^3$.
   (a) The plane through $(1, 2, 1)$ spanned by the vectors $(0, 1, 0)$ and $(1, 1, 4)$.
   (b) The plane through the three points $(1, 2, 1)$, $(0, 1, 0)$, and $(1, 1, 4)$.
4. A plane M in $\mathbf{R}^3$ has scalar parametric equations

$$x = 1 + s - 2t, \qquad y = 2 + s + 4t, \qquad z = 2s + t.$$

   (a) Determine which of the following points are on M: $(0, 0, 0)$, $(1, 2, 0)$, $(2, -3, -3)$.
   (b) Find vectors P, A, and B such that $M = \{P + sA + tB\}$.
5. Let M be the plane in $\mathbf{R}^3$ determined by three points P, Q, R, not on the same line.
   (a) If p, q, r are three scalars whose sum is 1, prove that $pP + qQ + rR$ is on M.
   (b) Prove that every point on M has the form $pP + qQ + rR$ where $p + q + r = 1$.
6. Determine a linear Cartesian equation of the form $ax + by + cz = d$ for each of the following planes in $\mathbf{R}^3$.
   (a) The plane through $(2, 3, 1)$ spanned by $(3, 2, 1)$ and $(-1, -2, -3)$.
   (b) The plane through $(2, 3, 1)$, $(-2, -1, -3)$ and $(4, 3, -1)$.
   (c) The plane through $(2, 3, 1)$ parallel to the plane through the origin spanned by $(2, 0, -2)$ and $(1, 1, 1)$.
7. A plane M in $\mathbf{R}^3$ has the Cartesian equation $3x - 5y + z = 9$.
   (a) Determine which of the following points are on M: $(0, -2, -1)$, $(-1, -2, 2)$, $(3, 1, -5)$.
   (b) Find vectors P, A, B, such that $M = \{P + sA + tB\}$.
8. Given two planes $M = \{P + sA + tB\}$ and $M' = \{Q + sC + tD\}$ in $\mathbf{R}^3$, where $P = (1, 1, 1)$, $A = (2, -1, 3)$, $B = (-1, 0, 2)$, $Q = (2, 3, 1)$, $C = (1, 2, 3)$, and $D = (3, 2, 1)$. Find two distinct points on the intersection $M \cap M'$.
9. In $\mathbf{R}^3$, given a plane $M = \{P + sA + tB\}$, where $P = (2, 3, 1)$, $A = (1, 2, 3)$, and $B = (3, 2, 1)$, and another plane M' with Cartesian equation $x - 2y + z = 0$.
   (a) Determine whether M and M' are parallel.
   (b) If a third plane M'' has the Cartesian equation $x + 2y + z = 0$, find two points on the intersection of M' and M''.
10. In $\mathbf{R}^3$, let L be the line through $(1, 1, 1)$ parallel to the vector $(2, -1, 3)$, and let M be the plane through $(1, 1, -2)$ spanned by the vectors $(2, 1, 3)$ and $(0, 1, 1)$. Prove that there is one and only one point on the intersection $L \cap M$, and determine this point.
11. A line in $\mathbf{R}^3$ with direction vector X is said to be parallel to a plane M if X is parallel to M. Let L be the line through $(1, 1, 1)$ parallel to the vector $(2, -1, 3)$. Determine whether L is parallel to each of the following planes.

(a) The plane through $(1, 1, -2)$ spanned by $(2, 1, 3)$ and $(\frac{3}{4}, 1, 1)$.

(b) The plane through $(1, 1, -2)$, $(3, 5, 2)$, and $(2, 4, -1)$.

(c) The plane with Cartesian equation $x + 2y + 3z = -3$.

12. In $\mathbf{R}^n$, two distinct points P and Q lie on a plane M. Prove or disprove: Every point on the line through P and Q also lies on M.

13. In $\mathbf{R}^3$, given a line L through $(1, 2, 3)$ parallel to the vector $(1, 1, 1)$, and given a point $(2, 3, 5)$ that is not on L. Find a Cartesian equation for the plane M through $(2, 3, 5)$ that contains every point on L.

14. In $\mathbf{R}^n$, given a line L and a point P not on L. Prove or disprove: There is one and only one plane through P that contains every point on L.

2.10 The cross product of two vectors in $\mathbf{R}^3$

In many applications of vector algebra to problems in geometry and mechanics it is helpful to have an easy method for constructing a vector perpendicular to each of two given vectors A and B. This is accomplished by means of the cross product $A \times B$ (read "A cross B") which is defined as follows.

DEFINITION. *Let $A = (a_1, a_2, a_3)$ and $B = (b_1, b_2, b_3)$ be two vectors in 3-space. Their cross product $A \times B$ (in that order) is defined to be the vector*

$$A \times B = (a_2 b_3 - a_3 b_2, a_3 b_1 - a_1 b_3, a_1 b_2 - a_2 b_1).$$

The following properties are easily deduced from this definition:

THEOREM 2.12. *For all vectors A, B, C in 3-space and for all real c we have:*

(a) $A \times B = -(B \times A)$ *(skew symmetry),*

(b) $A \times (B + C) = (A \times B) + (A \times C)$ *(distributive law),*

(c) $c(A \times B) = (cA) \times B = A \times (cB)$,

(d) $A \cdot (A \times B) = 0$ *(orthogonality to A),*

(e) $B \cdot (A \times B) = 0$ *(orthogonality to B),*

(f) $\|A \times B\|^2 = \|A\|^2 \|B\|^2 - (A \cdot B)^2$ *(Lagrange's identity),*

(g) $A \times B = O$ *if and only if A and B are linearly dependent.*

Proof. Parts (a), (b), and (c) follow quickly from the definition and are left as exercises for the reader. To prove (d), we note that

$$A \cdot (A \times B) = a_1(a_2 b_3 - a_3 b_2) + a_2(a_3 b_1 - a_1 b_3) + a_3(a_1 b_2 - a_2 b_1) = 0.$$

Part (e) follows in the same way, or it can be deduced from (a) and (d). To prove (f) we write

$$\|A \times B\|^2 = (a_2 b_3 - a_3 b_2)^2 + (a_3 b_1 - a_1 b_3)^2 + (a_1 b_2 - a_2 b_1)^2$$

and

$$\|A\|^2 \|B\|^2 - (A \cdot B)^2 = \left(a_1^2 + a_2^2 + a_3^2\right)\left(b_1^2 + b_2^2 + b_3^2\right) - (a_1 b_1 + a_2 b_2 + a_3 b_3)^2$$

and then verify by brute force that the two right-hand members are identical.

Property (f) shows that $A \times B = O$ if and only if $(A \cdot B)^2 = \|A\|^2\|B\|^2$. By the Cauchy-Schwarz inequality (Theorem 1.3), this happens if and only if one of the vectors is a scalar multiple of the other. In other words, $A \times B = O$ if and only if A and B are linearly dependent, which proves (g).

EXAMPLES.   Both (a) and (g) show that $A \times A = O$. From the definition of cross product we find that the unit coordinate vectors satisfy

$$i \times j = k, \qquad j \times k = i, \qquad k \times i = j.$$

The cross product is *not* associative. For example, we have

$$i \times (i \times j) = i \times k = -j \qquad \text{but} \qquad (i \times i) \times j = O.$$

The next theorem describes two further properties of the cross product.

THEOREM 2.13.   *Let A and B be linearly independent vectors in 3-space. Then:*
(a) *The vectors A, B and $A \times B$ are linearly independent.*
(b) *Every vector N in 3-space orthogonal to both A and B is a scalar multiple of $A \times B$.*

Proof.   Let $C = A \times B$. Then $C \neq O$ because A and B are linearly independent. Suppose that $aA + bB + cC = O$. Take the dot product of each member with C and use the relations $A \cdot B = B \cdot C = 0$ to find $c = 0$. This gives $aA + bB = 0$, so $a = b = 0$ because A and B are independent. This proves (a).

To prove (b) let N be orthogonal to both A and B, and let $C = A \times B$. We shall prove that

$$(N \cdot C)^2 = (N \cdot N)(C \cdot C).$$

Then by the Cauchy-Schwarz inequality (Theorem 1.3) it follows that N is a scalar multiple of C.

Since A, B, C are independent, we know by Theorem 1.10(c) that they span $\mathbf{R}^3$. In particular, we can write

$$N = aA + bB + cC$$

for some scalars a, b, c. This gives us

$$N \cdot N = N \cdot (aA + bB + cC) = cN \cdot C$$

since $N \cdot A = N \cdot B = 0$. Also, since $C \cdot A = C \cdot B = 0$ we find

$$C \cdot N = C \cdot (aA + bB + cC) = cC \cdot C.$$

Therefore $(N \cdot N)(C \cdot C) = (cN \cdot C)(C \cdot C) = (N \cdot C)(cC \cdot C) = (N \cdot C)^2$, which completes the proof.

Theorem 2.12 helps us visualize the cross product geometrically. From properties (d) and (c), we know that $A \times B$ is perpendicular to both A and B. When the vector $A \times B$ is

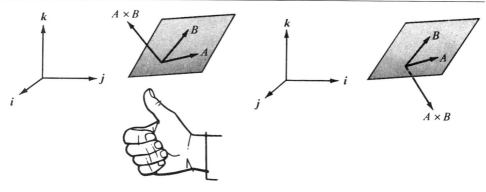

(a) A right-handed coordinate system   (b) A left-handed coordinate system

FIGURE 2.4   Illustrating the relative positions of A, B and $A \times B$.

represented geometrically by an arrow, the direction of the arrow depends on the relative positions of the three unit coordinate vectors. If i, j, and k are arranged as shown in Figure 2.4(a), they are said to form a *right-handed coordinate system*. In this case, the direction of $A \times B$ is determined by the "right-hand rule." That is to say, when A is rotated into B in such a way that the fingers of the right hand point in the direction of rotation, then the thumb indicates the direction of $A \times B$ (assuming, for the sake of the discussion, that the thumb is perpendicular to the other fingers). In a left-handed coordinate system, as shown in Figure 2.4(b), the direction of $A \times B$ is reversed and can be determined by a corresponding left-hand rule.

The length of $A \times B$ has an interesting geometric interpretation. If A and B are nonzero vectors making an angle θ with each other, where $0 \le \theta \le \pi$, we can write $A \cdot B = \|A\|\|B\| \cos \theta$ in property (f) of Theorem 2.12 to obtain

$$\|A \times B\|^2 = \|A\|^2\|B\|^2(1 - \cos^2 \theta) = \|A\|^2\|B\|^2 \sin^2 \theta,$$

from which we find

$$\|A \times B\| = \|A\|\|B\| \sin \theta.$$

Since $\|B\| \sin \theta$ is the altitude of the parallelogram determined by A and B (see Figure 2.5), we see that the length of $A \times B$ is equal to the area of this parallelogram.

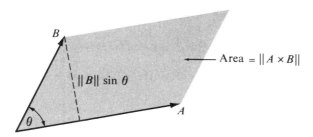

FIGURE 2.5   The length of $A \times B$ is the area of the parallelogram determined by A and B.

2.11   The cross product expressed as a determinant

The formula that defines the cross product can be put in a more compact form with the aid of determinants. If a, b, c, d are four numbers, the difference $ad - bc$ is often denoted by the symbol

$$\begin{vmatrix} a & b \\ c & d \end{vmatrix}$$

and is called a *determinant* (of order two). The numbers a, b, c, d are called its *elements* or *entries* and they are said to be arranged in two horizontal *rows*, a, b and c, d and in two vertical *columns* a, c and b, d. Note that an interchange of two rows or of two columns changes the sign of the determinant. For example, since $ad - bc = -(bc - ad)$ we have

$$\begin{vmatrix} a & b \\ c & d \end{vmatrix} = - \begin{vmatrix} b & a \\ d & c \end{vmatrix}.$$

We can express each of the components of the cross product as a determinant of order two, and the formula defining $A \times B$ becomes

$$A \times B = \left(\begin{vmatrix} a_2 & a_3 \\ b_2 & b_3 \end{vmatrix}, \begin{vmatrix} a_3 & a_1 \\ b_3 & b_1 \end{vmatrix}, \begin{vmatrix} a_1 & a_2 \\ b_1 & b_2 \end{vmatrix} \right).$$

This can also be expressed in terms of the unit coordinate vectors i, j, k as follows:

$$(2.7) \qquad A \times B = \begin{vmatrix} a_2 & a_3 \\ b_2 & b_3 \end{vmatrix} i + \begin{vmatrix} a_3 & a_1 \\ b_3 & b_1 \end{vmatrix} j + \begin{vmatrix} a_1 & a_2 \\ b_1 & b_2 \end{vmatrix} k.$$

Determinants of order three are written with three rows and three columns and they can be defined in terms of second-order determinants by the formula

$$(2.8) \qquad \begin{vmatrix} a_1 & a_2 & a_3 \\ b_1 & b_2 & b_3 \\ c_1 & c_2 & c_3 \end{vmatrix} = a_1 \begin{vmatrix} b_2 & b_3 \\ c_2 & c_3 \end{vmatrix} - a_2 \begin{vmatrix} b_1 & b_3 \\ c_1 & c_3 \end{vmatrix} + a_3 \begin{vmatrix} b_1 & b_2 \\ c_1 & c_2 \end{vmatrix}.$$

This is said to be an "expansion" of the determinant along its first row. Note that the determinant on the right that multiplies a_1 can be obtained from that on the left by deleting the row and column in which a_1 appears. The other determinants on the right are similarly obtained.

When written out explicitly in terms of the entries, the formula in (2.8) becomes

$$\begin{vmatrix} a_1 & a_2 & a_3 \\ b_1 & b_2 & b_3 \\ c_1 & c_2 & c_3 \end{vmatrix} = a_1 b_2 c_3 - a_1 b_3 c_2 + a_2 b_3 c_1 - a_2 b_1 c_3 + a_3 b_1 c_2 - a_3 b_2 c_1.$$

This formula shows that a determinant is unchanged if its rows and columns are interchanged. In other words,

$$\begin{vmatrix} a_1 & a_2 & a_3 \\ b_1 & b_2 & b_3 \\ c_1 & c_2 & c_3 \end{vmatrix} = \begin{vmatrix} a_1 & b_1 & c_1 \\ a_2 & b_2 & c_2 \\ a_3 & b_3 & c_3 \end{vmatrix}.$$

Determinants of order greater than three are discussed in Chapter 5. Our only purpose in introducing determinants of order two and three at this stage is to have a useful device for writing certain formulas in a compact form that makes them easier to remember.

Determinants are meaningful if the elements in the first row are vectors. For example, if we write the determinant

$$\begin{vmatrix} i & j & k \\ a_1 & a_2 & a_3 \\ b_1 & b_2 & b_3 \end{vmatrix}$$

and "expand" this according to the rule prescribed in (2.8), we find that the result is equal to the right member of (2.7). This enables us to write the definition of the cross product $A \times B$ in the following compact form:

$$A \times B = \begin{vmatrix} i & j & k \\ a_1 & a_2 & a_3 \\ b_1 & b_2 & b_3 \end{vmatrix}.$$

For example, to compute the cross product of $A = 2i - 8j + 3k$ and $B = 4j + 3k$, we write

$$A \times B = \begin{vmatrix} i & j & k \\ 2 & -8 & 3 \\ 0 & 4 & 3 \end{vmatrix} = \begin{vmatrix} -8 & 3 \\ 4 & 3 \end{vmatrix} i - \begin{vmatrix} 2 & 3 \\ 0 & 3 \end{vmatrix} j + \begin{vmatrix} 2 & -8 \\ 0 & 4 \end{vmatrix} k = -36i - 6j + 8k.$$

2.12 Exercises

1. Let $A = -i + 2k$, $B = 2i + j - k$, $C = i + 2j + 2k$. Express each of the following vectors in terms of i, j, k:
   - (a) $A \times B$
   - (b) $B \times C$
   - (c) $C \times A$
   - (d) $A \times (C \times A)$
   - (e) $(A \times B) \times C$
   - (f) $A \times (B \times C)$
   - (g) $(A \times C) \times B$
   - (h) $(A + B) \times (A - C)$
   - (i) $(A \times B) \times (A \times C)$.

2. In each case find a vector of length 1 in $\mathbf{R}^3$ orthogonal to both A and B.
   - (a) $A = i + j + k$, $\quad B = 2i + 3j - k$.
   - (b) $A = 2i - 3j + 4k$, $\quad B = -i + 5j + 7k$.
   - (c) $A = i - 2j + 3k$, $\quad B = -3i + 2j - k$.

3. In each case use the cross product to compute the area of the triangle with vertices A, B, C.
   - (a) $A = (0, 2, 2)$, $\quad B = (2, 0, -1)$, $\quad C = (3, 4, 0)$.
   - (b) $A = (-2, 3, 1)$, $\quad B = (1, -3, 4)$, $\quad C = (1, 2, 1)$.
   - (c) $A = (0, 0, 0)$, $\quad B = (0, 1, 1)$, $\quad C = (1, 0, 1)$.

4. If $A = 2i + 5j + 3k$, $B = 2i + 7j + 4k$, and $C = 3i + 3j + 6k$, express the cross product $(A - C) \times (B - A)$ in terms of i, j, k.

5. Prove that $\|A \times B\| = \|A\|\|B\|$ if and only if A and B are orthogonal.

6. Given two linearly independent vectors A and B in $\mathbf{R}^3$. Let $C = (B \times A) - B$.
   - (a) Prove that A is orthogonal to $B + C$.
   - (b) Prove that the angle θ between B and C satisfies $\frac{1}{2}\pi < \theta < \pi$.
   - (c) If $\|B\| = 1$ and $\|B \times A\| = 2$, compute the length of C.

7. Let A and B be two orthogonal vectors in $\mathbf{R}^3$, each having length 1.
   - (a) Prove that $A, B, A \times B$ is an orthonormal basis for $\mathbf{R}^3$.
   - (b) Let $C = (A \times B) \times A$. Prove that $\|C\| = 1$.

(c) Draw a figure showing the geometric relation between A, B, and $A \times B$, and use this figure to obtain the relations

$$(A \times B) \times A = B, \qquad (A \times B) \times B = -A.$$

(d) Prove the relations in part (c) algebraically.

8. (a) If $A \times B = O$ and $A \cdot B = 0$, then at least one of A or B is zero. Prove this statement and give its geometric interpretation.

(b) Given $A \neq O$. If $A \times B = A \times C$ and $A \cdot B = A \cdot C$, prove that $B = C$.

9. Let $A = 2i - j + 2k$, and $C = 3i + 4j - k$.

(a) Find a vector B such that $A \times B = C$. Is there more than one solution?

(b) Find a vector B such that $A \times B = C$ and $A \cdot B = 1$. Is there more than one solution?

10. Given a nonzero vector A and a vector C orthogonal to A, both vectors in $\mathbf{R}^3$. Prove that there is exactly one vector B such that $A \times B = C$ and $A \cdot B = 1$.

11. Three vertices of a parallelogram are at points $A = (1, 0, 1)$, $B = (-1, 1, 1)$, $C = (2, -1, 2)$.

(a) Find all possible points D that can be the fourth vertex of the parallelogram.

(b) Compute the area of triangle ABC.

12. Given two nonparallel vectors A and B in $\mathbf{R}^3$ with $A \cdot B = 2$, $\|A\| = 1$, $\|B\| = 4$. Let $C = 2(A \times B) - 3B$. Compute $A \cdot (B + C)$, $\|C\|$, and the cosine of the angle between B and C.

13. Given two linearly independent vectors A and B in $\mathbf{R}^3$. Determine whether each of the following statements is always true or sometimes false.

(a) $A + B$, $A - B$, $A \times B$ are linearly independent.

(b) $A + B$, $A + (A \times B)$, $B + (A \times B)$ are linearly independent.

(c) A, B, $(A + B) \times (A - B)$ are linearly independent.

14. (a) Prove that three vectors A, B, C in $\mathbf{R}^3$ lie on a line if and only if $(B - A) \times (C - A) = O$.

(b) If $A \neq B$, prove that the line through A and B consists of the set of all vectors P such that $(P - A) \times (P - B) = O$.

15. Given two orthogonal vectors A, B in $\mathbf{R}^3$, each of length 1. Let P be a vector satisfying the equation $P \times B = A - P$. Prove each of the following statements.

(a) P is orthogonal to B and has length $\frac{1}{2}\sqrt{2}$.

(b) P, B, $P \times B$ form a basis for $\mathbf{R}^3$.

(c) $(P \times B) \times B = -P$.

(d) $P = \frac{1}{2}A - \frac{1}{2}(A \times B)$.

2.13   The scalar triple product

The dot and cross products can be combined to form the scalar triple product $A \cdot B \times C$, which can only mean $A \cdot (B \times C)$. Because this is a dot product of two vectors, its value is a scalar. We can compute this scalar by means of determinants. Write $A = (a_1, a_2, a_3)$, $B = (b_1, b_2, b_3)$, and $C = (c_1, c_2, c_3)$ and express $B \times C$ according to Equation (2.7). Forming the dot product with A, we obtain

$$A \cdot B \times C = a_1 \begin{vmatrix} b_2 & b_3 \\ c_2 & c_3 \end{vmatrix} + a_2 \begin{vmatrix} b_3 & b_1 \\ c_3 & c_1 \end{vmatrix} + a_3 \begin{vmatrix} b_1 & b_2 \\ c_1 & c_2 \end{vmatrix} = \begin{vmatrix} a_1 & a_2 & a_3 \\ b_1 & b_2 & b_3 \\ c_1 & c_2 & c_3 \end{vmatrix}.$$

Thus, the scalar triple product $A \cdot B \times C$ is equal to the determinant whose rows are the components of the factors A, B, C.

In Theorem 2.12 we found that two vectors A and B in $\mathbf{R}^3$ are linearly dependent if and only if their cross product $A \times B$ is the zero vector. The next theorem gives a corresponding criterion for linear dependence of three vectors in $\mathbf{R}^3$.

THEOREM 2.14.   *Three vectors A, B, C in* $\mathbf{R}^3$ *are linearly dependent if and only if*

$$A \cdot B \times C = 0.$$

Proof.   Assume first that A, B, C are dependent. If B and C are dependent, then $B \times C = O$, and hence $A \cdot B \times C = 0$. Supppose, then, that B and C are independent. Since all three are dependent, there exist scalars a, b, c, not all zero, such that $aA + bB + cC = O$. We must have $a \neq 0$ in this relation, otherwise B and C would be dependent. Therefore we can divide by a and express A as a linear combination of B and C, say $A = tB + sC$. Take the dot product of each member with $B \times C$ to obtain

$$A \cdot (B \times C) = tB \cdot B \times C + sC \cdot B \times C = 0,$$

since each of B and C is orthogonal to $B \times C$. Therefore dependence of A, B, C implies that the scalar triple product $A \cdot B \times C$ is zero.

To prove the converse, assume that $A \cdot B \times C = 0$. If B and C are dependent, then so are A, B, and C, and there is nothing more to prove. Assume, then, that B and C are linearly independent. Then, by Theorem 2.13, the three vectors B, C and $B \times C$ are linearly independent, hence they span A, so we can write

$$A = aB + bC + c(B \times C)$$

for some scalars a, b, c. Take the dot product of each member with $B \times C$ and use the fact that $A \cdot B \times C = 0$ to find $c = 0$, so $A = aB + bC$. This proves that A, B, C are linearly dependent.

EXAMPLE.   To determine whether the three vectors $(2, 3, -1)$, $(3, -7, 5)$, and $(1, -5, 2)$ are dependent, we calculate their scalar triple product, expressing it as the determinant

$$\begin{vmatrix} 2 & 3 & -1 \\ 3 & -7 & 5 \\ 1 & -5 & 2 \end{vmatrix} = 2(-14 + 25) - 3(6 - 5) - 1(-15 + 7) = 27.$$

Since the scalar triple product is nonzero, the vectors are linearly independent.

The scalar triple product has an interesting geometric interpretation. Figure 2.6 shows a parallelepiped determined by three geometric vectors A, B, C not in the same plane. Its

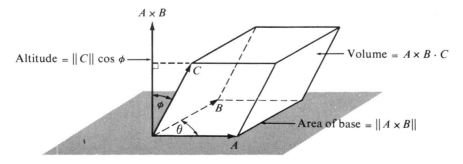

FIGURE 2.6   Geometric interpretation of the scalar triple product as the volume of a parallelepiped.

altitude is $\|C\| \cos \phi$, where ϕ is the angle between $A \times B$ and C. In this figure, $\cos \phi$ is positive because ϕ lies in the interval $0 \leq \phi < \frac{1}{2}\pi$. The area of the parallelogram that forms the base is $\|A \times B\|$. Multiplying the area of the base by the altitude we find that the volume of the parallelepiped is

$$\|A \times B\| \left(\|C\| \cos \phi \right) = (A \times B) \cdot C.$$

In other words, the scalar triple product $A \times B \cdot C$ is equal to the volume of the parallelepiped determined by A, B, C. When ϕ lies in the interval $\frac{1}{2}\pi < \phi \leq \pi$, $\cos \phi$ is negative and the product $A \times B \cdot C$ is the negative of the volume. So for any value of ϕ the volume of the parallelepiped is the absolute value of the scalar triple product $A \times B \cdot C$. If A, B, C are on a plane through the origin, they are linearly dependent and their scalar triple product is zero. In this case, the parallelepiped degenerates and has zero volume.

The geometric interpretation of the scalar triple product suggests certain algebraic properties of this product. For example, a cyclic permutation of the vectors A, B, C leaves the scalar triple product unchanged. By this we mean that

$$(2.9) \qquad A \times B \cdot C = B \times C \cdot A = C \times A \cdot B.$$

An algebraic proof of this property is outlined in Exercise 8 of Section 2.15. This property implies that the dot and cross are interchangeable in a scalar triple product. In fact, commutativity of the dot product implies $(B \times C) \cdot A = A \cdot (B \times C)$, and when this is combined with the first equation in (2.9) we find that

$$(2.10) \qquad A \times B \cdot C = A \cdot B \times C.$$

The scalar triple product $A \cdot B \times C$ is often denoted by the symbol $[ABC]$ without indicating the dot or cross. Because of Equation (2.10) there is no ambiguity in this notation—the product depends only on the order of the factors A, B, C and not on the positions of the dot and cross.

2.14   Cramer's rule for solving a system of three linear equations

The scalar triple product can be used to solve a system of three simultaneous linear equations in three unknowns x, y, z. Suppose the system is written in the form

$$
\begin{aligned}
a_1 x + b_1 y + c_1 z &= d_1 \\
(2.11) \qquad a_2 x + b_2 y + c_2 z &= d_2 \\
a_3 x + b_3 y + c_3 z &= d_3.
\end{aligned}
$$

Let A be the vector with components a_1, a_2, a_3 and define B, C, and D similarly. Then the three equations in (2.11) are equivalent to the single vector equation

$$(2.12) \qquad xA + yB + zC = D.$$

If we dot multiply both sides of this equation with $B \times C$, writing $[ABC]$ for $A \cdot B \times C$, we find

$$x[ABC] + y[BBC] + z[CBC] = [DBC].$$

Now $[BBC] = [CBC] = 0$, so the coefficients of x and y drop out and we can solve for x, giving us

$$(2.13) \qquad\qquad x = \frac{[DBC]}{[ABC]} \quad\text{if}\quad [ABC] \neq 0.$$

A similar argument yields analogous formulas for y and z. Thus we have

$$(2.14) \qquad y = \frac{[ADC]}{[ABC]} \quad\text{and}\quad z = \frac{[ABD]}{[ABC]} \quad\text{if}\quad [ABC] \neq 0.$$

The condition $[ABC] \neq 0$ means that the three vectors A, B, C are linearly independent. In this case, Eq. (2.12) shows that every vector D in 3-space is spanned by A, B, C; and the multipliers x, y, z are uniquely determined by the formulas in (2.13) and (2.14). When the scalar triple products in these formulas are written as determinants, the result is known as *Cramer's rule* for solving the system in (2.11):

$$x = \frac{\begin{vmatrix} d_1 & b_1 & c_1 \\ d_2 & b_2 & c_2 \\ d_3 & b_3 & c_3 \end{vmatrix}}{\begin{vmatrix} a_1 & b_1 & c_1 \\ a_2 & b_2 & c_2 \\ a_3 & b_3 & c_3 \end{vmatrix}}, \qquad y = \frac{\begin{vmatrix} a_1 & d_1 & c_1 \\ a_2 & d_2 & c_2 \\ a_3 & d_3 & c_3 \end{vmatrix}}{\begin{vmatrix} a_1 & b_1 & c_1 \\ a_2 & b_2 & c_2 \\ a_3 & b_3 & c_3 \end{vmatrix}}, \qquad z = \frac{\begin{vmatrix} a_1 & b_1 & d_1 \\ a_2 & b_2 & d_2 \\ a_3 & b_3 & d_3 \end{vmatrix}}{\begin{vmatrix} a_1 & b_1 & c_1 \\ a_2 & b_2 & c_2 \\ a_3 & b_3 & c_3 \end{vmatrix}}.$$

If $[ABC] = 0$, then A, B, C lie on a plane through the origin and the system has no solution unless D lies in the same plane. In this latter case it is easy to show that there are infinitely many solutions for the system. In fact, the vectors A, B, C are linearly dependent so there exist scalars u, v, w not all zero such that $uA + vB + wC = O$. If the triple (x, y, z) satisfies (2.12), then so does the triple $(x + tu, y + tv, z + tw)$ for all real t, because we have

$$(x + tu)A + (y + tv)B + (z + tw)C = xA + yB + zC + t(uA + vB + wC) = xA + yB + zC.$$

2.15 Exercises

1. Compute the scalar triple product $A \cdot B \times C$ in each case.
   (a) $A = (3, 0, 0)$, $B = (0, 4, 0)$, $C = (0, 0, 8)$.
   (b) $A = (2, 3, -1)$, $B = (3, -7, 5)$, $C = (1, -5, 2)$.
   (c) $A = (2, 1, 3)$, $B = (-3, 0, 6)$, $C = (4, 5, -1)$.
2. Find all real t for which the three vectors $(1, t, 1)$, $(t, 1, 0)$, $(0, 1, t)$ are linearly dependent.
3. Compute the volume of the parallelepiped detemined by the vectors $i + j$, $j + k$, $k + i$.
4. Prove that $A \times B = A \cdot (B \times i)i + A \cdot (B \times j)j + A \cdot (B \times k)k$.
5. Prove that $i \times (A \times i) + j \times (A \times j) + k \times (A \times k) = 2A$.
6. (a) Find all vectors $ai + bj + ck$ that satisfy the relation

$$(ai + bj + ck) \cdot k \times (6i + 3j + 4k) = 3.$$

   (b) Find that vector $ai + bj + ck$ of shortest length that satisfies the relation in (a).
7. Use algebraic properties of the dot and cross products to derive the following properties of the scalar triple product.
   (a) $(A + B) \cdot (A + B) \times C = 0$.
   (b) $A \cdot B \times C = -B \cdot A \times C$. (Switching the first two vectors reverses the sign.) [*Hint:* Use part (a) and the distributive laws.]

(c) $A \cdot B \times C = -A \cdot C \times B$. (Switching the second and third vectors reverses the sign.) [*Hint:* Use skew-symmetry.]

(d) $A \cdot B \times C = -C \cdot B \times A$. (Switching the first and third vectors reverses the sign.)

8. Use parts (b), (c), (d) of Exercise 7 to deduce that

$$A \cdot B \times C = B \cdot C \times A = C \cdot A \times B.$$

This shows that a cyclic permutation of A, B, C leaves their scalar triple product unchanged.

9. This exercise outlines a proof of the vector identity

$$(2.15) \qquad\qquad A \times (B \times C) = (C \cdot A)B - (B \cdot A)C,$$

sometimes referred to as the "cab minus bac" formula. Let $B = (b_1, b_2, b_3)$, $C = (c_1, c_2, c_3)$ and prove that

$$i \times (B \times C) = c_1 B - b_1 C.$$

This proves (2.15) in the special case $A = i$. Prove corresponding formulas for the special cases $A = j$ and $A = k$, then combine them to obtain (2.15).

10. Use the formula (2.15) to derive the following vector identities:
(a) $(A \times B) \times (C \times D) = (A \times B \cdot D)C - (A \times B \cdot C)D$.
(b) $A \times (B \times C) + B \times (C \times A) + C \times (A \times B) = O$.
(c) $A \times (B \times C) = (A \times B) \times C$ if and only if $B \times (C \times A) = O$.
(d) $(A \times B) \cdot (C \times D) = (B \cdot D)(A \cdot C) - (B \cdot C)(A \cdot D)$.

11. Four vectors A, B, C, D in $\mathbf{R}^3$ satisfy the relations $A \times C \cdot B = 5$, $A \times D \cdot B = 3$, $C + D = i + 2j + k$, $C - D = i - k$. Compute $(A \times B) \times (C \times D)$ in terms of i, j, k.

12. Prove that $(A \times B) \cdot (B \times C) \times (C \times A) = (A \cdot B \times C)^2$.

13. Prove or disprove the formula $A \times [A \times (A \times B)] \cdot C = -\|A\|^2 A \cdot (B \times C)$.

14. (a) Prove that the volume of the tetrahedron whose vertices are A, B, C, D is

$$\frac{1}{6}|(B - A) \cdot (C - A) \times (D - A)|.$$

(b) Compute this volume when $A = (1, 1, 1)$, $B = (0, 0, 2)$, $C = (0, 3, 0)$, and $D = (4, 0, 0)$.

15. (a) If $B \neq C$, prove that the perpendicular distance from A to the line through B and C is

$$\|(A - B) \times (C - B)\|/\|B - C\|.$$

(b) Compute this distance when $A = (1, -2, -5)$, $B = (-1, 1, 1)$, and $C = (4, 5, 1)$.

16. Hero of Alexandria gave a formula for computing the area S of a triangle whose sides have lengths a, b, c. It states that $S = \sqrt{s(s - a)(s - b)(s - c)}$, where $s = (a + b + c)/2$. This exercise outlines a proof of this formula by vector methods.

Assume the triangle has vertices at O, A, and B, with $\|A\| = a$, $\|B\| = b$, $\|B - A\| = c$.

(a) Combine the two identities

$$\|A \times B\|^2 = \|A\|^2 \|B\|^2 - (A \cdot B)^2 \qquad \text{and} \qquad -2A \cdot B = \|A - B\|^2 - \|A\|^2 - \|B\|^2$$

to obtain the formula

$$4S^2 = a^2 b^2 - \frac{1}{4}(c^2 - a^2 - b^2)^2 = \frac{1}{4}(2ab - c^2 + a^2 + b^2)(2ab + c^2 - a^2 - b^2).$$

(b) Rewrite the formula in part (a) to obtain

$$S^2 = \frac{1}{16}(a + b + c)(a + b - c)(c - a + b)(c + a - b),$$

and thereby deduce Hero's formula.

Use Cramer's rule to solve the system of equations in each of Exercises 17, 18, and 19.

17. $x + 2y + 3z = 5,$ $\quad 2x - y + 4z = 11,$ $\quad -y + z = 3.$
18. $\quad x + y + 2z = 4,$ $\quad 3x - y - z = 2,$ $\quad 2x + 5y + 3z = 3.$
19. $\quad\quad x + y = 5,$ $\quad\quad x + z = 2,$ $\quad\quad y + z = 5.$
20. If $P = (1, 1, 1)$ and $A = (2, 1, -1)$, prove that each point (x, y, z) on the line $\{P + tA\}$ satisfies the system of linear equations $x - y + z = 1, x + y + 3z = 5, 3x + y + 7z = 11.$

2.16 Normal vectors to planes in **R**^{3}

In Section 2.6 a plane in n-space was defined as a set of the form $\{P + sA + tB\}$, where A and B are linearly independent vectors. Now we show that planes in 3-space can be described in an entirely different way, using the concept of a normal vector.

DEFINITION. *Let $M = \{P + sA + tB\}$ be the plane through P spanned by A and B. A vector N in* **R**^{3} *is said to be perpendicular to M if N is perpendicular to both A and B. If, in addition, N is nonzero, then N is called a normal vector to the plane.*

Note. If $N \cdot A = N \cdot B = 0$, then $N \cdot (sA + tB) = 0$, so a vector perpendicular to both A and B is perpendicular to every vector in the linear span of A and B. Also, if N is normal to a plane, so is tN for every real $t \neq 0$.

THEOREM 2.15. *In* **R**^{3}, *for a given plane $M = \{P + sA + tB\}$ through P spanned by A and B, let $N = A \times B$. Then we have the following:*
(a) *N is a normal vector to M.*
(b) *M is the set of all X in* **R**^{3} *satisfying the equation*

$$(2.16) \qquad (X - P) \cdot N = 0.$$

Proof. Since M is a plane, A and B are linearly independent, so $A \times B \neq O$. This proves (a) because $A \times B$ is orthogonal to both A and B.

To prove (b), let M' be the set of all X in **R**^{3} satisfying Equation (2.16). If $X \in M$, then $X - P$ is in the linear span of A and B, so $X - P$ is orthogonal to N. Therefore $X \in M'$, so $M \subseteq M'$. Conversely, suppose $X \in M'$. Then X satisfies (2.16). Because A, B, N are linearly independent (Theorem 2.13), they span every vector in **R**^{3} so, in particular, we have

$$X - P = sA + tB + uN$$

for some scalars s, t, u. Taking the dot product of each member with N, we find $u = 0$, so $X - P = sA + tB$. This shows that $X \in M$. Hence $M' \subseteq M$, and this completes the proof of (b).

The geometric meaning of Theorem 2.15 is shown in Figure 2.7. The points P and X are on the plane, and the normal vector N is orthogonal to $X - P$. This figure also suggests the following theorem:

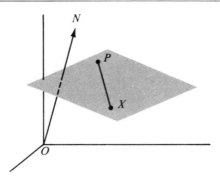

FIGURE 2.7   A plane through P and X with normal vector N.

THEOREM 2.16.   *Given a plane M through a point P, and a nonzero vector N normal to M. Let*

$$(2.17) \qquad d = \frac{|P \cdot N|}{\|N\|}.$$

Then every X on M has length $\|X\| \geq d$. Moreover, we have $\|X\| = d$ if and only if X is the projection of P along N; that is, if and only if

$$X = tN, \qquad where \qquad t = \frac{P \cdot N}{N \cdot N}.$$

Proof.   The proof follows from the Cauchy-Schwarz inequality in exactly the same way that we proved Theorem 2.6, the corresponding result for lines in $\mathbf{R}^2$.

By the same argument we find that if Q is a point not on M, then among all points X on M the smallest length $\|X - Q\|$ occurs when $X - Q$ is the projection of $P - Q$ along N. This minimum length is $|(P - Q) \cdot N|/\|N\|$ and is called the *distance from Q to the plane*. The number d in Equation (2.17) is the distance from the origin to the plane.

2.17   Linear cartesian equations for planes in $\mathbf{R}^3$

The results of Theorems 2.15 and 2.16 can also be expressed in terms of components. If we write $N = (a, b, c)$, $P = (x_1, y_1, z_1)$, and $X = (x, y, z)$, Equation (2.16) becomes

$$(2.18) \qquad a(x - x_1) + b(y - y_1) + c(z - z_1) = 0.$$

This is called a *Cartesian equation for the plane*, and it is satisfied by those and only those points (x, y, z) that lie on the plane. The set of points satisfying (2.18) is not altered if we multiply each of a, b, c by a nonzero scalar t. This simply amounts to a different choice of normal vector in (2.16).

We can transpose the terms not involving x, y, z, and write (2.18) in the form

$$(2.19) \qquad ax + by + cz = d_1,$$

where $d_1 = ax_1 + by_1 + cz_1$. An equation of this type is said to be *linear* in x, y, and z. We have just shown that in 3-space every point (x, y, z) on a plane satisfies a linear Cartesian equation of the form (2.19) in which not all three of a, b, c are zero. Conversely, every linear equation with this property represents a plane in 3-space. (The reader should verify this as an exercise.)

The number d_1 in Eq. (2.19) bears a simple relation to the distance d of the plane from the origin. Since $d_1 = P \cdot N$, we have $|d_1| = |P \cdot N| = d\|N\|$. In particular, $|d_1| = d$ if the normal vector N has length 1. The plane passes through the origin if and only if $d_1 = 0$.

EXAMPLE. The Cartesian equation $2x + 6y + 3z = 6$ represents a plane with normal vector $N = 2\mathbf{i} + 6\mathbf{j} + 3\mathbf{k}$. The length of N is $(N \cdot N)^{1/2} = 7$. If we divide the Cartesian equation by 6 and rewrite it in the form

$$\frac{x}{3} + \frac{y}{1} + \frac{z}{2} = 1,$$

we see that the plane intersects the coordinate axes at the points $(3, 0, 0)$, $(0, 1, 0)$ and $(0, 0, 2)$. The numbers 3, 1, 2 are called, respectively, the x, y, and z *intercepts of the plane*. A knowledge of the intercepts makes it possible to sketch the plane quickly. A portion of the plane is shown in Figure 2.8. Its distance from the origin is $d = 6/\|N\| = 6/7$.

If two planes are parallel they have a common normal $N = (a, b, c)$, and the Cartesian equations of these planes can be written as follows:

$$ax + by + cz = d_1, \qquad ax + by + cz = d_2,$$

the only difference being in the right-hand members. The number $|d_1 - d_2|/\|N\|$ is called the *perpendicular distance* between the two planes, a definition suggested by Theorem 2.16.

Two planes are called perpendicular if a normal of one is perpendicular to a normal of the other. More generally, if the normals of two planes make an angle θ with each other, then we say that θ is the angle between the two planes.

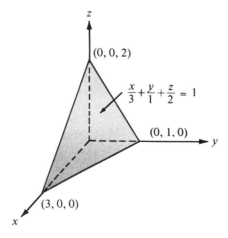

FIGURE 2.8   A plane with intercepts 3, 1, 2.

2.18　Exercises

1. Given vectors $A = 2i + 3j - 4k$ and $B = j + k$.
   (a) Find a nonzero vector N perpendicular to both A and B.
   (b) Find a Cartesian equation for the plane through the origin spanned by A and B.
   (c) Find a Cartesian equation for the plane through $(1, 2, 3)$ spanned by A and B.
2. A plane has Cartesian equation $x + 2y - 2z + 7 = 0$. Find the following:
   (a) A normal vector of unit length.　　　　　(c) The distance of the plane from the origin.
   (b) The intercepts of the plane.　　　　　　(d) The point Q on the plane nearest the origin.
3. Find a Cartesian equation of the plane that passes through $(1, 2, -3)$ and is parallel to the plane given by $3x - y + 2z = 4$. Find the perpendicular distance between the two planes.
4. Four planes have Cartesian equations $x + 2y - 2z = 5$, $3x - 6y + 3z = 2$, $2x + y + 2z = -1$, and $x - 2y + z = 7$.
   (a) Show that two of them are parallel and the other two are perpendicular.
   (b) Find the distance between the two parallel planes.
5. Three points $(1, 1, -1)$, $(3, 3, 2)$, and $(3, -1, -2)$ determine a plane. For this plane, find:
   (a) a normal vector;
   (b) a Cartesian equation;
   (c) its distance from the origin.
6. Find a Cartesian equation for the plane determined by $(1, 2, 3)$, $(2, 3, 4)$, and $(-1, 7, -2)$.
7. Determine an angle between the planes with Cartesian equations $x + y = 1$ and $y + z = 2$.
8. A line parallel to a nonzero vector N is said to be perpendicular to a plane M if N is normal to M. Find a Cartesian equation for the plane through $(2, 3, -7)$, given that the line through the points $(1, 2, 3)$ and $(2, 4, 12)$ is perpendicular to this plane.
9. Find a vector parametric equation for the line that contains the point $(2, 1, -3)$ and is perpendicular to the plane given by $4x - 3y + z = 5$.
10. A point moves in space in such a way that at time t its position is given by the vector $X(t) = (1 - t)i + (2 - 3t)j + (2t - 1)k$.
    (a) Prove that the point moves along a line. (Call it L.)
    (b) Find a vector N parallel to L.
    (c) At what time does the point strike the plane given by $2x + 3y + 2z + 1 = 0$?
    (d) Find a Cartesian equation for the plane that is parallel to the one in part (c) and contains the point $X(3)$.
11. Find a Cartesian equation for the plane through $(1, 1, 1)$ if a normal vector N makes angles $\frac{1}{3}\pi$, $\frac{1}{4}\pi$, $\frac{1}{3}\pi$ with i, j, k, respectively.
12. Compute the volume of the tetrahedron whose vertices are at the origin and at the points where the coordinate axes intersect the plane given by $x + 2y + 3z = 6$.
13. Find a vector A of length 1 perpendicular to $i + 2j - 3k$ and parallel to the plane with Cartesian equation $x - y + 5z = 1$.
14. Find a Cartesian equation of the plane that is parallel to both vectors $i + j$ and $j + k$ and intersects the x axis at $(2, 0, 0)$.
15. Find all points that lie on the intersection of the three planes given by $3x + y + z = 5$, $3x + y + 5z = 7$, and $x - y + 3z = 3$.
16. Prove that three planes with linearly independent normals intersect in one and only one point.
17. A line with direction vector A is said to be parallel to a plane M if A is parallel to M. A line containing $(1, 2, 3)$ is parallel to each of the planes given by $x + 2y + 3z = 4$, $2x + 3y + 4z = 5$. Find a vector parametric equation for the line.
18. In $\mathbf{R}^n$, given a plane M and a line L not parallel to M. Prove that the intersection $L \cap M$ contains at most one point. When $n = 3$ prove that $L \cap M$ contains exactly one point.
19. (a) Prove that the distance from the point (x_0, y_0, z_0) to the plane with Cartesian equation $ax + by + cz + d = 0$ is
$$\frac{|ax_0 + by_0 + cz_0 + d|}{(a^2 + b^2 + c^2)^{1/2}}.$$

    (b) Find the point P on the plane given by $5x - 14y + 2z + 9 = 0$ that is nearest to the point $Q = (-2, 15, -7)$.

20. Find a Cartesian equation for the plane parallel to the plane given by $2x - y + 2z + 4 = 0$ if the point $(3, 2, -1)$ is equidistant from both planes.

21. (a) If three points A, B, C determine a plane, prove that the distance from a point Q to this plane is $|(Q - A) \cdot (B - A) \times (C - A)| / \|(B - A) \times (C - A)\|$.
    (b) Compute this distance if $Q = (1, 0, 0)$, $A = (0, 1, 1)$, $B = (1, -1, 1)$, and $C = (2, 3, 4)$.

22. Prove that if two planes M and M' in $\mathbf{R}^3$ are not parallel, their intersection $M \cap M'$ is a line.

23. Find a Cartesian equation for the plane that is parallel to j and passes through the intersection of the planes described by the equations $x + 2y + 3z = 4$, and $2x + y + z = 2$.

24. Find a Cartesian equation for the plane parallel to the vector $3i - j + 2k$ if it contains every point on the line of intersection of the planes given by $x + y = 3$ and $2y + 3z = 4$.

2.19 The conic sections

A moving line G that intersects a fixed line A at a given point P, making a constant angle θ with A, where $0 < \theta < \frac{1}{2}\pi$, generates a surface in 3-space called a *right circular cone*. The line G is called a *generator* of the cone, A is its *axis*, and P its *vertex*. Each of the cones shown in Figure 2.9 has a vertical axis. The upper and lower portions of the cone that meet at the vertex are called *nappes* of the cone. The curves obtained by slicing the cone with a plane not passing through the vertex are called *conic sections*, or simply *conics*. If the cutting plane is parallel to a line of the cone through the vertex, the conic is called a *parabola*. Otherwise the intersection is called an *ellipse* or a *hyperbola*, according as the plane cuts just one or both nappes (See Figure 2.9.) The hyperbola consists of two "branches," one on each nappe.

Many important discoveries in both pure and applied mathematics have been related to the conic sections. Appolonius' treatment of conics as early as the third century B.C.E. was one of the most profound achievements of classical Greek geometry. Nearly 2000 years later, Galileo discovered that a projectile fired horizontally from the top of a tower falls to earth along a parabolic path (if air resistance is neglected and if the motion takes place above a part of the earth that can be regarded as a flat plane). One of the turning points in the

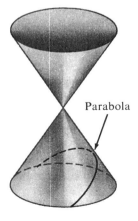
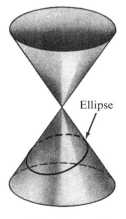
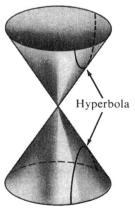

FIGURE 2.9   The conic sections.

history of astronomy occurred around 1600 when Johannes Kepler (1571–1630) suggested that all planets move in elliptical orbits. Some 80 years later, Isaac Newton (1642–1727) demonstrated that an elliptical planetary path implies an inverse-square law of universal gravitation, a result often referred to as one of the greatest scientific discoveries ever made. Conic sections appear not only as orbits of planets and satellites but also as trajectories of elementary atomic particles. They are used in the design of lenses and mirrors, and in architecture. These examples and many others show that the importance of the conic sections can hardly be overestimated.

There are other equivalent definitions of the conic sections. One of these refers to special points known as *foci* (singular: *focus*). An ellipse can be defined as the set of all points in a plane, the sum of whose distances d_1 and d_2 from two fixed points F_1 and F_2 (the foci) is constant (see Figure 2.10). If the foci coincide, the ellipse reduces to a circle.

A *hyperbola* is the set of all points for which the difference $|d_1 - d_2|$ is constant.

A *parabola* is the set of all points in a plane for which the distance to a fixed point F (called the *focus*) is equal to the distance to a given line (called the *directrix*).

There is a very simple and elegant argument showing that the focal property of an ellipse is a consequence of its definition as a section of a cone. This proof, which we refer to as the "ice-cream-cone proof," was discovered in 1822 by a Belgian mathematician, G. P. Dandelin (1794–1847), and makes use of two spheres S_1 and S_2 drawn so as to be tangent to the cutting plane and the cone, as illustrated in Figure 2.11. These spheres touch the cone along two parallel circles C_1 and C_2. We shall prove that the points F_1 and F_2, where the spheres contact the plane, can serve as foci of the ellipse.

Let P be an arbitrary point of the ellipse. We wish to prove that $\|\overrightarrow{PF_1}\| + \|\overrightarrow{PF_2}\|$ is constant, that is, independent of the choice of P. For this purpose, draw a line on the cone from the vertex O to P and let A_1 and A_2 be its intersections with the circles C_1 and C_2, respectively. Then $\overrightarrow{PF_1}$ and $\overrightarrow{PA_1}$ are two tangents to S_1 from P, and hence $\|\overrightarrow{PF_1}\| = \|\overrightarrow{PA_1}\|$. Similarly, $\|\overrightarrow{PF_2}\| = \|\overrightarrow{PA_2}\|$, and therefore we have

$$\|\overrightarrow{PF_1}\| + \|\overrightarrow{PF_2}\| = \|\overrightarrow{PA_1}\| + \|\overrightarrow{PA_2}\|.$$

But $\|\overrightarrow{PA_1}\| + \|\overrightarrow{PA_2}\| = \|\overrightarrow{A_1A_2}\|$, which is the distance between the parallel circles C_1 and C_2 measured along the surface of the cone, a number independent of P. The proves that F_1 and F_2 can serve as foci of the ellipse, as asserted.

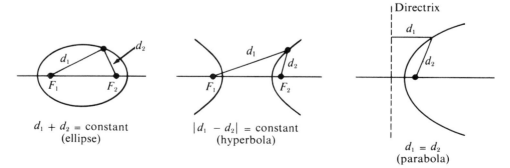

$d_1 + d_2 =$ constant
(ellipse)

$|d_1 - d_2| =$ constant
(hyperbola)

$d_1 = d_2$
(parabola)

FIGURE 2.10   Focal definitions of the conic sections.

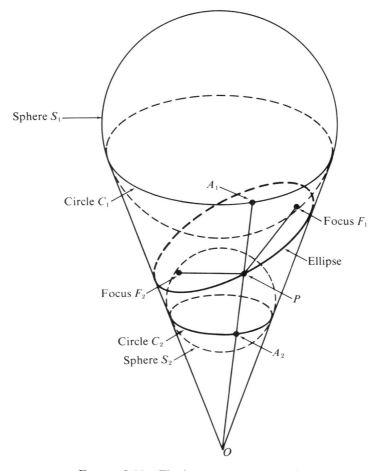

FIGURE 2.11   The ice-cream-cone proof.

Modifications of this proof work also for the hyperbola and for the parabola. In the case of the hyperbola, the proof employs one sphere in each portion of the cone. For the parabola, one sphere is used, tangent to the cutting plane at the focus F. This sphere touches the cone along a circle that lies in a plane whose intersection with the cutting plane is the directrix of the parabola. With these hints the reader should be able to show that the focal properties of the hyperbola and parabola can be deduced from their definitions as sections of a cone.

2.20   Eccentricity of conic sections

Another characteristic property of conic sections involves a concept called eccentricity. A conic section can be defined as a curve traced out by a point moving in a plane in such a way that the ratio of its distances from a fixed point and a fixed line is constant. This constant ratio is called the *eccentricity* of the curve and is denoted by e. (This should not be confused with the Euler number $e = 2.71828\ldots$.) The curve is an *ellipse* if $0 < e < 1$, a *parabola* if $e = 1$, and a *hyperbola* if $e > 1$. The fixed point is called a *focus* and the fixed line a *directrix*.

We shall adopt this definition as the basis for our study of conic sections because it permits a simultaneous treatment of all three types of conics and lends itself to the use of vector methods. In this definition it is understood that all points and lines lie in the same plane.

DEFINITION.   *Given a line L, a point F not on L, and a positive number e. Let $d(X, L)$ denote the distance from point X to L. The set of all X satisfying the relation*

$$(2.20) \qquad \qquad \|X - F\| = e\, d(X, L)$$

is called a conic section with eccentricity e. The conic is called an ellipse if $e < 1$, a parabola if $e = 1$, and a hyperbola if $e > 1$.

If N is a vector normal to L and if P is any point on L the distance $d(X, L)$ from any point X to L is given by the formula derived in Section 2.5:

$$d(X, L) = \frac{|(x - P) \cdot N|}{\|N\|}.$$

When N has length 1, this simplifies to $d(X, L) = |(X - P) \cdot N|$, and the basic equation (2.20) for the conic sections becomes

$$(2.21) \qquad \qquad \|X - F\| = e|(X - P) \cdot N|.$$

The line L separates the plane into two parts, which we shall arbitrarily label as "positive" and "negative" according to the choice of N. If $(X - P) \cdot N > 0$, we say that X is in the positive half-plane, and if $(X - P) \cdot N < 0$ we say that X is in the negative half-plane. On the line L itself we have $(X - P) \cdot N = 0$. In Figure 2.12 the choice of normal vector N dictates that points to the right of L are in the positive half-plane and those to left are in the negative half-plane.

Now we place the focus F in the negative half-plane, as indicated in Figure 2.12, and choose P to be that point on L nearest to F. Then $P - F = dN$, where $|d| = \|P - F\|$ is

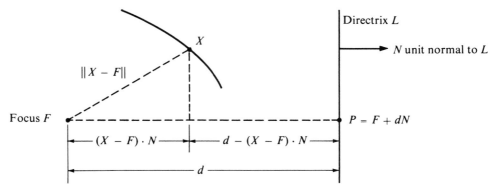

FIGURE 2.12   A conic section with eccentricity e is the set of all X satisfying $\|X - F\| = e|(X - F) \cdot N - d|$.

the distance from the focus to the directrix. Since F is in the negative half-plane, we have $(F - P) \cdot N = -d < 0$, so d is positive. Replacing P by $F + dN$ in (2.21), we obtain the following theorem, which is illustrated in Figure 2.12.

THEOREM 2.17. *Let C be a conic section with eccentricity e, focus F, and directrix L at a distance d from F. If N is a unit normal to L and if F is in the negative half-plane determined by N, then C consists of all points X satisfying the equation*

$$(2.22) \qquad \|X - F\| = e|(X - F) \cdot N - d|.$$

2.21 Polar equations for conic sections

The equation in Theorem 2.17 can be simplified if we place the focus in a special position. For example, if the focus is at the origin the equation becomes

$$(2.23) \qquad \|X\| = e|X \cdot N - d|.$$

This form is especially useful if we wish to express X in terms of polar coordinates. Take the directrix L to be vertical, as shown in Figure 2.13, and let $N = i$. If X has polar coordinates r and θ, we have $\|X\| = r$, and $X \cdot N = r \cos \theta$, so Equation (2.23) becomes

$$(2.24) \qquad r = e|r \cos \theta - d|.$$

If X lies to the left of the directrix, as in Figure 2.13(a), $r \cos \theta < d$, so $|r \cos \theta - d| = d - r \cos \theta$ and (2.24) becomes $r = e(d - r \cos \theta)$. Solving for r, we obtain

$$(2.25) \qquad r = \frac{ed}{e \cos \theta + 1}.$$

If X lies to the right of the directrix, we have $r \cos \theta > d$, so (2.24) becomes

$$r = e(r \cos \theta - d),$$

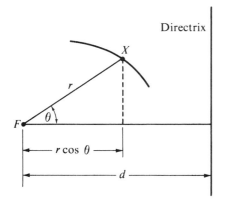

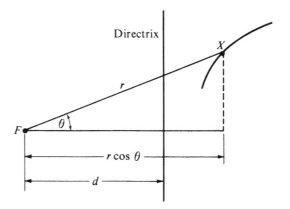

(a) $r \cos \theta < d$ on the ellipse, parabola, and left branch of the hyperbola

(b) $r \cos \theta > d$ on the right branch of the hyperbola

FIGURE 2.13 Conic sections with polar equation $r = e|r \cos \theta - d|$. The focus F is at the origin and lies to the left of the directrix.

which, when solved for r, gives us

$$(2.26) \qquad\qquad r = \frac{ed}{e \cos \theta - 1}.$$

Since $r > 0$, this last equation implies $e > 1$. In other words, there are points to the right of the directrix only for the hyperbola. Thus, we have proved the following theorem, which is illustrated in Figure 2.13.

THEOREM 2.18.   *Let C be a conic section with eccentricity e, with a focus F at the origin, and with a vertical directrix L at a distance d to the right of F. If $0 < e \le 1$, the conic C is an ellipse or a parabola; every point on C lies to the left of L and satisfies the polar equation*

$$(2.27) \qquad\qquad r = \frac{ed}{e \cos \theta + 1}.$$

If $e > 1$, the curve is a hyperbola with a branch on each side of L. Points on the left branch satisfy (2.27), and points on the right branch satisfy

$$(2.28) \qquad\qquad r = \frac{ed}{e \cos \theta - 1}.$$

Polar equations corresponding to other positions of the directrix are discussed in the next set of exercises.

2.22   Exercises

1. Prove that Equation (2.22) in Theorem 2.17 must be replaced by

$$\|X - F\| = e|(X - F) \cdot N + d|$$

   if F is in the positive half-plane determined by N.
2. Let C be a conic section with eccentricity e, with a focus at the origin, and with a vertical directrix L at a distance d to the left of F.
   (a) Prove that if C is an ellipse or parabola, every point of C lies to the right of L and satisfies the polar equation

$$r = \frac{ed}{1 - e \cos \theta}.$$

   (b) Prove that if C is a hyperbola, point on the right branch satisfy the equation in part (a), and points on the left branch satisfy $r = -ed/(1 + e \cos \theta)$. Note that $1 + e \cos \theta$ is negative in this case.
3. If a conic section has a horizontal directrix at a distance d above a focus at the origin, prove that its points satisfy the polar equations obtained from those in Theorem 2.18 by replacing $\cos \theta$ by $\sin \theta$. What are the corresponding polar equations if the directrix is horizontal and lies below the focus?

Each of Exercises 4 through 9 gives a polar equation for a conic section with a focus F at the origin and a vertical directrix lying to the right of F. In each case, determine the eccentricity e and the distance d from the focus to the directrix. Make a sketch showing the relation of the curve to its focus and directrix.

4. $r = \dfrac{2}{1 + \cos\theta}$.

5. $r = \dfrac{3}{1 + \frac{1}{2}\cos\theta}$.

6. $r = \dfrac{6}{3 + \cos\theta}$.

7. $r = \dfrac{1}{-\frac{1}{2} + \cos\theta}$.

8. $r = \dfrac{4}{1 + 2\cos\theta}$.

9. $r = \dfrac{4}{1 + \cos\theta}$.

In each of Exercises 10 through 12, a conic section of eccentricity e has a focus at the origin and a directrix with the given Cartesian equation. In each case, compute the distance d from the focus to the directrix and determine a polar equation for the conic section. For a hyperbola, give a polar equation for each branch. Make a sketch showing the relation of the curve to its focus and directrix.

10. $e = \frac{1}{2}$;   directrix: $3x + 4y = 25$.
11. $e = 1$;   directrix: $4x + 3y = 25$.
12. $e = 2$;   directrix: $x + y = 1$.
13. A comet moves in a parabolic orbit with the sun at one focus. When the comet is 10^8 miles from the sun, a vector from the focus to the comet makes an angle of $\pi/3$ radians with a unit vector N from the focus perpendicular to the directrix, the focus being in the negative half-plane determined by N.
    (a) Find a polar equation for the orbit, taking the origin at the focus, and compute the smallest distance from the comet to the sun.
    (b) Solve part (a) if the focus is in the positive half-plane determined by N.

2.23 Cartesian equation for a general conic

For a general conic of eccentricity e the basic equation (2.22) can be written as follows:

$$(2.29) \quad \|X - F\| = e|(X - F) \cdot N - d| = e|X \cdot N - F \cdot N - d| = |eX \cdot N - a|,$$

where $a = ed + eF \cdot N$. Squaring both members, we obtain

$$(2.30) \quad \|X\|^2 - 2X \cdot F + \|F\|^2 = e^2(X \cdot N)^2 - 2eaX \cdot N + a^2.$$

When we rewrite this equation in terms of the components of all vectors involved we obtain a Cartesian equation satisfied by every conic.

Let $X = (x, y)$, $F = (f_1, f_2)$, and $N = (n_1, n_2)$. Using these values in (2.30) and rearranging terms, remembering that $n_1^2 + n_2^2 = 1$ because $\|N\| = 1$, we find that (2.30) implies

$$x^2(1 - e^2 n_1^2) - 2e^2 n_1 n_2 xy + y^2(1 - e^2 n_2^2) + 2(aen_1 - f_1)x + 2(aen_2 - f_2)y + G = 0,$$

where $G = f_1^2 + f_2^2 - a^2$. This equation has the form

$$Ax^2 + Bxy + Cy^2 + Dx + Ey + G = 0,$$

where

$$(2.31) \quad A = 1 - e^2 n_1^2, \qquad B = -2e^2 n_1 n_2, \qquad C = 1 - e^2 n_2^2.$$

In other words, every conic is represented by a Cartesian equation that involves both quadratic and linear terms in x and y. Now we shall show that the *type* of conic depends only on the coefficients A, B, C of the quadratic part. The linear terms determine the location of the conic in the xy plane.

THEOREM 2.19.   *Every conic section has a Cartesian equation of the form*

$$(2.32) \qquad Ax^2 + Bxy + Cy^2 + Dx + Ey + G = 0.$$

The type of conic is determined by the algebraic sign of the number $\Delta = 4AC - B^2$, *called the discriminant, which is related to the eccentricity e by the equation*

$$(2.33) \qquad \Delta = 4(1 - e^2).$$

Consequently, the conic is an ellipse, parabola, or hyperbola according as Δ *is positive, zero, or negative.*

Proof.   From the formulas for A, B, C in (2.31) we find, after some algebra and use of the relation $n_1^2 + n_2^2 = 1$, that $4AC = 4(1 - e^2) + B^2$, so $4AC - B^2 = 4(1 - e^2)$. Therefore, if we introduce the discriminant $\Delta = 4AC - B^2$ we obtain Eq. (2.33). This shows that $\Delta > 0$ if $e < 1$, $\Delta = 0$ if $e = 1$, and $\Delta < 0$ if $e > 1$.

> *Note.* The converse of Theorem 2.19 would say that every Cartesian equation of the form (2.32) represents a conic section. This is true only if the definition of conic is extended to allow for degenerate cases. For example, the set of points satisfying $x^2 + y^2 + 1 = 0$ is the empty set; the set of points satisfying $x^2 + y^2 = 0$ consists of a single point $(0,0)$, yet for these examples the discriminant is positive. If all three of A, B, C are zero, the discriminant is zero but Equation (2.32) is linear in x and y and represents a line. If the equation factors into the product of two linear factors, the graph consists of a pair of lines.

2.24   Conic sections symmetric about the origin

A set of points is said to be *symmetric about the origin* if $-X$ is in the set whenever X is in the set. We show next that a focus of an ellipse or hyperbola can always be placed so the conic section will be symmetric about the origin. To do this we return to Equation (2.30):

$$(2.30) \qquad \|X\|^2 - 2X \cdot F + \|F\|^2 = e^2(X \cdot N)^2 - 2eaX \cdot N + a^2.$$

If we are to have symmetry about the origin, this equation must also be satisfied when X is replaced by $-X$, giving us

$$(2.34) \qquad \|X\|^2 + 2X \cdot F + \|F\|^2 = e^2(X \cdot N)^2 + 2eaX \cdot N + a^2.$$

Subtracting (2.34) from (2.30), we see that we have symmetry if and only if $F \cdot X = eaX \cdot N$, or

$$(F - eaN) \cdot X = 0.$$

This equation, in turn, is satisfied for all X on the curve if and only if the factor $(F - eaN) = O$, which means

$$(2.35) \qquad\qquad F = eaN, \qquad \text{where} \qquad a = ed + eF \cdot N.$$

The relation $F = eaN$ implies $F \cdot N = ea$, giving us $a = ed + e^2a$, or $a(1 - e^2) = ed$. If $e = 1$, this last equation cannot be satisfied because d, the distance from the focus to the directrix, is nonzero. This means there is no symmetry about the origin for a parabola. If $e \neq 1$, we can always satisfy the relations in (2.35) by taking

$$(2.36) \qquad\qquad a = \frac{ed}{1 - e^2}, \qquad \text{and} \qquad F = eaN.$$

Note that $a > 0$ if $e < 1$ and $a < 0$ if $e > 1$. Putting $F = eaN$ in (2.30) and rearranging terms we obtain the following:

THEOREM 2.20. *Let C be a conic section with eccentricity $e \neq 1$ and with a focus F at a distance d from a directrix L. If N is a unit normal to L and if $F = eaN$, where $a = ed/(1 - e^2)$, then C is the set of all points X satisfying the equation*

$$(2.37) \qquad\qquad \|X\|^2 - e^2(X \cdot N)^2 = a^2(1 - e^2).$$

This equation displays the symmetry about the origin because it remains unchanged when X is replaced by $-X$. Symmetry implies that the ellipse and hyperbola each have two foci, symmetrically located about the center at the points $\pm aeN$, and two directrices, also symmetrically located about the center. Because they have a center, the ellipse and hyperbola are called *central conics*.

Equation (2.37) is satisfied when $X = \pm aN$. These two points are called *vertices* of the central conic. The segment joining them is called the *major axis* if the conic is an ellipse, the *transverse axis* if the conic is a hyperbola. It is easy to show that the directrices intersect the line through the foci at the points $\pm(a/e)N$ (see Exercise 7 in Section 2.28).

Let N' be a unit vector orthogonal to N. If $X = bN'$, then $X \cdot N = 0$, so Equation (2.37) is satisfied by $X = bN'$ if and only if $b^2 + e^2a^2 = a^2$. Therefore $b^2 = a^2(1 - e^2)$, so $e < 1$. The segment joining the points $X = \pm bN'$, where $b = a\sqrt{1 - e^2}$, is called the *minor axis* of the ellipse.

> *Note.* If we put $e = 0$ in (2.37), it becomes $\|X\| = a$, the equation of a circle of radius a and center at the origin. In view of (2.36), we can consider such a circle as a limiting case of an ellipse in which $e \to 0$ as $d \to \infty$ in such a way that $ed \to a$.

2.25 Cartesian equations for the ellipse and the hyperbola in standard position

To obtain Cartesian equations for the ellipse and hyperbola, we simply write (2.37) in terms of the rectangular coordinates of X. If we choose $N = i$ the conics are said to be in *standard position*, which means the directrices are vertical. Now let $X = (x, y)$. Then $\|X\|^2 = x^2 + y^2$ and $X \cdot N = x$, so (2.37) becomes $x^2 + y^2 + e^2a^2 = e^2x^2 + a^2$, or $x^2(1 - e^2) + y^2 = a^2(1 - e^2)$, which gives us

$$(2.38) \qquad\qquad \frac{x^2}{a^2} + \frac{y^2}{a^2 \left(1 - e^2\right)} = 1.$$

This Cartesian equation represents both central conics, the ellipse (if $e < 1$) and the hyperbola (if $e > 1$) and is said to be in *standard form*. The foci are at the points $(ae, 0)$ and $(-ae, 0)$; the directrices are the vertical lines $x = a/e$ and $x = -a/e$. Recall that $a > 0$ if $e < 1$, and $a < 0$ if $e > 1$.

If $e < 1$, we let $b = a\sqrt{1 - e^2}$ and write the equation of the ellipse in the standard form

$$(2.39) \qquad \frac{x^2}{a^2} + \frac{y^2}{b^2} = 1.$$

Its foci are located at $(c, 0)$ and $(-c, 0)$, where $c = ae = \sqrt{a^2 - b^2}$. An example is shown in Figure 2.14(a).

If $e > 1$, we let $b = |a|\sqrt{e^2 - 1}$ and write the equation of the hyperbola in the standard form

$$(2.40) \qquad \frac{x^2}{a^2} - \frac{y^2}{b^2} = 1.$$

Its foci are at the points $(c, 0)$ and $(-c, 0)$, where $c = |a|e = \sqrt{a^2 + b^2}$. An example is shown in Figure 2.14(b).

Note. Solving for y in terms of x in (2.40), we obtain two relations

$$(2.41) \qquad y = \pm\frac{b}{|a|}\sqrt{x^2 - a^2}.$$

For large positive x, the number $\sqrt{x^2 - a^2}$ is nearly equal to x, so the right member of (2.41) is nearly $\pm bx/|a|$. It is easy to prove that the difference between the two values $y_1 = bx/|a|$ and

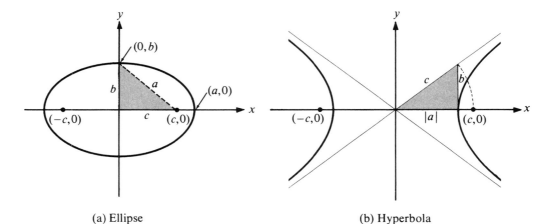

(a) Ellipse

$$\frac{x^2}{a^2} + \frac{y^2}{b^2} = 1; \quad b^2 = a^2 - c^2$$

(b) Hyperbola

$$\frac{x^2}{a^2} - \frac{y^2}{b^2} = 1; \quad b^2 = c^2 - a^2$$

FIGURE 2.14   Central conics of eccentricity $e \neq 1$. The foci are at $(\pm c, 0)$, where $c = |a|e$. The triangles relate a, b, c geometrically.

$y_2 = b\sqrt{x^2 - a^2}/|a|$ approaches 0 as $x \to \infty$. This difference is

$$y_1 - y_2 = \frac{b}{|a|}\left(x - \sqrt{x^2 - a^2}\right) = \frac{b}{|a|}\frac{x^2 - (x^2 - a^2)}{x + \sqrt{x^2 - a^2}} = \frac{|a|b}{x + \sqrt{x^2 - a^2}} < \frac{|a|b}{x},$$

so $y_1 - y_2 \to 0$ as $x \to +\infty$. The line $y = bx/|a|$ is called an *asymptote* of the hyperbola. The line $y = -bx/|a|$ is another asymptote. The hyperbola is said to approach these lines asymptotically. The asymptotes are shown in Figure 2.14(b).

The Cartesian equation for central conics will take a different form if the directrices are not vertical. For example, if the directrices are taken to be horizontal, we may take $N = j$ in Equation (2.37). Since $X \cdot N = X \cdot j = y$, we obtain a Cartesian equation like that in (2.38), except that x and y are interchanged. The standard form of a central conic in this case is

(2.42)
$$\frac{y^2}{a^2} + \frac{x^2}{a^2\left(1 - e^2\right)} = 1.$$

If the conic is translated by adding a vector $X_0 = (x_0, y_0)$ to each of its points, the center will be at (x_0, y_0) instead of at the origin. The corresponding Cartesian equation may be obtained from (2.35) or (2.39) by replacing x by $x - x_0$ and y by $y - y_0$.

2.26 Cartesian equations for the parabola

To obtain a Cartesian equation for the parabola, we return to the basic equation in (2.20) with $e = 1$. Take the directrix to be the vertical line $x = -c$ and place the focus at $(c, 0)$. If $X = (x, y)$, we have $X - F = (x - c, y)$, and Equation (2.20) gives us $(x - c)^2 + y^2 = |x + c|^2$. This simplifies to the standard form

(2.43)
$$y^2 = 4cx.$$

The point midway between the focus and directrix (the origin in Figure 2.15) is called the *vertex* of the parabola, and the line passing through the vertex and focus is the *axis* of the

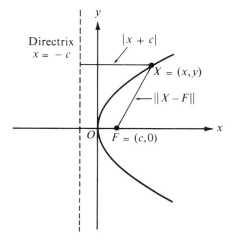

FIGURE 2.15   The parabola $y^2 = 4cx$.

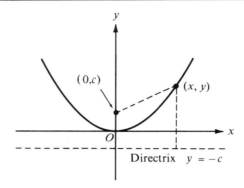

FIGURE 2.16   The parabola $x^2 = 4cy$.

parabola. The parabola is symmetric about its axis. If $c > 0$, the parabola lies to the right of the y axis, as in Figure 2.15. When $c < 0$, the curve lies to the left of the y axis.

If the axes are chosen so the focus is on the y axis at the $(0, c)$ and if the horizontal line $y = -c$ is taken as directrix, the standard form of the Cartesian equation becomes

$$x^2 = 4cy.$$

When $c > 0$ the parabola opens upward as shown in Figure 2.16. When $c < 0$, it opens downward.

If the parabola in Figure 2.15 is translated so that its vertex is at the point (x_0, y_0), the corresponding Cartesian equation becomes

$$(y - y_0)^2 = 4c(x - x_0).$$

Now the focus is at the point $(x_0 + c, y_0)$ and the directrix is the line $x = x_0 - c$. The axis of the parabola is the line $y = y_0$.

Similarly, a translation of the parabola in Figure 2.16 leads to the Cartesian equation

$$(x - x_0)^2 = 4c(y - y_0),$$

with focus at $(x_0, y_0 + c)$. The line $y = y_0 - c$ is its directrix, the line $x = x_0$ is its axis.

The reader may find it amusing to prove that a parabola does not have any asymptotes.

2.27   Exercises

Each of the equations in Exercises 1 through 6 represents an ellipse. Find the coordinates of the center, the foci, and the vertices, and sketch the curve. Also determine the eccentricity.

1. $\dfrac{x^2}{100} + \dfrac{y^2}{36} = 1.$ 4. $9x^2 + 25y^2 = 25.$

2. $\dfrac{y^2}{100} + \dfrac{x^2}{36} = 1.$ 5. $4y^2 + 3x^2 = 1.$

3. $\dfrac{(x - 2)^2}{16} + \dfrac{(y + 3)^2}{9} = 1.$ 6. $\dfrac{(x + 1)^2}{16} + \dfrac{(y + 2)^2}{25} = 1.$

In each of Exercises 7 through 12, find a Cartesian equation (in the appropriate standard form) for the ellipse that satisfies the conditions given. Sketch each curve.

7. Center at $(0,0)$, one focus at $(\frac{3}{4}, 0)$, one vertex at $(1,0)$.
8. Center at $(-3,4)$, semiaxes of lengths 4 and 3, major axis parallel to the x axis.
9. Same as Exercise 8, except with major axis parallel to the y axis.
10. Vertices at $(-1,2)$, $(-7,2)$, minor axis of length 2.
11. Vertices at $(3,-2)$, $(13,-2)$, foci at $(4,-2)$, $(12,-2)$.
12. Center at $(2,1)$, major axis parallel to the x axis, curve passes through points $(6,1)$ and $(2,3)$.

Each of the equations in Exercises 13 through 18 represents a hyperbola. Find the coordinates of the center, the foci, and the vertices. Sketch each curve and show the positions of the asymptotes. Also, compute the eccentricity.

13. $\dfrac{x^2}{100} - \dfrac{y^2}{64} = 1$.   16. $9x^2 - 16y^2 = 144$.

14. $\dfrac{y^2}{100} - \dfrac{x^2}{64} = 1$.   17. $4x^2 - 5y^2 + 20 = 0$.

15. $\dfrac{(x+3)^2}{4} - (y-3)^2 = 1$.   18. $\dfrac{(x-1)^2}{4} - \dfrac{(y+2)^2}{9} = 1$.

In each of Exercises 19 through 23, find a Cartesian equation (in the appropriate standard form) for the hyperbola that satisfies the conditions given. Sketch each curve and show the positions of its asymptotes.

19. Center at $(0,0)$, one focus at $(4,0)$, one vertex at $(2,0)$.
20. Foci at $(0, \pm\sqrt{2})$, vertices at $(0, \pm 1)$.
21. Vertices at $(\pm 2, 0)$, asymptotes $y = \pm 2x$.
22. Center at $(-1,4)$, one focus at $(-1,2)$, one vertex at $(-1,3)$.
23. Center at $(2,-3)$, transverse axis parallel to one of the coordinate axes, curve passes through the points $(3,-1)$ and $(-1,0)$.
24. The asymptotes of a hyperbola are the lines $2x - y = 0$ and $2x + y = 0$. Find a Cartesian equation for the hyperbola if it passes through the point $(3,-5)$.

Each of the equations in Exercises 25 through 30 represents a parabola. Find the coordinates of the vertex, an equation for the directrix, and an equation for the axis. Sketch each of the curves.

25. $y^2 = -8x$.   28. $x^2 = 6y$.
26. $y^2 = 3x$.   29. $x^2 + 8y = 0$.
27. $(y-1)^2 = 12x - 6$.   30. $(x+2)^2 = 4y + 9$.

In each of Exercises 31 through 36, find a Cartesian equation (in appropriate standard form) for the parabola that satisfies the conditions given, and sketch the curve.

31. Focus at $(0, -\frac{1}{4})$; equation of directrix, $y = \frac{1}{4}$.
32. Vertex at $(0,0)$; equation of directrix, $x = -2$.
33. Vertex at $(-4,3)$; focus at $(-4,1)$.
34. Focus at $(3,-1)$; equation of directrix, $x = \frac{1}{2}$.
35. Axis is parallel to the y axis; curve passes through $(0,1)$, $(1,0)$ and $(2,0)$.
36. Axis is parallel to the x axis; vertex at $(1,3)$; curve passes through $(-1,-1)$.
37. Proceeding directly from the focal definition, find a Cartesian equation for the parabola whose focus is at the origin and whose directrix is the line $2x + y = 10$.
38. Find a Cartesian equation for the parabola whose focus is at the origin and whose directrix is the line $x + y + 1 = 0$.
39. Find a Cartesian equation for the conic section consisting of all points (x, y) whose distance from the point $(0,2)$ is half the distance from the line $y = 8$.
40. Find a Cartesian equation for a hyperbola passing through the origin, given that its asymptotes are the lines $y = 2x + 1$ and $y = -2x + 3$.

2.28 Miscellaneous exercises on conic sections

1. (a) For each $p > 0$, the equation $px^2 + (p + 2)y^2 = p^2 + 2p$ represents an ellipse. Find (in terms of p) the eccentricity and the coordinates of the foci.
   (b) Find a Cartesian equation for the hyperbola that has the same foci as the ellipse of part (a) and that has eccentricity $\sqrt{3}$.

2. In Section 2.24 we proved that a conic section symmetric about the origin satisfies the equation $\|X - F\| = |eX \cdot N - a|$, where $a = ed + eF \cdot N$. Use this relation to prove that if the conic is an ellipse we have $\|X - F\| + \|X + F\| = 2a$. In other words, the sum of the distances from any point on an ellipse to its foci is constant.

3. See Exercise 2. Prove that on each branch of a hyperbola the difference $\|X - F\| - \|X + F\|$ is constant.

4. (a) Prove that a similarity transformation (replacing x by tx and y by ty) carries an ellipse with center at the origin into another ellipse with the same eccentricity. In other words, similar ellipses have the same eccentricity.
   (b) Prove the converse of (a). That is, if two concentric ellipses have the same eccentricity and major axes on the same line, then they are related by a similarity transformation.
   (c) State and prove results corresponding to (a) and (b) for hyperbolas.

5. (a) Prove that a similarity transformation (see Exercise 4) carries a parabola into a parabola.
   (b) Find all the parabolas similar to $y = x^2$.

6. Two points P and Q are said to be symmetric with respect to a circle if P and Q are collinear with the center, if the center is not between them, and if the product of their distances from the center is equal to the square of the radius. If point Q traces out the straight line $x + 2y = 5$, find the locus of the point P symmetric to Q with respect to the circle $x^2 + y^2 = 4$.

7. For a central conic with its center at the origin as described in Theorem 2.20, prove that the directrices intersect the line through the foci at the points $\pm(a/e)N$.

8. Prove that a parabola has no asymptotes.

9. A conic of eccentricity e with its focus on the x axis has a vertical directrix and passes through the origin. Show that the rectangular coordinates (x, y) of each point on the conic satisfy an equation of the form $y^2 = (e^2 - 1)x^2 - cx$, where c is a constant depending on the conic.

10. The point $(2, 2)$ is a focus and the line $x + y = 2$ is a directrix of the hyperbola with eccentricity $e = \sqrt{2}$. Show that the Cartesian equation of the hyperbola is $xy = 2$.

Exercises requiring a knowledge of calculus

11. For what constants C will the line $3x - 2y = C$ be tangent to the hyperbola $x^2 - 3y^2 = 1$?

12. A parabolic arch has a base of length b and altitude h. Determine the area of the region bounded by the arch and the base.

13. The region bounded by the parabola $y^2 = 8x$ and the line $x = 2$ is rotated about the x axis. Find the volume of the solid of revolution so generated.

14. Two parabolas with equations $y^2 = 2(x - 1)$ and $y^2 = 4(x - 2)$ enclose a plane region R.
    (a) Compute the area of R by integration.
    (b) Find the volume of the solid of revolution generated by rotating R about the x axis.
    (c) Same as (b), but rotate R about the y axis.

15. Show that the area of the region bounded by the ellipse $x^2/a^2 + y^2/b^2 = 1$ with a and b positive, is ab times the area of a circular disk of radius 1. *Note.* This statement can be proved from general properties of the integral, without performing any integration.

16. Find all positive numbers A and B, $A > B$, such that the area of the region enclosed by the ellipse $Ax^2 + By^2 = 3$ is equal to the area of the region enclosed by the ellipse

$$(A + B)x^2 + (A - B)y^2 = 3.$$

17. The line $x - y + 4 = 0$ is tangent to the parabola $y^2 = 16x$. Find the point of contact.

3

LINEAR SPACES

3.1 Introduction

Throughout mathematics we encounter many examples of mathematical objects that can be added to each other and multiplied by real numbers. First of all, real numbers themselves are such objects. Other examples are real-valued functions, the complex numbers, infinite series, vectors in n-space, and vector-valued functions. This chapter introduces a general mathematical concept, called a *linear space*, which includes all these examples and many others as special cases.

Briefly, a linear space is a set of elements of any kind on which certain operations (called *addition* and *multiplication by numbers*) can be performed. In defining a linear space, we do not specify the nature of the elements nor do we tell how the operations are to be performed on them. Instead, we require that the operations have certain properties that we take as axioms for a linear space. We turn now to a detailed description of these axioms.

3.2 Axiomatic definition of a linear space

Let V denote a nonempty set of objects, called *elements*. The set V is called a *linear space* if it satisfies the following ten axioms, which we separate into three groups.

Closure axioms

AXIOM 1. CLOSURE UNDER ADDITION. *For every pair of elements x and y in V there corresponds a unique element in V called the sum of x and y, denoted by $x + y$.*

AXIOM 2. CLOSURE UNDER MULTIPLICATION BY REAL NUMBERS. *For every x in V and every real number a there corresponds an element in V called the product of a and x, denoted by ax.*

Axioms for addition

AXIOM 3. COMMUTATIVE LAW. *For all x and y in V, we have $x + y = y + x$.*

AXIOM 4. ASSOCIATIVE LAW. *For all x, y, z in V, we have $(x + y) + z = x + (y + z)$.*

AXIOM 5.   EXISTENCE OF ZERO ELEMENT.   *There is an element in V, denoted by O, such that*

$$x + O = x \qquad \text{for all } x \text{ in } V.$$

AXIOM 6.   EXISTENCE OF NEGATIVES.   *For every x in V, the element* $(-1)x$ *has the property*

$$x + (-1)x = O.$$

Axioms for multiplication by numbers

AXIOM 7.   ASSOCIATIVE LAW.   *For every x in V and all real numbers a and b, we have*

$$a(bx) = (ab)x.$$

AXIOM 8.   DISTRIBUTIVE LAW FOR ADDITION IN V.   *For all x and y in V and all real numbers a, we have*

$$a(x + y) = ax + ay.$$

AXIOM 9.   DISTRIBUTIVE LAW FOR ADDITION OF NUMBERS.   *For all x in V and all real numbers a and b, we have*

$$(a + b)x = ax + bx.$$

AXIOM 10.   EXISTENCE OF IDENTITY.   *For every x in V, we have* $1x = x$.

Linear spaces, as just defined, are sometimes called *real* linear spaces to emphasize the fact that we are multiplying the elements of V by real numbers. The numbers used as multipliers are also called *scalars*. If *real number* is replaced by *complex number* in Axioms 2, 7, 8, and 9, the resulting structure is called a *complex linear space*. A real linear space has real numbers as scalars; a complex linear space has complex numbers as scalars. Although we shall deal primarily with examples of real linear spaces, all the theorems are valid for complex linear spaces as well. When we use the term linear space without further designation, it is understood that the space can be real or complex.

Some authors refer to a linear space as a *linear vector space* or simply as a *vector space*. In this book we reserve the term *vector space* for those examples in which the elements are vectors.

3.3   Examples of linear spaces

Because nothing is said about the nature of the elements, the axioms describe a structure that is called an *abstract* linear space. When we specify the set V and tell how to add its elements and how to multiply them by scalars, we get a concrete example of a linear space. The reader can easily verify that each of the following examples satisfies all the axioms for a real linear space.

EXAMPLE 1.   Let $V = \mathbf{R}$, the set of all real numbers, and let $x + y$ and ax be ordinary addition and multiplication of real numbers.

EXAMPLE 2.   Let $V = \mathbf{C}$, the set of all complex numbers, define $x + y$ to be ordinary addition of complex numbers, and define ax to be multiplication of the complex number x by the real number a. Even though the elements of V are complex numbers, this is a real linear space because the scalars are real.

EXAMPLE 3.   Let $V = \mathbf{R}^n$, the vector space of all n-tuples of real numbers, with addition and multiplication by scalars defined in the usual way in terms of components.

EXAMPLE 4.   Let V be the set of all vectors in $\mathbf{R}^n$ orthogonal to a given nonzero vector N. If $n = 2$, this linear space is a line through O with N as normal vector. If $n = 3$, it is a plane through O with N as normal vector.

The following examples are called *function spaces*. The elements of V are real-valued functions, with addition of two functions f and g defined in the usual way:

$$(f + g)(x) = f(x) + g(x)$$

for every real x in the intersection of the domains of f and g. Multiplication of a function f by a real scalar a is defined as follows: af is that function whose value at each x in the domain of f is $af(x)$. The zero element is the function whose values are everywhere zero. The reader can easily verify that each of the following sets is a function space.

EXAMPLE 5.   The set of all functions defined on a given interval.

EXAMPLE 6.   The set of all polynomials. Recall that a polynomial is a function f defined for all real x by a finite sum of the form

$$f(x) = c_0 + c_1 x + c_2 x^2 + \cdots + c_k x^k = \sum_{i=0}^{k} c_i x^i.$$

The numbers $c_0, c_1, \ldots, c_k$ that multiply the various powers of x are called coefficients. If x^k is the highest power of x with a nonzero coefficient, the polynomial is said to have degree k. If all the coefficients are zero except c_0 the polynomial is called a constant polynomial of degree 0. A constant polynomial with $c_0 = 0$ is called the zero polynomial and has no degree. The set of *all* polynomials consists of the zero polynomial, together with all polynomials of degree 0, of degree 1, of degree 2, etc.

EXAMPLE 7.   The set of all polynomials of degree $\leq n$, where n is fixed. (Whenever we consider this set it is understood that the zero polynomial is also included, even though it has no degree.) The set of all polynomials of degree *equal to n* is not a linear space because the closure axioms are not satisfied (the sum of two polynomials of degree n need not have degree n). For example, the polynomials $x^2 + x$ and $-x^2 + x$, each of degree 2, have sum $2x$, a polynomial of degree 1.

EXAMPLE 8.   The set of all functions f defined at 1 with $f(1) = 0$. The number 0 is essential in this example. If we replace 0 by a nonzero number c, we violate the closure axioms.

Examples 9 through 12 are for those readers acquainted with calculus.

EXAMPLE 9.   The set of all functions continuous on a given interval. If the interval is $[a, b]$, we denote this space by $C(a, b)$.

EXAMPLE 10.   The set of all functions differentiable at a given point.

EXAMPLE 11.   The set of all functions integrable on a given interval.

EXAMPLE 12.   The set of all solutions of the linear differential equation $y'' + ay' + by = 0$, where a and b are given constants. Here again, 0 is essential. If the right member is replaced by a nonzero quantity the set of solutions of the differential equation does not satisfy the closure axioms.

These examples and many others illustrate how the linear space concept permeates algebra, geometry, and analysis. When a theorem is deduced from the axioms of a linear space, we obtain, in one stroke, a result valid for each concrete example. By unifying diverse examples in this way we gain a deeper insight into each. Sometimes special knowledge of one particular example helps to anticipate or interpret results valid for other examples and reveals relationships that might otherwise escape notice.

3.4   Elementary consequences of the axioms

The following theorems are easily deduced from the axioms for a linear space.

THEOREM 3.1.   UNIQUENESS OF THE ZERO ELEMENT.   *In any linear space there is one and only one zero element.*

Proof.   Axiom 5 tells us that there is at least one zero element. Suppose there were two, say, O_1 and O_2. In Axiom 5 take $x = O_1$ and $O = O_2$ to obtain $O_1 + O_2 = O_1$. Similarly, if we take $x = O_2$ and $O = O_1$, we find $O_2 + O_1 = O_2$. But $O_1 + O_2 = O_2 + O_1$ because of the commutative law, so $O_1 = O_2$.

THEOREM 3.2.   UNIQUENESS OF NEGATIVE ELEMENTS.   *In any linear space every element has exactly one negative. That is, for every x there is one and only one y such that $x + y = O$.*

Proof.   Axiom 6 tells us that each x has at least one negative, namely $(-1)x$. Suppose x has two negatives y_1 and y_2. Then $x + y_1 = O$ and $x + y_2 = O$. Adding y_2 to both members of the first equation and using axioms 5, 4, and 3, we find that

$$y_2 + (x + y_1) = y_2 + O = y_2,$$

and

$$y_2 + (x + y_1) = (y_2 + x) + y_1 = O + y_1 = y_1.$$

Therefore $y_1 = y_2$, so x has exactly one negative, the element $(-1)x$.

Notation.   The negative of x is denoted by $-x$. The sum of y and $(-1)x$ is called the *difference $y - x$.*

The next theorem describes a number of properties that govern elementary algebraic manipulations in a linear space.

THEOREM 3.3. *In a given linear space, let x and y denote arbitrary elements, and let a and b denote arbitrary scalars. Then we have the following properties:*
(a) $0x = O$.
(b) $aO = O$.
(c) $(-a)x = -(ax) = a(-x)$.
(d) *If* $ax = O$, *then either* $a = 0$ *or* $x = O$.
(e) *If* $ax = ay$ *and* $a \neq 0$, *then* $x = y$.
(f) *If* $ax = bx$ *and* $x \neq O$, *then* $a = b$.
(g) $-(x + y) = (-x) + (-y) = -x - y$.
(h) $x + x = 2x$, $x + x + x = 3x$, *and, in general,* $\sum_{i=1}^{n} x = nx$.

We shall prove (a), (b), and (c), and leave the proofs of the other properties as exercises.

Proof of (a). Let $z = 0x$. We wish to prove that $z = O$. Adding z to itself and using Axiom 9, we find that

$$z + z = 0x + 0x = (0 + 0)x = 0x = z.$$

Now add $-z$ to both members to get $z = O$.

Proof of (b). Let $z = aO$, add z to itself, and use Axiom 8.

Proof of (c). Let $z = (-a)x$. Adding z to ax and using Axiom 9, we find

$$z + ax = (-a)x + ax = (-a + a)x = 0x = O,$$

so $z = -(ax)$, the negative of ax. Similarly, if we add $a(-x)$ to ax and use Axiom 8 and property (b), we find that $a(-x) = -(ax)$.

3.5 Exercises

In Exercises 1 through 20, determine whether each of the given sets is a linear space, if addition and multiplication by real scalars are defined in the usual way. For those that are not, tell which axioms fail to hold. The functions in Exercises 9 through 20 are real-valued. In Exercises 11, 12, and 13 each function has domain containing 0 and 1.

1. All vectors (x, y, z) in $\mathbf{R}^3$ with $z = 0$.
2. All vectors (x, y, z) in $\mathbf{R}^3$ with $x = 0$ or $y = 0$.
3. All vectors (x, y, z) in $\mathbf{R}^3$ with $y = 5x$.
4. All vectors (x, y, z) in $\mathbf{R}^3$ with $3x + 4y = 1$, $z = 0$.
5. All vectors (x, y, z) in $\mathbf{R}^3$ that are scalar multiples of $(1, 2, 3)$.
6. All vectors (x, y, z) in $\mathbf{R}^3$ whose components satisfy three linear equations of the form:

$$a_{11}x + a_{12}y + a_{13}z = 0, \qquad a_{21}x + a_{22}y + a_{23}z = 0, \qquad a_{31}x + a_{32}y + a_{33}z = 0.$$

7. All vectors (x, y, z) in $\mathbf{R}^3$ that are linear combinations of $(1, 2, 3)$ and $(2, 3, 4)$.
8. All vectors in $\mathbf{R}^n$ that are linear combinations of two given vectors A and B.
9. All rational functions (quotients of polynomials).

10. All rational functions f/g, with the degree of $f \leq$ the degree of g (including $f = 0$).
11. All functions f with $f(0) = f(1)$.
12. All functions f with $2f(0) = f(1)$.
13. All functions f with $f(1) = 1 + f(0)$.
14. All even functions: $f(-x) = f(x)$ for all x.
15. All odd functions: $f(-x) = -f(x)$ for all x.
16. All functions f satisfying $f(x) = f(1 - x)$ for all x.
17. All bounded functions: $|f(x)| \leq M$ for all x, where M is a constant depending on f.
18. All increasing functions: $f(x) \leq f(y)$ whenever $x \leq y$.
19. All functions with period 2π: $f(x + 2\pi) = f(x)$ for all x.
20. All linear combinations of $\sin x$ and $\cos x$.
21. Let $V = \mathbf{R}^+$, the set of positive real numbers. Define the "sum" of two elements x and y in V to be their product xy (in the usual sense), and define "multiplication" of an element x in V by a scalar c to be x^c. Prove that V is a linear space with 1 as zero element.
22. (a) Prove that Axiom 10 can be deduced from the other axioms.
    (b) Prove that Axiom 10 cannot be deduced from the other axioms if Axiom 6 is replaced by Axiom 6′: *For every x in V there is an element y in V such that $x + y = O$.*
23. Let S be the set of ordered pairs (x_1, x_2) of real numbers. In each case determine whether or not S is a linear space with the operations of addition and multiplication by scalars defined as indicated. If the set is not a linear space, tell which axioms are violated.
    (a) $(x_1, x_2) + (y_1, y_2) = (x_1 + y_1, x_2 + y_2)$, $a(x_1, x_2) = (ax_1, 0)$.
    (b) $(x_1, x_2) + (y_1, y_2) = (x_1 + y_1, 0)$, $a(x_1, x_2) = (ax_1, ax_2)$
    (c) $(x_1, x_2) + (y_1, y_2) = (x_1, x_2 + y_2)$, $a(x_1, x_2) = (ax_1, ax_2)$.
    (d) $(x_1, x_2) + (y_1, y_2) = (|x_1 + y_1|, |x_2 + y_2|)$, $a(x_1, x_2) = (|ax_1|, |ax_2|)$.
24. Prove parts (d) through (h) of Theorem 3.3.

The following exercise requires a knowledge of calculus.

25. Determine if the following sets of real-valued functions form a linear space, with addition and multiplication by scalars defined in the usual way.
    (a) All f integrable on $[0, 1]$ with $\int_0^1 f(x)\,dx = 0$.
    (b) All f integrable on $[0, 1]$ with $\int_0^1 f(x)\,dx \geq 0$.
    (c) All f with $f(x) \to 0$ as $x \to +\infty$.
    (d) All f satisfying a linear second-order differential equation $f'' + P(x)f' + Q(x)f = 0$, where P and Q are given functions, continuous everywhere.

3.6 Subspaces of a linear space

Given a linear space V, let S be a nonempty subset of V. If S is also a linear space, with the same operations of addition and multiplication by scalars, then S is called a *subspace of V*. To check whether or not an arbitrary set is a linear space we have to verify all ten axioms. But, as the next theorem shows, if the set is known to be a subset of a linear space, we need only check the closure axioms.

THEOREM 3.4.   *Let S be a nonempty subset of a linear space V. Then S is a subspace if and only if S satisfies the closure axioms.*

Proof.   If S is a subspace, it satisfies all the axioms for a linear space, and hence, in particular, it satisfies the closure axioms.

Now we show that if S satisfies the closure axioms it satisfies the others as well. The commutative and associative laws for addition (Axioms 3 and 4) and the axioms for multiplication by scalars (Axioms 7 through 10) are automatically satisfied in S because they hold for all elements of V. It remains to verify Axioms 5 and 6, the existence of a zero element in S, and the existence of a negative for each element in S.

Let x be any element of S. (S has at least one element because S is not empty.) By Axiom 2, ax is in S for every scalar a. Taking $a = 0$, it follows that $0x$ is in S. But $0x = O$, by Theorem 3.3(a), so $O \in S$, and Axiom 5 is satisfied. Now take $a = -1$, and we see that $(-1)x$ is in S. But the sum $x + (-1)x = O$ because both x and $(-1)x$ are in V, so Axiom 6 is satisfied in S. Therefore S is a subspace of V.

In the following examples, V is the linear space of all real functions defined on the real line, and S is a subset of V.

EXAMPLE 1.   If S is the set of all *even* functions, that is, functions satisfying $f(-x) = f(x)$ for all x, then S satisfies the closure axioms, so S is a subspace of V.

EXAMPLE 2.   If S is the set of all f such that $f(0) = 0$, then S satisfies the closure axioms, so S is a subspace of V.

EXAMPLE 3.   If S is the set of all f such that $f(0) = 1$, then S does not satisfy the closure axioms, so S is not a subspace of V.

EXAMPLE 4.   If S is the set of all functions of the form $f(x) = ae^x + be^{-x}$, where a and b are constants, then S satisfies the closure axioms, so S is a linear subspace of V. A function of the form $ae^x + be^{-x}$ is called a *linear combination* of e^x and e^{-x}.

DEFINITION.   *Let S be a nonempty subset of a linear space V. An element x in V of the form*

$$x = \sum_{i=1}^{k} c_i x_i,$$

where $x_1, \ldots, x_k$ are in S and $c_1, \ldots, c_k$ are scalars, is called a finite linear combination of elements of S. The set of all finite linear combinations of elements of S satisfies the closure axioms and hence is a subspace of V. We call this the subspace spanned by S, or the linear span of S, and denote it by $L(S)$. If S is empty, we define $L(S)$ to be $\{O\}$, the set consisting of the zero element alone.

Different sets may span the same subspace. For example, the space $\mathbf{R}^2$ is spanned by each of the following sets of vectors: $\{i, j\}$, $\{i, j, i + j\}$, $\{O, i, -i, j, -j, i + j\}$. The space of all polynomials $p(t)$ of degree $\leq n$ is spanned by the set of $n + 1$ polynomials

$$\{1, t, t^2, \ldots, t^n\},$$

and also by the set $\{1, t/2, t^2/3, \ldots, t^n/(n + 1)\}$, and by $\{1, (1 + t), (1 + t)^2, \ldots, (1 + t)^n\}$.

The space of all polynomials cannot be spanned by any finite set of polynomials, but it is spanned by the infinite set of polynomials $\{1, t, t^2, \ldots\}$.

A number of questions arise naturally this point. For example, which spaces can be spanned by a finite set of elements? If a space can be spanned by a finite set, what is the smallest number of elements required? To discuss these and related questions, we introduce the concepts of *dependence, independence, bases,* and *dimension.* These ideas were encountered in Chapter 1 in our study of the vector space $\mathbf{R}^n$, and now we extend them to general linear spaces.

3.7 Dependent and independent sets in a linear space

DEFINITION. *A set S of elements in a linear space V is called dependent if there is a finite set of distinct elements in S, say $x_1, \ldots, x_k$, and a corresponding set of scalars $c_1, \ldots, c_k$, not all zero, such that*

$$(3.1) \qquad\qquad \sum_{i=1}^{k} c_i x_i = O.$$

An equation like (3.1) with not all $c_i = 0$ is said to be a nontrivial representation of O.

The set S is called independent if it is not dependent. In this case, for all choices of distinct elements $x_1, \ldots, x_k$ in S and scalars $c_1, \ldots, c_k$,

$$\sum_{i=1}^{k} c_i x_i = O \qquad implies \qquad c_1 = c_2 = \cdots = c_k = 0.$$

Although dependence and independence are properties of sets of elements, we also apply these terms to the elements themselves. For example, the elements in an independent set are called independent elements.

If S is a finite set, the foregoing definition agrees with that given in Chapter 1 for the space $\mathbf{R}^n$. However, the present definition is not restricted to finite sets. An infinite set S is independent if every finite subset of S is independent.

EXAMPLE 1.   If $T \subseteq S$ and if T is dependent, then S is dependent. This is logically equivalent to the statement that every subset of an independent set is independent.

EXAMPLE 2.   If one element of S is a scalar multiple of another, then S is dependent.

EXAMPLE 3.   Any set containing the zero element is dependent.

EXAMPLE 4.   The empty set is independent.

Many examples of dependent and independent sets of vectors in n-space were encountered in Chapter 1. The following examples illustrate these concepts in function spaces. In each case the underlying linear space V is the set of all real-valued functions defined on the real line.

EXAMPLE 5.   Let $u_1(t) = \cos^2 t, u_2(t) = \sin^2 t, u_3(t) = 1$ for all real t. The Pythagorean identity $\cos^2 t + \sin^2 t = 1$ shows that $u_1 + u_2 - u_3 = O$, so the three functions u_1, u_2, u_3 are dependent.

EXAMPLE 6.   Let $u_k(t) = t^k$ for $k = 0, 1, 2, \ldots$, and t real. The infinite set $S = \{u_0, u_1, u_2, \ldots\}$ is independent. To prove this, it suffices to show that for each n the $n + 1$ polynomials $u_0, u_1, \ldots u_n$ are independent. If these polynomials span the zero polynomial we have

$$(3.2) \qquad \sum_{k=0}^{n} c_k t^k = 0$$

for all real t. Putting $t = 0$ in (3.2) we see that $c_0 = 0$. Now divide by t in (3.2) and put $t = 0$ again to find $c_1 = 0$. Repeating the process we find every c_k is 0, so $u_0, u_1, \ldots, u_n$ are independent.

EXAMPLE 7. (This example is needed in the study of linear differential equations.) If $a_0, a_1, \ldots, a_n$ are distinct real numbers, the n exponential functions

$$u_1(x) = e^{a_1 x}, \ldots, u_n(x) = e^{a_n x}$$

are independent. We prove this by induction on n. For $n = 1$ the result holds trivially. Therefore, assume it is true for $n - 1$ exponential functions and consider scalars $c_1, \ldots, c_n$ such that

$$(3.3) \qquad \sum_{k=1}^{n} c_k e^{a_k x} = 0.$$

Let a_M be the largest of the n numbers $a_1, \ldots, a_n$. Multiply each member of Eq. (3.3) by $e^{-a_M x}$ to obtain

$$(3.4) \qquad \sum_{k=1}^{n} c_k e^{(a_k - a_M)x} = 0.$$

Each factor $a_k - a_M$ in the exponent with $k \neq M$ is negative. Therefore, if we let $x \to +\infty$ in Equation (3.4), each term with $k \neq M$ tends to zero and we find $c_M = 0$. Now delete the Mth term from Eq. (3.3), apply the induction hypothesis, and we find that each of the remaining coefficients c_k is zero.

THEOREM 3.5. *Let $S = \{x_1, \ldots, x_k\}$ be an independent set consisting of k elements in a linear space V, and let $L(S)$ be the subspace spanned by S. Then every set of $k + 1$ elements in $L(S)$ is dependent.*

Proof. The proof is by induction on k, the number of elements in S. First suppose $k = 1$, so that $S = \{x_1\}$, where $x_1 \neq O$ because S is independent. Now take any two distinct elements y_1 and y_2 in $L(S)$. Then each is a scalar multiple of x_1, say, $y_1 = c_1 x_1$ and $y_2 = c_2 x_1$, where not both c_1 and c_2 are 0. Multiply y_1 by c_2 and y_2 by c_1 and subtract to obtain $c_2 y_1 - c_1 y_2 = O$. This is a nontrivial representation of O, so y_1 and y_2 are dependent. This proves the theorem when $k = 1$.

Now we assume the theorem is true for $k - 1$ and prove that it is also true for k. Take any set of $k + 1$ elements in $L(S)$, say $T = \{y_1, y_2, \ldots, y_{k+1}\}$. We wish to prove that T is dependent. Because each y_i is in $L(S)$ it can be written as a linear combination of the x_j, say,

$$(3.5) \qquad y_i = \sum_{j=1}^{k} a_{ij} x_j$$

for each $i = 1, 2, \ldots, k + 1$. We examine the scalars a_{i1} that multiply x_1 and split the proof into two cases, according to whether all these scalars are 0 or not.

CASE 1. $a_{i1} = 0$ *for every* $i = 1, 2, \ldots, k + 1$. In this case the sum in (3.5) does not involve x_1, so each y_i in T is in the linear span of the set $S' = \{x_2, \ldots, x_k\}$. But S' is independent and consists of $k - 1$ elements. By the induction hypothesis, the theorem is true for $k - 1$, so the set T is dependent. This proves the theorem in Case 1.

CASE 2. *Not all the scalars a_{i1} are zero.* Let us assume that $a_{11} \neq 0$. (If necessary, we can renumber the y_i to achieve this.) Taking $i = 1$ in Equation (3.5) and multiplying both members by the scalar $c_i = a_{i1}/a_{11}$, we get

$$c_i y_1 = a_{i1} x_1 + \sum_{j=2}^{k} c_i a_{ij} x_j.$$

Subtract Equation (3.5) from this to get

$$c_i y_1 - y_i = \sum_{j=2}^{k} (c_i a_{ij} - a_{ij}) x_j,$$

for $i = 2, \ldots, k + 1$. This equation expresses each of the k elements $c_i y_1 - y_i$ as a linear combination of the $k - 1$ elements $x_2, \ldots, x_k$. By the induction hypothesis, the k elements $c_i y_1 - y_i$ must be dependent. Hence, for some choice of scalars $t_2, \ldots, t_{k+1}$, not all zero, we have

$$\sum_{i=2}^{k+1} t_i (c_i y_1 - y_i) = O,$$

from which we find

$$\left(\sum_{i=2}^{k+1} t_i c_i \right) y_1 - \sum_{i=2}^{k+1} t_i y_i = O.$$

But this is a nontrivial linear combination of $y_1, y_2, \ldots, y_{k+1}$ that represents the zero element, so the elements $y_1, y_2, \ldots, y_{k+1}$ must be dependent. This completes the proof.

3.8   Bases and dimension

DEFINITION. *A finite set S of elements in a linear space V is called a finite basis for V if S is independent and spans V. The space V is called finite-dimensional if it has a finite basis, or if V consists of O alone. Otherwise, V is called infinite-dimensional.*

THEOREM 3.6. *Let V be a finite-dimensional linear space. Then every finite basis for V has the same number of elements.*

Proof. Let S and T be two finite bases for V. Suppose S consists of k elements and T of m elements. Since S is independent and spans V, Theorem 3.5 tells us that every set of $k + 1$ elements in V is dependent. Therefore, every set of more than k elements in V is dependent. But T is an independent set of m elements, so $m \leq k$. The same argument with S and T interchanged shows that $k \leq m$. Therefore $k = m$.

DEFINITION. *If a linear space V has a basis of n elements, the integer n is called the dimension of V, and we write $n = \dim V$. If $V = \{O\}$, we say V has dimension 0.*

EXAMPLE 1. The space $\mathbf{R}^n$ has dimension n. One basis is the set of n unit coordinate vectors.

EXAMPLE 2. The space of all polynomials $p(t)$ of degree $\leq n$ has dimension $n + 1$. One basis is the set of $n + 1$ polynomials $\{1, t, t^2, \ldots, t^n\}$. Every polynomial of degree $\leq n$ is a linear combination of these polynomials.

EXAMPLE 3. The space of *all* polynomials is infinite-dimensional. Although the infinite set of polynomials $\{1, t, t^2, \ldots\}$ is independent and spans this space, no *finite* set of polynomials spans this space, so it has no finite basis.

EXAMPLE 4. (This example requires a knowledge of differential equations.) The linear differential equation $y'' - 2y' - 3y = 0$ has two independent solutions $u_1(x) = e^{-x}$ and $u_2(x) = e^{3x}$, and every solution is a linear combination of these two. Therefore the space of solutions has a basis consisting of two elements, $\{u_1, u_2\}$, so this space has dimension 2.

THEOREM 3.7. *Let V be a finite-dimensional linear space with $\dim V = n$. Then we have:*
   (a) *Any set of independent elements in V is a subset of some basis for V.*
   (b) *Any set of n independent elements is a basis for V.*

Proof. To prove (a), let $S = \{x_1, \ldots, x_k\}$ be any independent set of elements in V. If $L(S) = V$, then S is a basis. If not, there is some element y in V that is not in $L(S)$. Adjoin this element to S and consider the new set $S' = \{x_1, \ldots, x_k, y\}$. We will show that S' is independent. If not, there would be scalars $c_1, \ldots, c_{k+1}$, not all zero, such that

$$\sum_{i=1}^{k} c_i x_i + c_{k+1} y = O.$$

But $c_{k+1} \neq 0$ because $x_1, \ldots, x_k$ are independent. Hence we could solve this equation for y and find that y is in the linear span of S, contradicting the fact that y is not in $L(S)$. Therefore the set S' is independent but contains $k + 1$ elements. If $L(S) = V$, then S' is a basis, and since S is a subset of S', part (a) is proved. But if S' is not a basis, we can argue with S' as we did with S, getting a new set S'' that contains $k + 2$ elements and is independent. If S'' is a basis, then part (a) is proved. If not, we repeat the process. We must arrive at a basis in a finite number of steps, otherwise we would eventually obtain an independent set with $n + 1$ elements, contradicting Theorem 3.5. Therefore part (a) is proved.

To prove (b), let S be any independent set consisting of n elements. By part (a), S is a subset of some basis, say B. But, by Theorem 3.6, basis B has exactly n elements, so $S = B$.

Note. Part (b) tells us that in a linear space of dimension n, every set of n independent elements spans the space.

3.9 Components

Let V be a linear space of dimension n and consider a basis whose elements $e_1, \ldots, e_n$ are taken in a given order. We denote such an ordered basis as an ordered n-tuple of elements $(e_1, \ldots, e_n)$. Every x in V can be expressed as a linear combination of these basis elements:

$$(3.6) \qquad\qquad x = \sum_{i=1}^{n} c_i e_i.$$

The coefficients in this equation comprise an n-tuple of numbers $(c_1, \ldots, c_n)$ that is uniquely determined by x. In fact, if we have another representation of x as a linear combination of the same basis elements, say $x = \sum_{i=1}^{n} d_i e_i$, then by subtraction from (3.6) we find that $\sum_{i=1}^{n} (c_i - d_i) e_i = O$. But the basis elements are independent, so $c_i - d_i = 0$ for each i, hence $(c_1, \ldots, c_n) = (d_1, \ldots, d_n)$.

The components of the ordered n-tuple $(c_1, \ldots, c_n)$ determined by Equation (3.6) are called the *components of x relative to the ordered basis* $(e_1, \ldots, e_n)$.

EXAMPLE 1. Let P_n denote the linear space of all real polynomials of degree $\leq n$, where n is fixed. This space has dimension $n + 1$ and one basis is the set $\{1, t, t^2, \ldots, t^n\}$. If a polynomial $f(t)$ has the form $f(t) = c_0 + c_1 t + \cdots + c_n t^n$, then the coefficients c_0, $c_1, \ldots, c_n$ are the components of f relative to the ordered basis $(1, t, t^2, \ldots, t^n)$.

EXAMPLE 2. The set of all linear combinations of the functions e^x and e^{-x} is a linear space of dimension 2 with basis $\{e^x, e^{-x}\}$. The hyperbolic functions $\cosh x = (e^x + e^{-x})/2$ and $\sinh x = (e^x - e^{-x})/2$ are elements of this space whose components relative to the ordered basis (e^x, e^{-x}) are $(\frac{1}{2}, \frac{1}{2})$ and $(\frac{1}{2}, -\frac{1}{2})$, respectively.

3.10 Exercises

In each of Exercises 1 through 10, let S denote the set of all vectors (x, y, z) in $\mathbf{R}^3$ whose components satisfy the condition given. Determine whether S is a subspace of $\mathbf{R}^3$. If S is a subspace, compute $\dim S$.

1. $x = 0$.	6. $x = y$ or $x = z$.
2. $x + y = 0$.	7. $x^2 - y^2 = 0$.
3. $x + y + z = 0$.	8. $x + y = 1$.
4. $x = y$.	9. $y = 2x$ and $z = 3x$.
5. $x = y = z$.	10. $x + y + z = 0$ and $x - y - z = 0$.

Let P_n denote the linear space of all real polynomials of degree $\leq n$, where n is fixed. In each of Exercises 11 through 22, let S denote the set of all polynomials f in P_n satisfying the condition given. Determine whether or not S is a subspace of P_n. If S is a subspace, compute $\dim S$. (Exercises 19 through 22 require a knowledge of calculus.)

11. $f(0) = 0$.	17. f has degree $\leq k$, where $k < n$, or $f = 0$.
12. $f(0) = f(2)$.	18. f has degree k, where $k < n$, or $f = 0$.
13. $f(0) + f(1) = 0$.	19. $f'(0) = 0$.
14. f is even.	20. $f'(0) = 0$ and $f(0) = 0$.
15. f is odd	21. $f''(0) = 0$.
16. $f(x) \geq 0$ for all x.	22. $f'(0) + f(0) = 0$.

23. In the linear space of all real polynomials $p(t)$, describe the subspace spanned by each of the following subsets of polynomials, and determine the dimension of the subspace.

(a) $\{1, t^2, t^4\}$ (b) $\{t, t^3, t^5\}$ (c) $\{t, t^2\}$ (d) $\{1 + t, (1 + t)^2\}$.

24. Let V be the linear space consisting of all real-valued functions defined on the real line. Determine whether each of the following subsets of V is dependent or independent. Compute the dimension of the subspace spanned by each set.

    (a) $\{1, e^{ax}, e^{bx}\}, a \neq b$.   (f) $\{\cos x, \sin x\}$.
    (b) $\{e^{ax}, xe^{ax}\}$.   (g) $\{\cos^2 x, \sin^2 x\}$.
    (c) $\{1, e^{ax}, xe^{ax}\}$.   (h) $\{1, \cos 2x, \sin^2 x\}$.
    (d) $\{e^{ax}, xe^{ax}, x^2 e^{ax}\}$.   (i) $\{\sin x, \sin 2x\}$.
    (e) $\{e^x, e^{-x}, \cosh x\}$.   (j) $\{e^x \cos x, e^{-x} \sin x\}$.

25. In this exercise, $L(S)$ denotes the subspace spanned by a subset S of a linear space V. Prove each of the statements (a) through (f).

    (a) $S \subseteq L(S)$.
    (b) If $S \subseteq T \subseteq V$ and if T is a subspace of V, then $L(S) \subseteq T$. This property is described by saying that $L(S)$ is the *smallest* subspace of V that contains S.
    (c) A subset S of V is a subspace of V if and only if $L(S) = S$.
    (d) If $S \subseteq T \subseteq V$, then $L(S) \subseteq L(T)$.
    (e) If S and T are subspaces of V, then so is $S \cap T$.
    (f) If S and T are subsets of V, then $L(S \cap T) \subseteq L(S) \cap L(T)$.
    (g) Give an example in which $L(S \cap T) \neq L(S) \cap L(T)$.

26. Let V be a finite-dimensional linear space, and let S be a subspace of V. Prove each of the following statements.

    (a) S is finite-dimensional and $\dim S \leq \dim V$.
    (b) $\dim S = \dim V$ if and only if $S = V$.
    (c) Every basis for S is part of a basis for V.
    (d) A basis for V need not contain a basis for S.

3.11 Inner products, Euclidean spaces. Norms

In ordinary Euclidean geometry, properties of geometric figures that rely on lengths of line segments and angles between lines are called *metric* properties. In our study of $\mathbf{R}^n$ we defined lengths and angles in terms of the dot product. Now we wish to extend these ideas to more general linear spaces. We do this by introducing first a generalization of the dot product, which we call an *inner product*, and then we define length and angle in terms of the inner product.

The dot product $x \cdot y$ of two vectors $x = (x_1, \ldots, x_n)$ and $y = (y_1, \ldots, y_n)$ in $\mathbf{R}^n$ was defined in Chapter 1 by the formula

$$(3.7) \qquad\qquad x \cdot y = \sum_{i=1}^{n} x_i y_i.$$

In a general linear space, we write (x, y) instead of $x \cdot y$ for inner products, and we define the product axiomatically rather than by a specific formula. That is, we list a number of properties we want inner products to satisfy, and we regard these properties as *axioms*.

DEFINITION. *A real linear space V is said to have an inner product if for each pair of elements x and y in V there corresponds a unique real number (x, y) satisfying the following axioms for all choices of x, y, z in V and all real scalars c:*

1. $(x, y) = (y, x)$ *(commutativity, or symmetry),*
2. $(x, y + z) = (x, y) + (x, z)$ *(distributivity, or linearity),*
3. $c(x, y) = (cx, y)$ *(associativity, or homogeneity),*
4. $(x, x) > 0$ if $x \neq O$ *(positivity).*

A real linear space with an inner product is called a *real Euclidean space*.

> *Note.* Taking $c = 0$ in (3) we find that $(O, y) = 0$ for all y.

In a complex linear space, an inner product (x, y) is a complex number satisfying the same axioms as those for a real inner product, except that the symmetry axiom is replaced by the relation

$$(1')\qquad\qquad (x, y) = \overline{(y, x)} \qquad \text{(Hermitian}^*\ \text{symmetry}),$$

where $\overline{(y, x)}$ denotes the complex conjugate of (y, x). In the homogeneity axiom (3), the scalar c can be any complex number. From (3) and $(1')$, we get the companion relation

$$(3')\qquad\qquad (x, cy) = \overline{(cy, x)} = \overline{c}\,\overline{(y, x)} = \overline{c}(x, y).$$

In other words, scalars removed from the second factor of an inner product must be conjugated.

A complex linear space with an inner product is called a *complex Euclidean space*. (The term *unitary space* is also used.) One example is complex space $\mathbf{C}^n$ discussed briefly in Section 1.16.

Although we are interested primarily in examples of real Euclidean spaces, the theorems of this chapter are valid for complex Euclidean spaces as well. When we use the term Euclidean space without further designation, it is to be understood that the space can be real or complex.

The reader should verify that each of the following examples satisfies all the axioms for a real-valued inner product.

EXAMPLE 1.   In $\mathbf{R}^n$, let $(x, y) = x \cdot y$, the usual dot product of x and y as defined in Eq. (3.7).

EXAMPLE 2.   In $\mathbf{R}^2$, define the inner product of any two vectors $x = (x_1, x_2)$ and $y = (y_1, y_2)$ by the formula

$$(x, y) = 2x_1 y_1 + x_1 y_2 + x_2 y_1 + x_2 y_2.$$

This example shows that there may be more than one inner product in a given linear space.

The next three examples are for readers with a knowledge of integral calculus.

EXAMPLE 3.   Let $C(a, b)$ denote the linear space of all real-valued functions continuous on an interval $[a, b]$. Define an inner product of two continuous functions f and g by the formula

$$(f, g) = \int_a^b f(t)g(t)\, dt.$$

This formula is analogous to Equation (3.7) for the dot product of two vectors in $\mathbf{R}^n$. The function values $f(t)$, $g(t)$ play the role of the components x_i, y_i, and integration takes the place of summation.

---

*In honor of Charles Hermite (1822–1901), a French mathematician who made many contributions to algebra and analysis.

EXAMPLE 4.   In the space $C(a, b)$ of Example 3, define

$$(f, g) = \int_a^b w(t) f(t) g(t) \, dt,$$

where w is a fixed positive function in $C(a, b)$. The function w is called a *weight function*. In Example 3 we have $w(t) = 1$ for all t.

EXAMPLE 5.   In the linear space of all real polynomials, define

$$(f, g) = \int_0^\infty e^{-t} f(t) g(t) \, dt.$$

Because of the exponential factor, this improper integral converges for all polynomials f and g.

THEOREM 3.8.   *In a Euclidean space V, every inner product satisfies the Cauchy-Schwarz inequality:*

$$|(x, y)|^2 \leq (x, x)(y, y) \qquad \text{for all } x \text{ and } y \text{ in } V.$$

Moreover, the equality sign holds if and only if x and y are dependent.

Proof.   If either $x = O$ or $y = O$ the result holds trivially, so we can assume that both x and y are nonzero. Let $z = ax + by$, where a and b are scalars to be specified later. We have the inequality $(z, z) \geq 0$ for all a and b. When we express this inequality in terms of x and y with an appropriate choice of a and b we will obtain the Cauchy-Schwarz inequality.
To express (z, z) in terms of x and y we use properties $(1')$, (2) and $(3')$ to obtain

$$(z, z) = (ax + by, ax + by) = (ax, ax) + (ax, by) + (by, ax) + (by, by)$$

$$= a\bar{a}(x, x) + a\bar{b}(x, y) + b\bar{a}(y, x) + b\bar{b}(y, y) \geq 0.$$

Taking $a = (y, y)$ and cancelling a positive factor (y, y) in the inequality we obtain

$$(y, y)(x, x) + \bar{b}(x, y) + b(y, x) + b\bar{b} \geq 0.$$

Now we take $b = -(x, y)$. Then $\bar{b} = -(y, x)$ and the last inequality simplifies to

$$(y, y)(x, x) \geq (x, y)(y, x).$$

Since $(x, y)(y, x) = |(x, y)|^2$ this proves the Cauchy-Schwarz inequality. The equality sign holds throughout the proof if and only if $z = O$. This holds, in turn, if and only if x and y are dependent.

EXAMPLE.   When we apply the Cauchy-Schwarz inequality to the inner product space in Example 3, we obtain the following inequality involving integrals:

$$\left(\int_a^b f(t) g(t) \, dt \right)^2 \leq \left(\int_a^b f^2(t) \, dt \right) \left(\int_a^b g^2(t) \, dt \right).$$

The inner product is used to introduce the metric concept of length in any Euclidean space.

DEFINITION.   *In a Euclidean space V, the nonnegative number* $(x, x)^{1/2}$ *is called the norm of the element x and is denoted by* $\|x\|$.

When the Cauchy-Schwarz inequality is expressed in terms of norms, it takes the form

$$|(x, y)| \leq \|x\|\|y\|.$$

Since it may be possible to define an inner product in many different ways, the norm of an element will depend on the choice of inner product. This lack of uniqueness is to be expected. It is analogous to the fact that we can assign different numbers to measure the length of a given line segment, depending on the choice of scale or unit of measurement. The next theorem gives fundamental properties of norms that are valid regardless of the choice of inner product.

THEOREM 3.9.   *In a Euclidean space, every norm has the following properties for all elements x and y and all scalars c:*
(a) $\|x\| = 0$ *if* $x = O$.
(b) $\|x\| > 0$ *if* $x \neq O$ *(positivity).*
(c) $\|cx\| = |c|\|x\|$ *(homogeneity).*
(d) $\|x + y\| \leq \|x\| + \|y\|$ *(triangle inequality).*
The equality sign holds in (d) *if* $x = O$, *if* $y = O$, *or if* $y = cx$ *for some* $c > 0$.

Proof.   Properties (a), (b), (c) follow at once from the axioms for an inner product. To prove (d), note that

$$\|x + y\|^2 = (x + y, x + y) = (x, x) + (y, y) + (x, y) + (y, x)$$
$$= \|x\|^2 + \|y\|^2 + (x, y) + \overline{(x, y)}.$$

The sum $(x, y) + \overline{(x, y)}$ is real. By the Cauchy-Schwarz inequality we have $|(x, y)| \leq \|x\|\|y\|$ and $|\overline{(x, y)}| \leq \|x\|\|y\|$, so

$$\|x + y\|^2 \leq \|x\|^2 + \|y\|^2 + 2\|x\|\|y\| = \left(\|x\| + \|y\|\right)^2.$$

This proves (d). When $y = cx$, where $c > 0$, we have

$$\|x + y\| = \|x + cx\| = (1 + c)\|x\| = \|x\| + \|cx\| = \|x\| + \|y\|.$$

DEFINITION.   *In a real Euclidean space V, the angle between two nonzero elements x and y is defined to be that number in the interval* $0 \leq \theta \leq \pi$ *that satisfies the equation*

(3.8) $$\cos \theta = \frac{(x, y)}{\|x\|\|y\|}.$$

Note.   The Cauchy-Schwarz inequality shows that the quotient on the right of (3.8) lies in the interval $[-1, 1]$, so there is exactly one θ in $[0, \pi]$ whose cosine is equal to this quotient.

3.12 Orthogonality in a Euclidean space

DEFINITION. *In a Euclidean space V, two elements x and y are called orthogonal if their inner product is zero. A subset S of V is called an orthogonal set if $(x, y) = 0$ for every pair of distinct elements x and y in S. An orthogonal set is called orthonormal if each of its elements has norm equal to 1.*

The element O is orthogonal to every element of V, and (by Axiom 4) is the only element orthogonal to itself.

Two vectors A and B in $\mathbf{R}^n$ are orthogonal if and only if they satisfy the Pythagorean identity $\|A + B\|^2 = \|A\|^2 + \|B\|^2$. The next theorem extends this result to any Euclidean space.

THEOREM 3.10. *In any Euclidean space V, two elements x and y are orthogonal if and only if they satisfy the Pythagorean identity*

$$\|x + y\|^2 = \|x\|^2 + \|y\|^2.$$

Proof. This follows immediately from the identity

$$\|x + y\|^2 = (x + y, x + y) = \|x\|^2 + \|y\|^2 + (x, y) + (y, x).$$

Next we relate orthogonality with independence.

THEOREM 3.11. *In a Euclidean space V, every orthogonal set of nonzero elements is independent. In particular, in a finite-dimensional Euclidean space with* dim $V = n$, *every orthogonal set consisting of n nonzero elements is a basis for V.*

Proof. Let S be an orthogonal set of nonzero elements in V, and suppose some finite linear combination of elements of S is zero, say,

$$\sum_{i=1}^{k} c_i x_i = O,$$

where each $x_i \in S$. Taking the inner product of each member of this equation with x_1 and using the fact that $(x_1, x_i) = 0$ if $i \neq 1$, we find that $c_1(x_1, x_1) = 0$. But $(x_1, x_1) \neq 0$ because $x_1 \neq O$, so $c_1 = 0$. Repeating the argument with x_1 replaced by x_j we find that each $c_j = 0$. This proves that S is independent. If dim $V = n$ and if S consists of n elements, Theorem 3.7(b) shows that S is a basis for V.

EXAMPLE 1. (For readers having acquaintance with integral calculus.) In the real linear space $C(0, 2\pi)$ with the inner product $(f, g) = \int_0^{2\pi} f(x)g(x)\,dx$, let S be the set of trigonometric functions $\{u_0, u_1, u_2, \ldots\}$ given by

$$u_0(x) = 1, \qquad u_{2n-1}(x) = \cos nx, \qquad u_{2n}(x) = \sin nx, \qquad \text{for } n = 1, 2, \ldots.$$

It is known that

$$\int_0^{2\pi} u_n(x)u_m(x)\,dx = 0 \qquad \text{if } m \neq n.$$

(These are called orthogonality relations for $\cos nx$ and $\sin mx$.) Therefore, S is an orthogonal set and, since no member of S is the zero element, S is independent.

The norm of each element of S is easily calculated. We have $(u_0, u_0) = \int_0^{2\pi} dx = 2\pi$ and, for $n \geq 1$, we have

$$(u_{2n-1}, u_{2n-1}) = \int_0^{2\pi} \cos^2 nx \, dx = \pi, \qquad (u_{2n}, u_{2n}) = \int_0^{2\pi} \sin^2 nx \, dx = \pi.$$

Therefore, $\|u_0\| = \sqrt{2\pi}$ and $\|u_n\| = \sqrt{\pi}$ for $n \geq 1$. Dividing each u_n by its norm, we obtain an orthonormal set $\{\varphi_0, \varphi_1, \varphi_2, \ldots\}$, where $\varphi_n = u_n / \|u_n\|$. Thus, we have

$$\varphi_0(x) = \frac{1}{\sqrt{2\pi}}, \qquad \varphi_{2n-1}(x) = \frac{\cos nx}{\sqrt{\pi}}, \qquad \varphi_{2n}(x) = \frac{\sin nx}{\sqrt{\pi}}, \qquad \text{for } n \geq 1.$$

In Section 3.14 we will prove that every finite-dimensional Euclidean space has an orthogonal basis. The next theorem shows how to compute the components of an element relative to this basis.

THEOREM 3.12. *Let V be a finite-dimensional Euclidean space with* $\dim V = n$, *and assume that $S = \{e_1, \ldots, e_n\}$ is an orthogonal basis for V. If an element x in S is expressed as a linear combination of the basis elements, say*

$$(3.9) \qquad x = \sum_{i=1}^{k} c_i e_i,$$

then its components relative to the ordered basis $(e_1, \ldots, e_n)$ are given by the formulas

$$(3.10) \qquad c_j = \frac{(x, e_j)}{(e_j, e_j)} \qquad \text{for } j = 1, 2, \ldots, n.$$

In particular, if S is an orthonormal basis, each c_j is given by

$$(3.11) \qquad c_j = (x, e_j).$$

Proof. Taking the inner product of each member of (3.9) with e_j, we obtain

$$(x, e_j) = \sum_{i=1}^{k} c_i(e_i, e_j) = c_j(e_j, e_j)$$

since $(e_i, e_j) = 0$ if $i \neq j$. This implies (3.10), and when $(e_j, e_j) = 1$ we obtain (3.11).

If $\{e_1, \ldots, e_n\}$ is an orthonormal basis, Equation (3.9) can be written in the form

$$(3.12) \qquad x = \sum_{i=1}^{n} (x, e_i) e_i.$$

The next theorem shows that in a finite-dimensional Euclidean space with an orthonormal basis the inner product of two elements can be computed in terms of their components by a formula analogous to that used in $\mathbf{R}^n$ for dot products.

THEOREM 3.13. *Let V be a finite-dimensional Euclidean space of dimension n, and assume that $\{e_1, \ldots, e_n\}$ is an orthonormal basis for V. Then the inner product of any two elements x and y in V is given by*

$$(3.13) \qquad (x, y) = \sum_{i=1}^{n} (x, e_i)\overline{(y, e_i)} \qquad (Parseval's\ formula).$$

In particular, when $x = y$ we have

$$(3.14) \qquad \|x\|^2 = \sum_{i=1}^{n} |(x, e_i)|^2.$$

Proof. Taking the inner product of both members of Equation (3.12) with y and using the linearity property of the inner product, we obtain (3.13). When $x = y$, Eq. (3.13) reduces to (3.14).

Note. Equation (3.13) is named in honor of M. A. Parseval (circa 1776–1836), who obtained this type of formula in a special function space. Equation (3.14), a special case of Parseval's formula, can be regarded as a generalization of the theorem of Pythagoras.

3.13 Exercises

1. Let $x = (x_1, \ldots, x_n)$ and $y = (y_1, \ldots, y_n)$ be arbitrary vectors in $\mathbf{R}^n$. In each case, determine whether (x, y) is an inner product for $\mathbf{R}^n$ if (x, y) is defined by the formula given. In case (x, y) is not an inner product, tell which axioms are not satisfied.
   (a) $(x, y) = \sum_{i=1}^{n} x_i |y_i|$.
   (b) $(x, y) = \left| \sum_{i=1}^{n} x_i y_i \right|$.
   (c) $(x, y) = \sum_{i=1}^{n} x_i \sum_{j=1}^{n} y_j$.
   (d) $(x, y) = \left(\sum_{i=1}^{n} x_i^2 y_i^2 \right)^{1/2}$.
   (e) $(x, y) = \sum_{i=1}^{n} (x_i + y_i)^2 - \sum_{i=1}^{n} x_i^2 - \sum_{i=1}^{n} y_i^2$.
2. Suppose we retain the first three axioms for a real inner product (symmetry, linearity, and homogeneity) but replace the fourth axiom by a new axiom $4'$: $(x, x) = 0$ *if and only if $x = O$*. Prove that either $(x, x) > 0$ for all $x \neq O$ or else $(x, x) < 0$ for all $x \neq O$.
   [*Hint:* Assume $(x, x) > 0$ for some x and $(y, y) < 0$ for some y. In the space spanned by $\{x, y\}$, find an element $z \neq O$ with $(z, z) = 0$.]

Prove that each of the statements in Exercises 3 through 7 is valid for all elements x and y in a real Euclidean space.

3. $(x, y) = 0$ if and only if $\|x + y\| = \|x - y\|$.
4. $(x, y) = 0$ if and only if $\|x + y\|^2 = \|x\|^2 + \|y\|^2$.
5. $(x, y) = 0$ if and only if $\|x + cy\| \geq \|x\|$ for all real c.
6. $(x + y, x - y) = 0$ if and only if $\|x\| = \|y\|$.
7. If x and y are nonzero elements making an angle θ with each other, then

$$\|x - y\|^2 = \|x\|^2 + \|y\|^2 - 2\|x\|\|y\| \cos \theta.$$

8. Prove that the following identities are valid in every Euclidean space.
   (a) $\|x + y\|^2 = \|x\|^2 + \|y\|^2 + (x, y) + (y, x)$.
   (b) $\|x + y\|^2 - \|x - y\|^2 = 2(x, y) + 2(y, x)$.
   (c) $\|x + y\|^2 + \|x - y\|^2 = 2\|x\|^2 + 2\|y\|^2$.

9. In a complex Euclidean space, prove that the inner product has the following properties for all elements x, y, z and all complex a and b.
   (a) $(ax, by) = a\bar{b}(x, y)$. (b) $(x, ay + bz) = \bar{a}(x, y) + \bar{b}(x, z)$.
10. In the linear space P_n of all polynomials of degree $\leq n$, define

$$(f, g) = \sum_{k=0}^{n} f\left(\frac{k}{n}\right) g\left(\frac{k}{n}\right).$$

   (a) Prove that (f, g) is an inner product for P_n.
   (b) Compute (f, g) when $f(t) = t$ and $g(t) = at + b$.
   (c) If $f(t) = t$, find all linear polynomials g orthogonal to f.

Exercises 11 through 17 require a knowledge of calculus.

11. In the real linear space $C(1, e)$, define an inner product by the equation

$$(f, g) = \int_1^e (\log x) f(x) g(x) \, dx.$$

   (a) If $f(x) = \sqrt{x}$, compute $\|f\|$.
   (b) Find a linear polynomial $g(x) = a + bx$ orthogonal to the constant function $f(x) = 1$.
12. In the real linear space $C(-1, 1)$, let $(f, g) = \int_{-1}^{1} f(t) g(t) \, dt$. Consider the three functions

$$u_1(t) = 1, \qquad u_2(t) = t, \qquad u_3(t) = 1 + t.$$

   Prove that two of them are orthogonal, two make an angle $\pi/3$ with each other, and two make an angle $\pi/6$ with each other.
13. In the linear space of all real polynomials, define $(f, g) = \int_0^\infty e^{-t} f(t) g(t) \, dt$.
   (a) Prove that this improper integral converges absolutely for all polynomials f and g.
   (b) If $x_n(t) = t^n$ for $n = 0, 1, 2, \ldots$, prove that $(x_n, x_m) = (m + n)!$.
   (c) Compute (f, g) when $f(t) = (t + 1)^2$ and $g(t) = t^2 + 1$.
   (d) Find all linear polynomials $g(t) = a + bt$ orthogonal to $f(t) = 1 + t$.
14. In the linear space of all real polynomials, determine whether or not (f, g) is an inner product if (f, g) is defined by the formula given. In case (f, g) is not an inner product, indicate which axioms are violated. In (c), f' and g' denote derivatives.
   (a) $(f, g) = f(1)g(1)$. (c) $(f, g) = \int_0^1 f'(t)g'(t) \, dt$.
   (b) $(f, g) = |\int_0^1 f(t)g(t) \, dt|$. (d) $(f, g) = (\int_0^1 f(t) \, dt)(\int_0^1 g(t) \, dt)$.
15. Let V be the set of all real functions f continuous on $[0, +\infty)$ and such that the integral $\int_0^\infty e^{-t} f^2(t) \, dt$ converges. Define $(f, g) = \int_0^\infty e^{-t} f(t) g(t) \, dt$.
   (a) Prove that the integral for (f, g) converges absolutely for each pair f and g in V.
      [*Hint:* Use the Cauchy-Schwarz inequality to estimate the integral $\int_0^M e^{-t} |f(t)g(t)| \, dt$.]
   (b) Prove that V is a linear space with (f, g) as an inner product.
   (c) Compute (f, g) if $f(t) = e^{-t}$ and $g(t) = t^n$, where $n = 0, 1, 2, \ldots$.
16. Let V consist of all sequences $\{x_n\}$ of real numbers for which the series $\sum_{n=1}^{\infty} x_n^2$ converges. If $x = \{x_n\}$ and $y = \{y_n\}$ are two elements of V, define

$$(x, y) = \sum_{n=1}^{\infty} x_n y_n.$$

   (a) Prove that this series converges absolutely.
      [*Hint:* Use the Cauchy-Schwarz inequality to estimate the sum $\sum_{n=1}^{M} |x_n y_n|$.]
   (b) Prove that V is a linear space with (x, y) as an inner product.
   (c) Compute (x, y) if $x_n = 1/n$ and $y_n = 1/(n + 1)$ for $n \geq 1$.
   (d) Compute (x, y) if $x_n = 2^{-n}$ and $y_n = 1/n!$ for $n \geq 1$.

17. Prove that the space of all complex-valued functions continuous on an interval $[a, b]$ becomes a complex Euclidean space if we define an inner product by the formula

$$(f, g) = \int_a^b w(t) f(t) \overline{g(t)} \, dt,$$

where w is a fixed positive function, continuous on $[a, b]$.

3.14 Construction of orthogonal sets. The Gram-Schmidt process

Every finite-dimensional linear space has a finite basis. If the space is Euclidean, we can always construct an *orthogonal* basis. This result will be deduced as a consequence of a general theorem whose proof shows how to construct orthogonal sets in any Euclidean space, finite- or infinite-dimensional. The construction is called the *Gram-Schmidt orthogonalization process*, in honor of J. P. Gram (1850–1916) and E. Schmidt (1845–1921).

THEOREM 3.14.   ORTHOGONALIZATION THEOREM.   *Let* $x_1, x_2, \ldots$, *be a finite or infinite sequence of elements in a Euclidean space* V, *and let* $L(x_1, \ldots, x_k)$ *denote the subspace spanned by the first* k *of these elements. Then there is a corresponding sequence of elements* $y_1, y_2, \ldots$, *in* V *that has the following properties for each positive integer* k:
   (a) *The element* y_k *is orthogonal to every element in the subspace* $L(y_1, \ldots, y_{k-1})$.
   (b) *The subspace spanned by* $y_1, \ldots, y_k$ *is the same as that spanned by* $x_1, \ldots, x_k$:

$$L(y_1, \ldots, y_k) = L(x_1, \ldots, x_k).$$

   (c) *The sequence* $y_1, y_2, \ldots$, *is unique, except for scalar factors. That is, if* $z_1, z_2, \ldots$, *is another sequence of elements in* V *satisfying properties* (a) *and* (b) *for all* k, *then for each* k *there is a scalar* c_k *such that* $z_k = c_k y_k$.

Proof.   We construct the elements $y_1, y_2, \ldots$, inductively. To start the process, we take $y_1 = x_1$. Now assume we have constructed $y_1, \ldots, y_r$ so that (a) and (b) are satisfied when $k = r$. Then we define y_{r+1} by the equation

(3.15)
$$y_{r+1} = x_{r+1} - \sum_{i=1}^r a_i y_i,$$

where the scalars a_i will be chosen presently. For $j \leq r$, the inner product of y_{r+1} with y_j is given by

$$(y_{r+1}, y_j) = (x_{r+1}, y_j) - \sum_{i=1}^r a_i (y_i, y_j) = (x_{r+1}, y_j) - a_j (y_j, y_j),$$

since $(y_i, y_j) = 0$ if $i \neq j$. If $y_j \neq O$, we can make y_{r+1} orthogonal to y_j by taking

(3.16)
$$a_j = \frac{(x_{r+1}, y_j)}{(y_j, y_j)}.$$

If $y_j = O$, then y_{r+1} is orthogonal to y_j for any choice of a_j, and in this case we choose $a_j = 0$. Thus, the element y_{r+1} is well defined and is orthogonal to each of the earlier elements $y_1, \ldots, y_r$. Therefore it is orthogonal to every element in the subspace

$$L(y_1, \ldots, y_r).$$

This proves (a) when $k = r + 1$.

To prove (b) when $k = r + 1$, we must show that $L(y_1, \ldots, y_{r+1}) = L(x_1, \ldots, x_{r+1})$, given that $L(y_1, \ldots, y_r) = L(x_1, \ldots, x_r)$. The first r elements $y_1, \ldots, y_r$ are in $L(x_1, \ldots, x_r)$ and hence they are also in the larger subspace $L(x_1, \ldots, x_{r+1})$. The new element y_{r+1} given by (3.15) is a difference of two elements in $L(x_1, \ldots, x_{r+1})$ so it, too, is in $L(x_1, \ldots, x_{r+1})$. This proves that

$$L(y_1, \ldots, y_{r+1}) \subseteq L(x_1, \ldots, x_{r+1}).$$

Equation (3.15) shows that x_{r+1} is the sum of two elements in $L(y_1, \ldots, y_{r+1})$ so a similar argument gives inclusion in the other direction:

$$L(x_1, \ldots, x_{r+1}) \subseteq L(y_1, \ldots, y_{r+1}).$$

This proves (b) when $k = r + 1$. Therefore both (a) and (b) are proved by induction on k.

Finally we prove (c) by induction on k. The case $k = 1$ is trivial. Therefore, assume (c) is true for $k = r$ and consider the element z_{r+1}. Because of (b), this element is in the subspace

$$L(y_1, \ldots, y_{r+1})$$

so we can write

$$z_{r+1} = \sum_{i=1}^{r+1} c_i y_i = v_r + c_{r+1} y_{r+1},$$

where $v_r \in L(y_1, \ldots, y_r)$. We wish to prove that $v_r = O$. By property (a), both z_{r+1} and $c_{r+1} y_{r+1}$ are orthogonal to v_r. Therefore, their difference, v_r, is orthogonal to v_r. In other words, v_r is orthogonal to itself, so $v_r = O$. This completes the proof of the orthogonalization theorem.

In the foregoing construction, suppose we have $y_{r+1} = O$ for some r. Then (3.15) shows that x_{r+1} is a linear combination of $y_1, \ldots, y_r$, and hence of $x_1, \ldots, x_r$, so the elements $x_1, \ldots, x_{r+1}$ are dependent. In other words, if the first k elements $x_1, \ldots, x_k$ are independent, then the corresponding elements $y_1, \ldots, y_k$ are *nonzero*. In this case the coefficients a_i in (3.15) are given by (3.16), and the formulas defining $y_1, \ldots, y_k$ become

$$(3.17) \quad y_1 = x_1, \qquad y_{r+1} = x_{r+1} - \sum_{i=1}^{r} \frac{(x_{r+1}, y_i)}{(y_i, y_i)} y_i \qquad \text{for} \qquad r = 1, 2, \ldots, k - 1.$$

These formulas describe the Gram-Schmidt process for constructing an orthogonal set of nonzero elements $y_1, \ldots, y_k$ that span the same subspace as a given independent set $x_1, \ldots, x_k$. In particular, if $x_1, \ldots, x_k$ is a basis for a finite-dimensisonal Euclidean space, then $y_1, \ldots, y_k$ is an orthogonal basis for the same space. And we can convert this to an orthonormal basis by *normalizing* each element y_i, that is, by dividing it by its norm. Therefore, as a corollary of Theorem 3.14 we have the following:

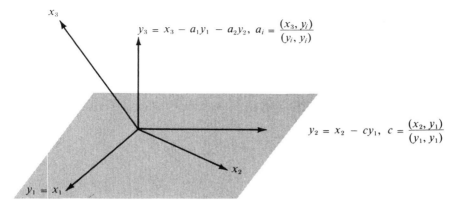

FIGURE 3.1   The Gram-Schmidt process in $\mathbf{R}^3$. An orthogonal set $\{y_1, y_2, y_3\}$ is constructed from a given independent set $\{x_1, x_2, x_3\}$.

THEOREM 3.15.   *Every finite-dimensional Euclidean space has an orthonormal basis.*

DEFINITION.   *If x and y are elements in a Euclidean space, with $y \neq O$, the element*

$$\frac{(x, y)}{(y, y)}y$$

is called the projection of x along y.

The geometric meaning of projection in $\mathbf{R}^2$ is shown in Figure 1.11, p. 38. In the Gram-Schmidt process (3.17), we construct the element y_{r+1} by subtracting from x_{r+1} the projection of x_{r+1} along each of the earlier elements $y_1, \ldots, y_r$. Figure 3.1 illustrates the construction geometrically in the vector space $\mathbf{R}^3$.

EXAMPLE 1.   In $\mathbf{R}^4$, find an orthonormal basis for the subspace spanned by the three vectors $x_1 = (1, -1, 1, -1)$, $x_2 = (5, 1, 1, 1)$, and $x_3 = (-3, -3, 1, -3)$.

Solution. Applying the Gram-Schmidt process, we find

$$y_1 = x_1 = (1, -1, 1, -1),$$

$$y_2 = x_2 - \frac{(x_2, y_1)}{(y_1, y_1)}y_1 = x_2 - y_1 = (4, 2, 0, 2),$$

$$y_3 = x_3 - \frac{(x_3, y_1)}{(y_1, y_1)}y_1 - \frac{(x_3, y_2)}{(y_2, y_2)}y_2 = x_3 - y_1 + y_2 = (0, 0, 0, 0).$$

Since $y_3 = O$, the three vectors x_1, x_2, x_3 must be dependent. But since y_1 and y_2 are nonzero, the vectors x_1 and x_2 are independent. Therefore $L(x_1, x_2, x_3)$ is a subspace of dimension 2. The set $\{y_1, y_2\}$ is an orthogonal basis for this subspace. Dividing each of y_1 and y_2 by its norm we get an orthonormal basis consisting of the two vectors

$$\frac{y_1}{\|y_1\|} = \frac{1}{2}(1, -1, 1, -1) \qquad \text{and} \qquad \frac{y_2}{\|y_2\|} = \frac{1}{\sqrt{6}}(2, 1, 0, 1).$$

EXAMPLE 2. *The Legendre polynomials.* (For readers acquainted with calculus.) In the linear space of all polynomials, with the inner product $(x, y) = \int_{-1}^{1} x(t)y(t)\,dt$, consider the infinite sequence $x_0, x_1, x_2, \ldots$, where $x_n(t) = t^n$. When the orthogonalization theorem is applied to this sequence it yields another sequence of polynomials $y_0, y_1, y_2, \ldots$, first encountered by the French mathematician A. M. Legendre (1752–1833) in his work on potential theory. The first few polynomials are easily calculated by the Gram-Schmidt process. First of all, we have $y_0(t) = x_0(t) = 1$. Since

$$(y_0, y_0) = \int_{-1}^{1} dt = 2 \quad \text{and} \quad (x_1, y_0) = \int_{-1}^{1} t\,dt = 0,$$

we find that

$$y_1(t) = x_1(t) - \frac{(x_1, y_0)}{(y_0, y_0)} y_0(t) = x_1(t) = t.$$

Next, we use the relations

$$(x_2, y_0) = \int_{-1}^{1} t^2\,dt = \frac{2}{3}, \quad (x_2, y_1) = \int_{-1}^{1} t^3\,dt = 0, \quad (y_1, y_1) = \int_{-1}^{1} t^2\,dt = \frac{2}{3},$$

to obtain

$$y_2(t) = x_2(t) - \frac{(x_2, y_0)}{(y_0, y_0)} y_0(t) - \frac{(x_2, y_1)}{(y_1, y_1)} y_1(t) = t^2 - \frac{1}{3}.$$

Similarly, we find that

$$y_3(t) = t^3 - \frac{3}{5}t, \quad y_4(t) = t^4 - \frac{6}{7}t^2 + \frac{3}{35}, \quad y_5(t) = t^5 - \frac{10}{9}t^3 + \frac{5}{21}t.$$

These polynomials occur in the study of differential equations, where it is shown that

$$y_n(t) = \frac{n!}{(2n)!} \frac{d^n}{dt^n}(t^2 - 1)^n.$$

The polynomials P_n given by

$$P_n(t) = \frac{(2n)!}{2^n(n!)^2} y_n(t) = \frac{1}{2^n n!} \frac{d^n}{dt^n}(t^2 - 1)^n$$

are known as *Legendre polynomials*. The polynomials in the corresponding orthonormal sequence $\varphi_0, \varphi_1, \varphi_2, \ldots$, given by $\varphi_n = y_n/\|y_n\|$ are called the *normalized Legendre polynomials*. From the formulas for $y_0, \ldots, y_5$ given above, we find the first few normalized polynomials:

$$\varphi_0(t) = \sqrt{\tfrac{1}{2}}, \quad \varphi_1(t) = \sqrt{\tfrac{3}{2}}t, \quad \varphi_2(t) = \tfrac{1}{2}\sqrt{\tfrac{5}{2}}(3t^2 - 1), \quad \varphi_3(t) = \tfrac{1}{2}\sqrt{\tfrac{7}{2}}(5t^3 - 3t),$$

$$\varphi_4(t) = \tfrac{1}{8}\sqrt{\tfrac{9}{2}}(35t^4 - 30t^2 + 3), \quad \varphi_5(t) = \tfrac{1}{8}\sqrt{\tfrac{11}{2}}(63t^5 - 70t^3 + 15t).$$

3.15 Orthogonal complements. Projections

Let V be a Euclidean space and let S be a finite-dimensional subspace. We turn now to the following approximation problem: *Given an element x in V, determine an element in S whose distance from x is as small as possible.* The distance between two elements x and y in V is defined to be the norm $\|x - y\|$.

Before discussing this problem in its general form, we consider a special case, illustrated in Figure 3.2. Here $V = \mathbf{R}^3$ and S is a two-dimensional subspace, a plane through the origin. Given x in V, the problem is to find, in the plane S, that point s nearest to x.

If $x \in S$, then clearly $s = x$ is the solution. If x is not in S, then the nearest point s is obtained by dropping a perpendicular from x to the plane. This simple example suggests an approach to the general approximation problem and motivates the discussion that follows.

DEFINITION. *Let S be a subset of a Euclidean space V. An element in V is said to be orthogonal to S if it is orthogonal to every element of S. The set of all elements orthogonal to S is denoted by $S^\perp$ and is called "S perpendicular."*

It is a simple exercise to verify that $S^\perp$ is a subspace of V, whether or not S itself is one. In case S is a subspace, then $S^\perp$ is called the *orthogonal complement* of S.

EXAMPLE. If S is a plane through the origin, as shown in Figure 3.2, then $S^\perp$ is a line through the origin perpendicular to this plane.

DEFINITION. *In any Euclidean space V, let S be a finite-dimensional subspace with an orthonormal basis $\{e_1, \ldots, e_n\}$. If $x \in V$, the element $p(x)$ in S defined by the equation*

(3.18)
$$p(x) = \sum_{i=1}^{n} (x, e_i) e_i$$

is called the projection of x on the subspace S.

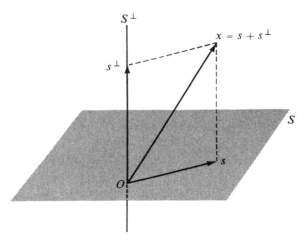

FIGURE 3.2 The orthogonal complement of a plane through the origin in $\mathbf{R}^3$ is a line perpendicular to the plane. Here $s = p(x)$ and $s^\perp = x - p(x)$.

Note that $(p(x), e_k) = (x, e_k)$, hence $(x - p(x), e_k) = 0$ for each basis element e_k, so $x - p(x) \in S^\perp$. We prove next that the projection $p(x)$ is the solution to the approximation problem stated at the beginning of this section.

3.16   Best approximation of elements in a Euclidean space by elements in a finite-dimensional subspace

THEOREM 3.16.   APPROXIMATION THEOREM.   *Let S be a finite-dimensional subspace of a Euclidean space V, and let x be any element of V. Then $p(x)$, the projection of x on S, is nearer to x than any other element of S. That is, we have*

$$\|x - p(x)\| \le \|x - t\| \qquad \text{for all } t \text{ in } S.$$

The equality sign holds if and only if $t = p(x)$.

Proof.   If $t \in S$ write $x - t = \{x - p(x)\} + \{p(x) - t\}$. Now $x - p(x) \in S^\perp$ and $p(x) - t \in S$, so $x - t$ is written as the sum of two orthogonal elements. By the Pythgorean formula we have

$$\|x - t\|^2 = \|x - p(x)\|^2 + \|p(x) - t\|^2.$$

But $\|p(x) - t\|^2 \ge 0$ so $\|x - t\|^2 \ge \|x - p(x)\|^2$, with equality holding if and only if $t = p(x)$.

Note.   Because x is the sum of two orthogonal elements $p(x)$ and $x - p(x)$, the Pythagorean formula tells us that $\|x\|^2 = \|p(x)\|^2 + \|x - p(x)\|^2$, so the minimal distance from x to the subspace S can be calculated from the relation

(3.19) $$\|x - p(x)\|^2 = \|x\|^2 - \|p(x)\|^2.$$

The easiest way to calculate $\|p(x)\|$ is to use the definition of $p(x)$ in (3.18), which gives us

$$\|p(x)\|^2 = \sum_{i=1}^{n} |(x, e_i)|^2.$$

We illustrate the approximation theorem with two examples involving calculus.

EXAMPLE 1.   *Approximation of continuous functions on $[0, 2\pi]$ by trigonometric polynomials.*   Let $V = C(0, 2\pi)$ be the linear space of all real functions continuous on the interval $[0, 2\pi]$, and define an inner product by the usual integral formula $(f, g) = \int_0^{2\pi} f(x)g(x)\,dx$. In Section 3.12 we exhibited an orthonormal set of trigonometric functions $\varphi_0, \varphi_1, \varphi_2, \dots,$ where

(3.20) $$\varphi_0(x) = \frac{1}{\sqrt{2\pi}}, \qquad \varphi_{2k-1}(x) = \frac{\cos kx}{\sqrt{\pi}}, \qquad \varphi_{2k}(x) = \frac{\sin kx}{\sqrt{\pi}}, \qquad \text{for } k \ge 1.$$

The $2n + 1$ elements $\varphi_0, \varphi_1, \dots, \varphi_{2n}$ span a subspace S of dimension $2n + 1$. The elements of S are called *trigonometric polynomials*.

If $f \in C(0, 2\pi)$, let f_n denote the projection of f on the subspace S. Then we have

$$(3.21) \qquad f_n = \sum_{k=0}^{2n} (f, \varphi_k)\varphi_k, \qquad \text{where} \qquad (f, \varphi_k) = \int_0^{2\pi} f(x)\varphi_k(x)\,dx.$$

The numbers (f, φ_k) are called *Fourier coefficients of* f. Using the formulas in (3.20) we can rewrite (3.21) in the form

$$(3.22) \qquad f_n(x) = \frac{1}{2}a_0 + \sum_{k=1}^{n} (a_k \cos kx + b_k \sin kx),$$

where

$$a_k = \frac{1}{\pi}\int_0^{2\pi} f(x) \cos kx\,dx, \qquad b_k = \frac{1}{\pi}\int_0^{2\pi} f(x) \sin kx\,dx,$$

for $k = 0, 1, 2, \ldots, n$. The approximation theorem tells us that the trigonometric polynomials in (3.22) approximate f better than any other trigonometric polynomial in S, in the sense that the norm $\|f - f_n\|$ is as small as possible.

EXAMPLE 2. *Approximation of continuous functions on the interval* $[-1, 1]$ *by polynomials of degree* $\leq n$. Let $V = C(-1, 1)$ be the space of real continuous functions on $[-1, 1]$, and define an inner product by the equation $(f, g) = \int_{-1}^{1} f(x)g(x)\,dx$. The $n + 1$ normalized Legendre polynomials $\varphi_0, \varphi_1, \ldots, \varphi_n$ introduced in Section 3.14 span a subspace S of dimension $n + 1$ consisting of all polynomials of degree $\leq n$. If $f \in C(-1, 1)$, let f_n denote the projection of f on S. Then we have

$$f_n = \sum_{k=0}^{n} (f, \varphi_k)\varphi_k, \qquad \text{where} \qquad (f, \varphi_k) = \int_{-1}^{1} f(t)\varphi_k(t)\,dt.$$

This is the polynomial of degree $\leq n$ for which the norm $\|f - f_n\|$ is smallest. For example, when $f(x) = \sin \pi x$, the coefficients (f, φ_k) are given by

$$(f, \varphi_k) = \int_{-1}^{1} \sin \pi t\ \varphi_k(t)\,dt.$$

In particular, we have $(f, \varphi_0) = 0$ and

$$(f, \varphi_1) = \int_{-1}^{1} \sqrt{\frac{3}{2}}\,t \sin \pi t\,dt = \sqrt{\frac{3}{2}}\frac{2}{\pi}.$$

Therefore the linear polynomial $f_1(t)$ that is nearest to $\sin \pi t$ on the interval $[-1, 1]$ is

$$f_1(t) = \sqrt{\frac{3}{2}}\frac{2}{\pi}\varphi_1(t) = \frac{3}{\pi}t.$$

Since $(f, \varphi_2) = 0$, this is also the nearest quadratic approximation.

3.17   Exercises

1. In each case find an orthonormal basis for the subspace of $\mathbf{R}^3$ spanned by the given vectors.
   (a) $x_1 = (1, 1, 1)$, $x_2 = (1, 0, 1)$, $x_3 = (3, 2, 3)$.
   (b) $x_1 = (1, 1, 1)$, $x_2 = (-1, 1, -1)$, $x_3 = (1, 0, 1)$.
2. In each case find an orthonormal basis for the subspace of $\mathbf{R}^4$ spanned by the given vectors.
   (a) $x_1 = (1, 1, 0, 0)$, $x_2 = (0, 1, 1, 0)$, $x_3 = (0, 0, 1, 1)$, $x_4 = (1, 0, 0, 1)$.
   (b) $x_1 = (1, 1, 0, 1)$, $x_2 = (1, 0, 2, 1)$, $x_3 = (1, 2, -2, 1)$.

Exercises 3 through 10 require a knowledge of calculus.

3. In the real linear space $C(0, \pi)$ with inner product $(x, y) = \int_0^\pi x(t)y(t)\,dt$, let $x_n(t) = \cos nt$ for $n = 0, 1, 2, \ldots$. Prove that the functions $y_0, y_1, y_2, \ldots$, given by

$$y_0(t) = \frac{1}{\sqrt{\pi}} \qquad \text{and} \qquad y_n(t) = \sqrt{\frac{2}{\pi}} \cos nt \qquad \text{for } n \geq 1,$$

   form an orthonormal set spanning the same subspace as $x_0, x_1, x_2, \ldots$.
4. In the linear space of all real polynomials, with inner product $(x, y) = \int_0^1 x(t)y(t)\,dt$, let $x_n(t) = t^n$ for $n = 0, 1, 2, \ldots$. Prove that the functions

$$y_0(t) = 1, \qquad y_1(t) = \sqrt{3}(2t - 1), \qquad y_2(t) = \sqrt{5}(6t^2 - 6t + 1),$$

   form an orthonormal set spanning the same subspace as $\{x_0, x_1, x_2\}$.
5. Let V be the linear space of all real functions f continuous on $[0, +\infty)$ such that the integral $\int_0^\infty e^{-t} f^2(t)\,dt$ converges. Define $(f, g) = \int_0^\infty e^{-t} f(t)g(t)\,dt$, and let $y_0, y_1, y_2, \ldots$, be the set obtained by applying the Gram-Schmidt process to $x_0, x_1, x_2, \ldots$, where $x_n(t) = t^n$ for $n \geq 0$. Prove that $y_0(t) = 1$, $y_1(t) = t - 1$, $y_2(t) = t^2 - 4t + 2$, $y_3(t) = t^3 - 9t^2 + 18t - 6$.
6. In the real linear space $C(1, 3)$ with inner product $(f, g) = \int_1^3 f(x)g(x)\,dx$, let $f(x) = 1/x$ and show that the constant polynomial g nearest to f is $g = \frac{1}{2} \log 3$. Compute $\|g - f\|^2$ for this g.
7. In the real linear space $C(0, 2)$ with inner product $(f, g) = \int_0^2 f(x)g(x)\,dx$, let $f(x) = e^x$ and show that the constant polynomial g nearest to f is $g = \frac{1}{2}(e^2 - 1)$. Compute $\|g - f\|^2$ for this g.
8. In the real linear space $C(-1, 1)$ with inner product $(f, g) = \int_{-1}^1 f(x)g(x)\,dx$, let $f(x) = e^x$ and find the linear polynomial g nearest to f. Compute $\|g - f\|^2$ for this g.
9. In the real linear space $C(0, 2\pi)$ with inner product $(f, g) = \int_0^{2\pi} f(x)g(x)\,dx$, let $f(x) = x$. In the subspace spanned by $u_0(x) = 1$, $u_1(x) = \cos x$, $u_2(x) = \sin x$, find the trigonometric polynomial nearest to f.
10. In the linear space V of Exercise 5, let $f(x) = e^{-x}$, and find the linear polynomial nearest to f.

4

LINEAR TRANSFORMATIONS AND MATRICES

4.1 Linear transformations

One of the ultimate goals of mathematical analysis is a comprehensive study of functions whose domains and ranges are subsets of linear spaces. Such functions are called *transformations*, *mappings*, or *operators*. This chapter treats the simplest examples, called *linear transformations*, which occur in all branches of mathematics. Properties of more general transformations are often obtained by approximating them by linear transformations.

First we introduce some notation and terminology concerning arbitrary functions. Let V and W be two sets. We use the symbol

$$T : V \to W$$

to indicate that T is a function whose domain is V and whose values are in W. For each x in V, the element $T(x)$ in W is called the *image of x under T*, and we say that T *maps x onto* $T(x)$. If A is any subset of V, the set of all images $T(x)$ obtained as x runs through the elements of A is called the *image of A under T* and is denoted by $T(A)$. The image $T(V)$ of the domain V is the *range* of T.

Now we assume that V and W are linear spaces having the same set of scalars, and we define a linear transformation as follows.

DEFINITION. *If V and W are linear spaces, a function* $T : V \to W$ *is called a linear transformation of V into W if it has the following two properties:*
  (a) $T(x + y) = T(x) + T(y)$ *for all x and y in V,*
  (b) $T(cx) = cT(x)$ *for all x in V and all scalars c.*

These two properties are verbalized by saying that T preserves addition and multiplication by scalars. The two properties can be combined into one formula which states that

$$T(ax + by) = aT(x) + bT(y)$$

for all x, y in V and all scalars a and b. By induction, we also have the more general relation

$$T\left(\sum_{i=1}^{n} a_i x_i \right) = \sum_{i=1}^{n} a_i T(x_i)$$

for any n elements $x_1, \ldots, x_n$ in V and any n scalars $a_1, \ldots, a_n$.

The reader can easily verify that the following examples are linear transformations.

EXAMPLE 1. *The identity transformation.* The transformation $T : V \rightarrow V$ given by $T(x) = x$ for each x in V is called the identity transformation and is denoted by I or by I_V. The identity maps every element of V onto itself.

EXAMPLE 2. *The zero transformation.* The transformation $T : V \rightarrow V$ that maps each element of V onto O is called the *zero transformation* and is denoted by O.

EXAMPLE 3. *Multiplication by a fixed scalar c.* Here we have $T : V \rightarrow V$, where $T(x) = cx$ for all x in V. When $c = 1$ this is the identity transformation, and when $c = 0$ it is the zero transformation.

EXAMPLE 4. *Linear equations.* Let $V = \mathbf{R}^n$ and let $W = \mathbf{R}^m$. Given mn real numbers a_{ik}, where $i = 1, 2, \ldots m$ and $k = 1, 2, \ldots, n$, define $T : \mathbf{R}^n \rightarrow \mathbf{R}^m$ as follows: T maps each vector $x = (x_1, \ldots, x_n)$ in $\mathbf{R}^n$ onto the vector $y = (y_1, \ldots, y_m)$ in $\mathbf{R}^m$ according to the equations

$$y_i = \sum_{k=1}^{n} a_{ik} x_k \qquad \text{for } i = 1, 2, \ldots, m.$$

EXAMPLE 5. *Inner product with a fixed element.* Let V be a real Euclidean space. For a fixed element z in V define $T : V \rightarrow \mathbf{R}$ as follows: If $x \in V$, then $T(x) = (x, z)$, the inner product of x with z.

EXAMPLE 6. *Projection on a subspace.* Let V be a Euclidean space and let S be a finite-dimensional subspace of V with an orthonormal basis $e_1, \ldots, e_n$. Define $T : V \rightarrow S$ as follows: If $x \in V$, then $T(x) = p(x)$, the projection of x on the subspace S, given by

$$p(x) = \sum_{i=1}^{n} (x, e_i) e_i.$$

The next two examples involve calculus concepts.

EXAMPLE 7. *The differentiation operator.* Let V be the linear space of all real functions f differentiable on an open interval (a, b). The linear transformation that maps each function f in V onto its derivative f' is called the *differentiation operator* and is denoted by D. Thus, we have $D : V \rightarrow W$, where $D(f) = f'$ for each f in V. The space W consists of all derivatives f'.

EXAMPLE 8. *The integration operator.* Let $V = C(a, b)$ be the linear space of all real functions continuous on an interval $[a, b]$. If $f \in V$, define $g = T(f)$ to be that function in V given by

$$g(x) = \int_a^x f(t)\, dt \qquad \text{if } a \le x \le b.$$

This transformation T is called the *integration operator.*

4.2 Null space and range

In this section, T denotes a linear transformation of a linear space V into a linear space W.

THEOREM 4.1. *The set $T(V)$ (the range of T) is a subspace of W. Moreover, T maps the zero element of V onto the zero element of W.*

Proof. To show that $T(V)$ is a subspace of W we need only verify the closure axioms. Take any two elements of $T(V)$, say $T(x)$ and $T(y)$. Then $T(x) + T(y) = T(x + y)$, so $T(x) + T(y)$ is an element of $T(V)$. And for any scalar c we have $cT(x) = T(cx)$, so $cT(x)$ is an element of $T(V)$. Hence $T(V)$ is a subspace of W. Taking $c = 0$ in the relation $T(cx) = cT(x)$, we find that $T(O) = O$.

DEFINITION. *The set of all elements in V that T maps onto O is called the null space of T and is denoted by $N(T)$. Thus, we have*

$$N(T) = \{x : x \in V \quad and \quad T(x) = O\}.$$

The null space is sometimes called the kernel of T.

THEOREM 4.2. *The null space of T is a subspace of V.*

Proof. If x and y are in $N(T)$, then so are $x + y$ and cx for all scalars c, because

$$T(x + y) = T(x) + T(y) = O \qquad and \qquad T(cx) = cT(x) = O.$$

The following examples describe the null spaces of the linear transformations given above in Section 4.1.

EXAMPLE 1. *Identity transformation.* The null space is $\{O\}$, the subspace consisting of O.

EXAMPLE 2. *Zero transformation.* Every element is mapped onto zero, so the null space is V.

EXAMPLE 3. *Multiplication by a fixed scalar c.* If $c \neq 0$, the null space is $\{O\}$. If $c = 0$, the null space is V.

EXAMPLE 4. *Linear equations.* The null space consists of all vectors $(x_1, \ldots, x_n)$ in $\mathbf{R}^n$ for which

$$\sum_{k=1}^{n} a_{ik} x_k = 0 \qquad for \qquad i = 1, 2, \ldots, m.$$

EXAMPLE 5. *Inner product with a fixed element z.* The null space consists of all elements in V orthogonal to z.

EXAMPLE 6. *Projection on a subspace S.* If $x \in V$, the projection $p(x)$ is equal to $\sum_{i=1}^{n}(x, e_i)e_i$, where $e_1, \ldots, e_n$ is an orthonormal basis for S. Therefore $p(x) = O$ if and only if $(x, e_i) = 0$ for each basis element e_i, so the null space is $S^{\perp}$, the orthogonal complement of S.

EXAMPLE 7. *Differentiation operator.* The null space consists of all functions that are constant on the given interval.

EXAMPLE 8. *Integration operator.* The null space contains only the zero function.

4.3 Nullity and rank

Again in this section T denotes a linear transformation of a linear space V into a linear space W. We explore the relation between the dimensionalities of the domain V, the null space $N(T)$, and the range $T(V)$.

DEFINITION. *The dimension of the null space $N(T)$ is called the nullity of T. The dimension of the range $T(V)$ is called the rank of T.*

If V is finite-dimensional, then so is the null space, because it is a subspace of V. The next theorem shows that in this case the range is also finite-dimensional.

THEOREM 4.3. NULLITY PLUS RANK THEOREM. *If V is finite-dimensional, then $T(V)$ is also finite-dimensional, and we have*

$$(4.1) \qquad \dim N(T) + \dim T(V) = \dim V.$$

In other words, the nullity plus the rank is equal to the dimension of the domain.

Proof. Let $n = \dim V$ and let $e_1, \ldots, e_k$ be a basis for the null space. Then $k = \dim N(T) \le n$. By Theorem 7(a), these elements are part of some basis for V, say, the basis

$$(4.2) \qquad e_1, \ldots, e_k, e_{k+1}, \ldots, e_{k+r},$$

where $k + r = n$. We shall prove that the r elements

$$(4.3) \qquad T(e_{k+1}), \ldots, T(e_{k+r})$$

form a basis for $T(V)$, thus proving that $\dim T(V) = r$. Since $k + r = n$, this also proves (4.1).

First we show that the r elements in (4.3) span $T(V)$. If $y \in T(V)$ we have $y = T(x)$ for some x in V, so we can write $x = c_1 e_1 + \cdots + c_{k+r} e_{k+r}$ for some choice of scalars $c_1, \ldots, c_{k+r}$. Therefore

$$y = T(x) = \sum_{i=1}^{k+r} c_i T(e_i) = \sum_{i=1}^{k} c_i T(e_i) + \sum_{i=k+1}^{k+r} c_i T(e_i) = \sum_{i=k+1}^{k+r} c_i T(e_i)$$

because $T(e_1) = \cdots = T(e_k) = O$. This shows that the elements in (4.3) span the range $T(V)$.

Now we show that these elements are independent. Suppose that

$$\sum_{i=k+1}^{k+r} a_i T(e_i) = O$$

for some choice of scalars $a_{k+1}, \ldots, a_{k+r}$. This implies that

$$T\left(\sum_{i=k+1}^{k+r} a_i e_i\right) = O,$$

so the element $z = a_{k+1}e_{k+1} + \cdots + a_{k+r}e_{k+r}$ is in the null space $N(T)$. But $e_1, \ldots, e_k$ is a basis for the null space, so z can also be expressed in the form $z = a_1 e_1 + \cdots + a_k e_k$ for some choice of scalars $a_1, \ldots, a_r$. Hence we have

$$z - z = \sum_{i=1}^{k} a_i e_i - \sum_{i=k+1}^{k+r} a_i e_i = O.$$

But the elements in (4.2) are independent, so all the scalar multipliers a_i must be zero. Therefore the elements in (4.3) are independent and constitute a basis for $T(V)$.

Note. If at least one of $N(T)$ or $T(V)$ is infinite-dimensional, Theorem 4.3 implies that V is also infinite-dimensional. Conversely, if V is infinite-dimensional, then at least one of $N(T)$ or $T(V)$ is infinite-dimensional. A proof of this last statement is outlined in Exercise 24 of Section 4.4.

4.4 Exercises

In each of Exercises 1 through 10, a transformation $T : \mathbf{R}^2 \to \mathbf{R}^2$ is defined by the formula given for $T(x, y)$, where (x, y) is an arbitrary point in $\mathbf{R}^2$. In each case determine whether T is linear. If T is linear, describe its null space and its range, and compute its nullity and rank.

1. $T(x, y) = (y, x)$.
2. $T(x, y) = (x, -y)$.
3. $T(x, y) = (x, 0)$.
4. $T(x, y) = (x, x)$.
5. $T(x, y) = (x^2, y^2)$.
6. $T(x, y) = (e^x, e^y)$.
7. $T(x, y) = (x, 1)$.
8. $T(x, y) = (x + 1, y + 1)$.
9. $T(x, y) = (x - y, x + y)$.
10. $T(x, y) = (2x - y, x + y)$.

Do the same for each of Exercises 11 through 15 if $T : \mathbf{R}^2 \to \mathbf{R}^2$ is described as indicated.

11. T rotates every point through the same angle φ about the origin. That is, T maps a point with polar coordinates (r, θ) onto the point with polar coordinates $(r, \theta + \varphi)$, where φ is fixed. Also T maps O onto itself.
12. T maps each point onto its reflection with respect to a fixed line through the origin.
13. T maps every point onto the point $(1, 1)$.
14. T maps each point with polar coordinates (r, θ) onto the point with polar coordinates $(2r, \theta)$. Also T maps O onto itself.
15. T maps each point with polar coordinates (r, θ) onto the point with polar coordinates $(r, 2\theta)$. Also T maps O onto itself.

Do the same as above in each of Exercises 16 through 23 if a transformation $T : \mathbf{R}^3 \to \mathbf{R}^3$ is defined by the formula given for $T(x, y, z)$, where (x, y, z) is an arbitrary point in $\mathbf{R}^3$.

16. $T(x, y, z) = (z, y, x)$.
17. $T(x, y, z) = (x, y, 0)$.
18. $T(x, y, z) = (x, 2y, 3z)$.
19. $T(x, y, z) = (x, y, 1)$.
20. $T(x, y, z) = (x + 1, y + 1, z - 1)$.
21. $T(x, y, z) = (x + 1, y + 2, z + 3)$.
22. $T(x, y, z) = (x, y^2, z^3)$.
23. $T(x, y, z) = (x + z, 0, x + y)$.

24. Let $T : V \rightarrow W$ be a linear transformation of a linear space V into a linear space W. If V is infinite-dimensional, prove that at least one of $T(V)$ or $N(T)$ is infinite-dimensional.

    [*Hint:* Assume $\dim N(T) = k$, $\dim T(V) = r$, let $e_1, \ldots, e_k$ be a basis for $N(T)$ and let $e_1, \ldots, e_k, e_{k+1}, \ldots, e_{k+n}$ be independent elements in V, where $n > r$. The elements $T(e_{k+1}), \ldots, T(e_{k+n})$ are dependent since $n > r$. Use this fact to get a contradiction.]

In each of Exercises 25 through 29, a transformation $T : V \rightarrow V$ is described as indicated. In each case, determine whether T is linear. If T is linear, describe its null space and range, and compute the nullity and rank.

25. Let V be the linear space of all real polynomials $p(x)$ of degree $\leq n$. If $p \in V, q = T(p)$ means that $q(x) = p(x + 1)$ for all real x.

Exercises 26 through 30 require a knowledge of calculus.

26. Let V be the linear space of all real functions differentiable on the open interval $(-1, 1)$. If $f \in V$, $g = T(f)$ means that $g(x) = xf'(x)$ for all x in $(-1, 1)$.
27. Let V be the linear space of all real functions twice differentiable on an open interval (a, b). If $y \in V$, define $T(y) = y'' + Ay' + By$, where A and B are fixed constants.
28. Let V be the linear space of all real functions continuous on $[a, b]$. If $f \in V, g = T(f)$ means that

$$g(x) = \int_a^b f(t) \sin(x - t) \, dt \qquad \text{for } a \leq x \leq b.$$

29. Let V be the linear space of all real convergent sequences $\{x_n\}$. Define a transformation $T : V \rightarrow V$ as follows: If $x = \{x_n\}$ is a convergent sequence with limit a, let $T(x) = \{y_n\}$, where $y_n = a - x_n$ for $n \geq 1$.
30. Let V be the linear space of all real functions continuous on the interval $[-\pi, \pi]$. Let S be that subset of V consisting of all f satisfying the three equations

$$\int_{-\pi}^{\pi} f(t) \, dt = 0, \qquad \int_{-\pi}^{\pi} f(t) \cos t \, dt = 0, \qquad \int_{-\pi}^{\pi} f(t) \sin t \, dt = 0.$$

    (a) Prove that S is a subspace of V.
    (b) Prove that S contains the functions $f(x) = \cos nx$ and $f(x) = \sin nx$ for each $n = 2, 3, \ldots$.
    (c) Prove that S is infinite-dimensional.
        Let $T : V \rightarrow V$ be the linear transformation defined as follows: If $f \in V, g = T(f)$ means that

$$g(x) = \int_{-\pi}^{\pi} \{1 + \cos(x - t)\} f(t) \, dt.$$

    (d) Prove that $T(V)$, the range of T, is finite-dimensional, and find a basis for $T(V)$.
    (e) Determine the null space of T.
    (f) Find all real $c \neq 0$ and all nonzero f in V such that $T(f) = cf$. (Note that such an f lies in the range of T.)

4.5 Algebraic operations on linear transformations

Functions whose values lie in a linear space W can be added to each other and can be multiplied by the scalars in W according to the following definition.

DEFINITION. *Let $S : V \rightarrow W$ and $T : V \rightarrow W$ be two functions with a common domain V and with values in a linear space W. If c is any scalar in W, we define the sum $S + T$ and the product cT by the equations*

(4.4) $$(S + T)(x) = S(x) + T(x), \qquad (cT)(x) = cT(x)$$

for all x in V.

We are especially interested in the case where V is also a linear space having the same scalars as W. In this case we denote by $\mathcal{L}(V, W)$ the set of all linear transformations of V into W.

If S and T are two linear transformations in $\mathcal{L}(V, W)$, it is an easy exercise to verify that $S + T$ and cT are also linear transformations in $\mathcal{L}(V, W)$. More than this is true. With the operations just defined, the set $\mathcal{L}(V, W)$ itself becomes a new linear space. The zero transformation serves as the zero element of this space, and the transformation $(-1)T$ is the negative of T. It is a straightforward matter to verify that all ten axioms for a linear space are satisfied. Therefore, we have the following.

THEOREM 4.4.   *The set $\mathcal{L}(V, W)$ is a linear space with the operations of addition and multiplication by scalars defined as in (4.4).*

A more interesting algebraic operation on linear transformations is *composition* or *multiplication* of transformations. This operation makes no use of the algebraic structure of a linear space and can be defined quite generally as follows.

DEFINITION.   COMPOSITION OF FUNCTIONS.   *Let U, V, W be sets. Let $T : U \rightarrow V$ be a function with domain U and values in V, and let $S : V \rightarrow W$ be another function with domain V and values in W. Then the composition ST is the function $ST : U \rightarrow W$ defined by the equation*

$$(ST)(x) = S[T(x)]   \text{for every } x \text{ in } U.$$

Thus, to map x by the composition ST, we first map x by T and then map $T(x)$ by S. This is illustrated by the diagram in Figure 4.1.

Composition of real-valued functions is encountered in the study of elementary functions where we learn that, in general, composition is not commutative. For example $\cos(x^2)$ means first square x and then take the cosine of x^2. This is not always the same $(\cos x)^2$, which means first take the cosine of x and then square $\cos x$. However, composition always satisfies an associative law.

THEOREM 4.5.   *If $T : U \rightarrow V$, $S : V \rightarrow W$, and $R : W \rightarrow X$ are three functions, then we have*

$$R(ST) = (RS)T.$$

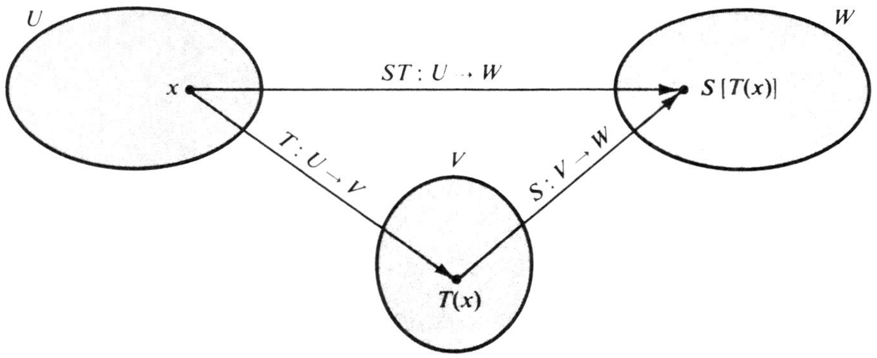

FIGURE 4.1   Illustrating the composition of two transformations.

Proof. Both $R(ST)$ and $(RS)T$ have domain U and values in X. For each x in U we have

$$[R(ST)](x) = R[(ST)(x)] = R[S\{T(x)\}] \quad \text{and} \quad [(RS)T](x) = (RS)[T(x)] = R[S\{T(x)\}],$$

which proves that $R(ST) = (RS)T$.

DEFINITION. *Let $T : V \to V$ be a function that maps V into itself. We define integral powers of T inductively as follows:*

$$T^0 = I, \qquad T^n = TT^{n-1} \qquad \text{for } n \geq 1.$$

Here I is the identity transformation. The reader can easily verify that the associative law implies the law of exponents $T^m T^n = T^{m+n}$ for all nonnegative integers m and n.

The next theorem shows that the composition of *linear* transformations is again linear.

THEOREM 4.6. *Let U, V, W be linear spaces with the same scalars. If $T : U \to V$ and $S : V \to W$ are linear transformations, then the composition $ST : U \to W$ is linear.*

Proof. For all x, y in U and all scalars a and b, we have

$$(ST)(ax + by) = S[T(ax + by)] = S[aT(x) + bT(y)] = aST(x) + bST(y).$$

Composition can be combined with the linear space operations addition and multiplication by scalars in $\mathcal{L}(V, W)$ to give us the following:

THEOREM 4.7. *Let U, V, W be linear spaces with the same scalars, assume S and T are in $\mathcal{L}(V, W)$, and let c be any scalar.*
(a) *For any function R with values in V, we have*

$$(S + T)R = SR + TR \qquad \text{and} \qquad (cS)R = c(SR).$$

(b) *For any linear transformation $R : W \to U$, we have*

$$R(S + T) = RS + RT \qquad \text{and} \qquad R(cS) = c(RS).$$

The proof is a simple application of the definition of composition and is left as an exercise.

4.6 Inverses

In the study of elementary functions, it is shown that monotonic functions can be inverted to construct new functions. For example, the natural logarithm function $f(x) = \log x$ is increasing for $x > 0$ and its inverse is the exponential function. That is, if $x > 0$, then $y = \log x$ if and only if $x = e^y$. This section extends the process of inversion to a more general class of functions.

Given any function T (not necessarily linear), our goal is to find, if possible, another function S whose composition with T is the identity transformation. Since composition is, in general, not commutative, we need to distinguish between ST and TS. Therefore we introduce two kinds of inverses, which we call left and right inverses.

DEFINITION. *Given two sets V and W and a function $T : V \to W$. A function $S :$ $T(V) \to V$ is called a left inverse of T if $S[T(x)] = x$ for all x in V; that is, if*

$$ST = I_V,$$

where I_V is the identity transformation on V.

A function $R : T(V) \to V$ is called a right inverse of T if $T[R(y)] = y$ for all y in $T(V)$; that is, if

$$TR = I_{T(V)},$$

where $I_{T(V)}$ is the identity transformation on $T(V)$.

EXAMPLE. *A function with no left inverse but with two right inverses.* Let $V = \{1, 2\}$ and let $W = \{0\}$. Define $T : V \to W$ as follows: $T(1) = T(2) = 0$. This function has two right inverses R and R' given by

$$R(0) = 1, \qquad R'(0) = 2.$$

It cannot have a left inverse S because this would require

$$1 = S[T(1)] = S(0) \qquad \text{and} \qquad 2 = S[T(2)] = S(0),$$

contradicting the fact that the function value $S(0)$ must be uniquely determined. This simple example shows that left inverses need not exist and right inverses need not be unique.

It is easy to show that every function $T : V \to W$ has at least one right inverse. In fact, each y in $T(V)$ has the form $y = T(x)$ for at least one x in V. If we select one such x and define $R(y) = x$, then $T[R(y)] = T(x) = y$ for each y in $T(V)$, so R is a right inverse. Nonuniqueness may occur because there may be more than one x in V that maps onto a given y in $T(V)$. We shall prove presently (in Theorem 4.9) that if each y in $T(V)$ is the image of *exactly one* x in V, then right inverses are unique.

First we prove that if a left inverse exists it is unique and, at the same time, is a right inverse.

THEOREM 4.8. *A function $T : V \to W$ can have at most one left inverse. If T has a left inverse S, then S is also a right inverse.*

Proof. Assume T has two left inverses, $S : T(V) \to V$ and $S' : T(V) \to V$. Choose any y in $T(V)$. We shall prove that $S(y) = S'(y)$. Now $y = T(x)$ for some x in V, so we have

$$S[T(x)] = x \qquad \text{and} \qquad S'[T(x)] = x,$$

because both S and S' are left inverses. This says that $S(y) = x$ and $S'(y) = x$, so $S(y) = S'(y)$ for all y in $T(V)$. Therefore $S = S'$, so if a left inverse exists there is only one.

Now we prove that every left inverse S is also a right inverse. Choose any y in $T(V)$. We will prove that $T[S(y)] = y$. Since $y \in T(V)$, $y = T(x)$ for some x in V. But S is a left inverse, so

$$x = S[T(x)] = S(y).$$

Applying T, we get $T(x) = T[S(y)]$. But $T(x) = y$, so $y = T[S(y)]$, which completes the proof.

Now it is a simple matter to characterize all functions having left inverses. First we introduce the following terminology.

DEFINITION. *A function $T : V \to W$ is said to be one-to-one on V if T maps distinct elements of V onto distinct elements of W; that is, T is called one-to-one on V if for all x and y in V we have*

(4.5) $x \neq y$ *implies* $T(x) \neq T(y)$.

A statement equivalent to (4.5) *is*

(4.6) $T(x) = T(y)$ *implies* $x = y$.

THEOREM 4.9.   *A function $T : V \to W$ has a left inverse if and only if T is one-to-one on V.*

Proof.   Assume T has a left inverse S, and suppose that $T(x) = T(y)$ for two elements x and y in V. Applying S to each member we find that $x = y$, so T is one-to-one on V.
  Conversely, assume T is one-to-one on V. We shall exhibit a function $S : T(V) \to V$ that is a left inverse for T. If $y \in T(V)$, then $y = T(x)$ for *exactly* one x in V because T is one-to-one. Define $S(y)$ to be this x. That is, we define S on $T(V)$ as follows:

$$S(y) = x   \text{ means that }   T(x) = y.$$

Then we have $S[T(x)] = S(y) = x$ for each x in V, so S is a left inverse of T.

DEFINITION.   INVERTIBLE FUNCTION.   *Let $T : V \to W$ be one-to-one on V. The unique left inverse of T (which is also a right inverse) is denoted by T^{-1}. We say that T is invertible, and we call T^{-1} the inverse of T.*

The results obtained in this section refer to arbitrary functions. The next section applies these ideas to *linear* transformations.

4.7   One-to-one linear transformations

In this section, V and W denote linear spaces with the same scalars, and $T : V \to W$ denotes a linear transformation in $\mathcal{L}(V, W)$. Linearity of T enables us to express the one-to-one property in several equivalent forms.

THEOREM 4.10.   *Let $T : V \to W$ be a linear transformation in $\mathcal{L}(V, W)$. Then the following statements are equivalent:*

(a) *T is one-to-one on V.*

(b) *T is invertible, and the inverse $T^{-1} : T(V) \to V$ is linear.*

(c) *$\dim N(T) = 0$. In other words, the null space $N(T)$ contains only the zero element.*

Proof.   We shall prove that (a) implies (b), (b) implies (c), and (c) implies (a). First assume (a) holds. Then T^{-1} exists (by Theorem 4.9) and we must show that it is linear. Take any two elements u and v in $T(V)$. Then $u = T(x)$ and $v = T(y)$ for some x and y in V. Because T is linear we have

$$au + bv = aT(x) + bT(y) = T(ax + by)$$

for all scalars a and b. Applying T^{-1} we have

$$T^{-1}(au + bv) = ax + by = aT^{-1}(u) + bT^{-1}(v),$$

so T^{-1} is linear. Therefore (a) implies (b).

Next assume that (b) holds. Take any x in V for which $T(x) = O$. Applying T^{-1} we find that $x = T^{-1}(O) = O$, since T^{-1} is linear. Therefore (b) implies (c).

Finally, assume (c) holds. Take any two elements x and y in V with $T(x) = T(y)$. By linearity, we have $T(x - y) = T(x) - T(y) = O$, so $x - y = O$ and $x = y$. Therefore T is one-to-one on V. Hence (c) implies (a), and the proof is complete.

When V is finite-dimensional, the one-to-one property can be formulated in terms of independence and dimensionality, as indicated by the next theorem.

THEOREM 4.11.   *Let $T : V \to W$ be a linear transformation in $\mathcal{L}(V, W)$, and assume that V is finite-dimensional, say $\dim V = n$. Then the following statements are equivalent:*

(a) *T is one-to-one on V.*

(b) *If $v_1, \ldots, v_p$ are independent elements in V, then $T(v_1), \ldots, T(v_p)$ are independent elements in $T(V)$.*

(c) *$\dim T(V) = \dim V$.*

(d) *If $\{v_1, \ldots, v_n\}$ is a basis for V, then $\{T(v_1), \ldots, T(v_n)\}$ is a basis for $T(V)$.*

Proof.   We shall prove that (a) implies (b), (b) implies (c), (c) implies (d), and (d) implies (a). First assume (a) holds. Let $v_1, \ldots, v_p$ be independent elements of V and consider the elements $T(v_1), \ldots, T(v_p)$ in $T(V)$. Suppose that $\sum_{i=1}^{p} c_i T(v_i) = O$ for certain scalars $c_1, \ldots, c_p$. Because T is linear and one-to-one, we have

$$T\left(\sum_{i=1}^{p} c_i v_i\right) = O, \qquad \text{and hence} \qquad \sum_{i=1}^{p} c_i v_i = O.$$

But $v_1, \ldots, v_p$ are independent, so $c_1 = \cdots = c_p = 0$. Therefore (a) implies (b).

Now assume (b) holds. If $\{v_1, \ldots, v_n\}$ is a basis for V then, by (b), the n elements $T(v_1), \ldots, T(v_n)$ in $T(V)$ are independent. Therefore $\dim T(V) \geq n$. On the other hand, by Theorem 3 we have $\dim T(V) \leq n$, so $\dim T(V) = n$. Hence (b) implies (c).

Next, assume (c) holds and let $\{v_1, \ldots, v_n\}$ be a basis for V. Take any element y in $T(V)$. Then $y = T(x)$ for some x in V, so we can write

$$x = \sum_{i=1}^{n} c_i v_i \qquad \text{and hence} \qquad y = T(x) = \sum_{i=1}^{n} c_i T(v_i).$$

Therefore the set of elements $\{T(v_1), \ldots, T(v_n)\}$ spans $T(V)$. But we are assuming $\dim T(V) = n$, so $\{T(v_1), \ldots, T(v_n)\}$ is a basis for $T(V)$. Therefore (c) implies (d).

Finally, assume (d) holds. We will show that $T(x) = O$ implies $x = O$. Let $\{v_1, \ldots, v_n\}$ be a basis for V. If $x \in V$ we may write

$$x = \sum_{i=1}^{n} c_i v_i \qquad \text{and hence} \qquad T(x) = \sum_{i=1}^{n} c_i T(v_i).$$

If $T(x) = O$, then $c_1 = \cdots = c_n = 0$, because the elements $T(v_1), \ldots, T(v_n)$ are independent. Therefore $x = O$, so T is one-to-one on V. Thus, (d) implies (a), and the proof is complete.

4.8 Exercises

1. Let $V = \{0, 1\}$. Describe all functions $T : V \to V$. There are four altogether. Label them as T_1, T_2, T_3, T_4, and make a multiplication table showing the composition of each pair. Indicate which functions are one-to-one on V, and give their inverses.

2. Let $V = \{0, 1, 2\}$. Describe all functions $T : V \to V$ for which $T(V) = V$. There are six altogether. Label them as $T_1, \ldots, T_6$, and make a multiplication table showing the composition of each pair. Indicate which functions are one-to-one on V, and give their inverses.

In each of Exercises 3 through 12, a function $T : \mathbf{R}^2 \to \mathbf{R}^2$ is defined by the formula given for $T(x, y)$, where (x, y) is an arbitrary point in $\mathbf{R}^2$. In each case determine whether T is one-to-one on $\mathbf{R}^2$. If it is, let $(x, y) = T^{-1}(u, v)$ and give formulas for determining x and y in terms of u and v.

3. $T(x, y) = (y, x)$.
4. $T(x, y) = (x, -y)$.
5. $T(x, y) = (x, 0)$.
6. $T(x, y) = (x, x)$.
7. $T(x, y) = (x^2, y^2)$.

8. $T(x, y) = (e^x, e^y)$.
9. $T(x, y) = (x, 1)$.
10. $T(x, y) = (x + 1, y + 1)$.
11. $T(x, y) = (x - y, x + y)$.
12. $T(x, y) = (2x - y, x + y)$.

In each of Exercises 13 through 20, a function $T : \mathbf{R}^3 \to \mathbf{R}^3$ is defined by the formula given for $T(x, y, z)$, where (x, y, z) is an arbitrary point in $\mathbf{R}^3$. In each case determine whether T is one-to-one on $\mathbf{R}^3$. If it is, let $(x, y, z) = T^{-1}(u, v, w)$ and give formulas for determining x, y, and z in terms of u, v, and w.

13. $T(x, y, z) = (z, y, x)$.
14. $T(x, y, z) = (x, y, 0)$.
15. $T(x, y, z) = (x, 2y, 3z)$.
16. $T(x, y, z) = (x, y, x + y + z)$.

17. $T(x, y, z) = (x + 1, y + 1, z - 1)$.
18. $T(x, y, z) = (x + 1, y + 2, z + 3)$.
19. $T(x, y, z) = (x, x + y, x + y + z)$.
20. $T(x, y, z) = (x + y, y + z, x + z)$.

21. Let $T : V \to V$ be a function that maps V onto itself. Powers are defined inductively by the formulas $T^0 = I$, $T^n = TT^{n-1}$ for $n \geq 1$. Prove that the associative law of composition implies the law of exponents: $T^m T^n = T^{m+n}$. If T is invertible, prove that T^n is also invertible and that $(T^n)^{-1} = (T^{-1})^n$.

In Exercises 22 through 25, S and T denote functions with domain V and values in V. In general, $ST \neq TS$. If $ST = TS$, we say that S and T *commute*.

22. If S and T commute, prove that $(ST)^n = S^n T^n$ for all integers $n \geq 0$.
23. If S and T are invertible, prove that ST is also invertible and that $(ST)^{-1} = T^{-1}S^{-1}$. In other words, the inverse of the composition ST is the composition of inverses, taken in reverse order.
24. If S and T are invertible and commute, prove that their inverses also commute.
25. Let V be a linear space. If S and T commute, prove that

$$(S + T)^2 = S^2 + 2ST + T^2 \qquad \text{and} \qquad (S + T)^3 = S^3 + 3S^2T + 3ST^2 + T^3.$$

Indicate how these formulas must be altered if $ST \neq TS$.
26. Let S and T be in $\mathcal{L}(V, V)$ and assume that $ST - TS = I$. Prove that $ST^n - T^nS = nT^{n-1}$ for all $n \geq 1$.
27. Let S and T be linear transformations of $\mathbf{R}^3$ into $\mathbf{R}^3$ defined by the formulas $S(x, y, z) = (z, y, x)$ and $T(x, y, z) = (x, x + y, x + y + z)$, where (x, y, z) is an arbitrary point of $\mathbf{R}^3$.
    (a) Determine the image of (x, y, z) under each of the following transformations: $ST, TS, ST - TS, S^2, T^2, (ST)^2, (TS)^2, (ST - TS)^2$.
    (b) Prove that S and T are one-to-one on $\mathbf{R}^3$, and find the image of (u, v, w) under each of the following transformations: $S^{-1}, T^{-1}, (ST)^{-1}, (TS)^{-1}$.
    (c) Find the image of (x, y, z) under $(T - I)^n$ for each $n \geq 1$.
28. Let V be the linear space of all real polynomials $p(x)$. Let R, S, T be the functions that map an arbitrary polynomial $p(x) = c_0 + c_1 x + \cdots + c_n x^n$ onto the polynomials $r(x)$, $s(x)$, and $t(x)$, respectively, where

$$r(x) = p(0), \qquad s(x) = \sum_{k=1}^{n} c_k x^{k-1}, \qquad t(x) = \sum_{k=0}^{n} c_k x^{k+1}.$$

    (a) Let $p(x) = 2 + 3x - x^2 + x^3$, and determine the image of p under each of the following transformations: $R, S, T, ST, TS, (TS)^2, T^2 S^2, S^2 T^2, TRS, RST$.
    (b) Prove that R, S, T are linear, and determine the null space and range of each.
    (c) Prove that T is one-to-one on V and determine its inverse.
    (d) If $n \geq 1$, express $(TS)^n$ and $S^n T^n$ in terms of I and R.

Exercises 29 through 32 require a knowledge of calculus.

29. Let V be the linear space of all real polynomials $p(x)$. Let D denote the differentiation operator, and let T be the linear transformation that maps $p(x)$ onto $xp'(x)$.
    (a) Let $p(x) = 2 + 3x - x^2 + 4x^3$, and determine the image of p under each of the following transformations: $D, T, DT, TD, DT - TD, T^2 D^2 - D^2 T^2$.
    (b) Determine those p in V for which $T(p) = p$.
    (c) Determine those p in V for which $(DT - 2D)(p) = O$.
    (d) Determine those p in V for which $(DT - TD)^n(p) = D^n(p)$.
30. Let V and D be as in Exercise 29, but let T be the linear transformation that maps $p(x)$ onto $xp(x)$. Prove that $DT - TD = I$ and that $DT^n - T^n D = nT^{n-1}$ for $n \geq 2$.
31. Let V be the linear space of all real polynomials $p(x)$. Let D denote the differentiation operator, and let T denote the integration operator that maps each polynomial p onto the polynomial q given by $q(x) = \int_0^x p(t)\,dt$. Prove that $DT = I_V$ but that $TD \neq I_V$. Describe the null space and range of TD.
32. Refer to Exercise 29 of Section 4.4. Determine whether T is one-to-one on V. If it is, describe its inverse.

4.9 Linear transformations with prescribed values at the elements of a basis

The next theorem shows that if V is finite-dimensional, a linear transformation $T : V \to W$ is completely determined by its action on the basis elements of V.

THEOREM 4.12.   *Let $v_1, \ldots, v_n$ be a basis for an n-dimensional linear space V. Let $u_1, \ldots, u_n$ be n arbitrary elements in a linear space W. Then there is one and only one* *linear transformation $T : V \to W$ such that*

(4.7)
$$T(v_k) = u_k \quad for \quad k = 1, 2, \ldots, n.$$

This T maps an arbitrary element x in V as follows:

(4.8)
$$If \; x = \sum_{k=1}^{n} x_k v_k \quad then \quad T(x) = \sum_{k=1}^{n} x_k u_k.$$

Proof. Every x in V is a linear combination of $v_1, \ldots, v_n$, the multipliers $x_1, \ldots, x_n$ being the components of x relative to the ordered basis $(v_1, \ldots, v_n)$. If we define T by (4.8) it is easy to verify that T is linear. If $x = v_k$ for some k, then all components of x are 0 except the kth, which is equal to 1, so (4.8) gives $T(v_k) = u_k$, as required.

To prove that there is only one linear transformation satisfying (4.7), let T' be another, and compute $T'(x)$. We find that

$$T'(x) = T'\left(\sum_{k=1}^{n} x_k v_k\right) = \sum_{k=1}^{n} x_k T'(v_k) = \sum_{k=1}^{n} x_k u_k = T(x).$$

Since $T'(x) = T(x)$ for all x in V, we have $T' = T$, which completes the proof.

EXAMPLE. Determine the linear transformation $T : \mathbf{R}^2 \to \mathbf{R}^2$ that maps the basis elements $i = (1, 0)$ and $j = (0, 1)$ as follows:

$$T(i) = i + j, \qquad T(j) = 2i - j.$$

Solution. If $x = x_1 i + x_2 j$ is an arbitrary element of $\mathbf{R}^2$, then $T(x)$ is given by

$$T(x) = x_1 T(i) + x_2 T(j) = x_1(i + j) + x_2(2i - j) = (x_1 + 2x_2)i + (x_1 - x_2)j.$$

4.10 Matrix representations of linear transformations

Theorem 4.12 shows that a linear transformation $T : V \to W$ of a finite-dimensional linear space V is completely determined by its action on a given basis $v_1, \ldots, v_n$. Now suppose the space W is also finite-dimensional, say, $\dim W = m$, and let $w_1, \ldots, w_m$ be a basis for W. (The dimensions n and m may or may not be equal.) Since T has values in W, each element $T(v_k)$ can be uniquely expressed as a linear combination of the basis elements $w_1, \ldots, w_m$, say,

$$T(v_k) = \sum_{i=1}^{m} t_{ik} w_i,$$

where $t_{1k}, \ldots, t_{mk}$ are the components of $T(v_k)$ relative to the ordered basis $(w_1, \ldots, w_m)$. We shall display the m-tuple $(t_{1k}, \ldots, t_{mk})$ vertically, as follows:

(4.9)
$$\begin{bmatrix} t_{1k} \\ t_{2k} \\ \vdots \\ t_{mk} \end{bmatrix}.$$

This array is called a *column vector* or a *column matrix*. We have such a column vector for each of the n elements $T(v_1), \ldots, T(v_n)$. We place them side by side and enclose them in one pair of brackets to obtain the following rectangular array:

$$\begin{bmatrix} t_{11} & t_{12} & \cdots & t_{1n} \\ t_{21} & t_{22} & \cdots & t_{2n} \\ \vdots & \vdots & & \vdots \\ t_{m1} & t_{m2} & \cdots & t_{mn} \end{bmatrix}.$$

This array is called a *matrix* consisting of m *rows* and n *columns*. We call it an $m \times n$ matrix (read: m by n matrix). The first row is the $1 \times n$ matrix $(t_{11}, t_{12}, \ldots, t_{1n})$. The $m \times 1$ matrix displayed in (4.9) is the kth column. The scalars t_{ik} are indexed so the first subscript i indicates the *row*, and the second subscript k indicates the *column* in which t_{ik} occurs. We call t_{ik} the *ik-entry* or the *ik-element* of the matrix. The more compact notations

$$(t_{ik}) \qquad \text{or} \qquad (t_{ik})_{i,k=1}^{m,n}$$

are often used to denote the matrix whose ik-entry is t_k. Square brackets $[t_{ik}]$ are also used.

Thus, every linear transformation T of an n-dimensional space V into an m-dimensional space W gives rise to an $m \times n$ matrix (t_{ik}) whose columns consist of the components of $T(v_1), \ldots, T(v_n)$ relative to the basis $w_1, \ldots, w_m$. We call this the *matrix representation* of T relative to the given choice of ordered bases $(v_1, \ldots, v_n)$ for V and $(w_1, \ldots, w_m)$ for W. Once we know the matrix (t_{ik}), the components of any element $T(x)$ relative to the basis $(w_1, \ldots, w_m)$ can be determined as described in the next theorem.

THEOREM 4.13. *Let T be a linear transformation in $\mathcal{L}(V, W)$, where* $\dim V = n$ *and* $\dim W = m$. *Let* $(v_1, \ldots, v_n)$ *and* $(w_1, \ldots, w_m)$ *be ordered bases for V and W, respectively, and let* (t_{ik}) *be the $m \times n$ matrix whose entries are determined by the equations*

$$(4.10) \qquad T(v_k) = \sum_{i=1}^{m} t_{ik} w_i, \qquad \text{for } k = 1, 2, \ldots, n.$$

Then an arbitrary element

$$(4.11) \qquad x = \sum_{k=1}^{n} x_k v_k$$

in V with components $(x_1, \ldots, x_n)$ relative to $(v_1, \ldots, v_n)$ is mapped by T onto the element

$$(4.12) \qquad T(x) = \sum_{i=1}^{m} y_i w_i$$

in W with components $(y_1, \ldots, y_m)$ relative to $(w_1, \ldots, w_m)$. The y_i are related to the components of x by the linear equations

$$(4.13) \qquad y_i = \sum_{k=1}^{n} t_{ik} x_k \qquad \text{for } i = 1, 2, \ldots, m.$$

Proof.   Applying T to each member of (4.11) and using (4.10), we obtain

$$T(x) = \sum_{k=1}^{n} x_k T(v_k) = \sum_{k=1}^{n} x_k \sum_{i=1}^{m} t_{ik} w_i = \sum_{i=1}^{m} \left(\sum_{k=1}^{n} t_{ik} x_k \right) w_i = \sum_{i=1}^{m} y_i w_i,$$

where each y_i is given by (4.13). This completes the proof.

Once we choose a pair of bases $(v_1, \ldots, v_n)$ and $(w_1, \ldots, w_m)$ for V and W, respectively, every linear transformation $T : V \rightarrow W$ has a well-defined matrix representation (t_{ik}). Conversely, if we start with any mn scalars arranged as a rectangular matrix (t_{ik}) and choose a pair of ordered bases for V and W, then it is easy to prove that there is exactly one linear transformation $T : V \rightarrow W$ having this matrix representation. We simply define T at the basis elements of V by the equations in (4.10). Then, by Theorem 12, there is one and only one linear transformation $T : V \rightarrow W$ with these prescribed values. The image $T(x)$ of an arbitrary element x in V is then given by Equations (4.12) and (4.13).

EXAMPLE 1.   *Construction of a linear transformation from a given matrix.*   Suppose we start with the 2×3 matrix

$$\begin{bmatrix} 3 & 1 & -2 \\ 1 & 0 & 4 \end{bmatrix}.$$

Choose the usual bases of unit coordinate vectors for $\mathbf{R}^3$ and $\mathbf{R}^2$. Then the given matrix represents a linear transformation $T : \mathbf{R}^3 \rightarrow \mathbf{R}^2$ that maps an arbitrary vector (x_1, x_2, x_3) in $\mathbf{R}^3$ onto the vector (y_1, y_2) in $\mathbf{R}^2$ according to the linear equations

$$y_1 = 3x_1 + x_2 - 2x_3,$$
$$y_2 = x_1 + 0x_2 + 4x_3.$$

EXAMPLE 2.   *Determining a matrix representation of a given linear transformation.* Let $V = W = \mathbf{R}^2$ with the usual basis of unit coordinate vectors. Let T be the linear transformation that maps each vector (x_1, x_2) onto the vector (y_1, y_2) according to the linear equations

$$y_1 = 2x_1 + 3x_2,$$
$$y_2 = 4x_1 - 5x_2.$$

This transformation can be represented by the 2×2 matrix

$$\begin{bmatrix} 2 & 3 \\ 4 & -5 \end{bmatrix}.$$

EXAMPLE 3.   Find a matrix representation relative to the basis (i, j) of the linear transformation $T : \mathbf{R}^2 \rightarrow \mathbf{R}^2$ that maps $(1, 1)$ onto $\left(1, \frac{1}{2}\right)$, and maps $(1, -1)$ onto $\left(\frac{1}{2}, 1\right)$.

Solution. From the given information we can determine the action of T on the basis elements i and j. The given information tells us that

$$T(i + j) = i + \frac{1}{2}j \quad \text{and} \quad T(i - j) = \frac{1}{2}i + j.$$

By linearity we have $T(i + j) = T(i) + T(j)$ and $T(i - j) = T(i) - T(j)$, so the given information implies

$$T(i) + T(j) = i + \frac{1}{2}j$$

and

$$T(i) - T(j) = \frac{1}{2}i + j.$$

By adding and then subtracting these two equations we find

$$T(i) = \frac{3}{4}i + \frac{3}{4}j$$

and

$$T(j) = \frac{1}{4}i - \frac{1}{4}j.$$

Using the coefficients of i and j as columns, we find that the matrix of T is

$$\begin{bmatrix} \frac{3}{4} & \frac{1}{4} \\ \frac{3}{4} & -\frac{1}{4} \end{bmatrix},$$

and the corresponding linear equations describing T are

$$y_1 = \frac{3}{4}x_1 + \frac{1}{4}x_2$$

$$y_2 = \frac{3}{4}x_1 - \frac{1}{4}x_2.$$

The foregoing examples show the relation between a matrix representation of a linear transformation and the linear equations relating the components. In the next two examples the linear transformation is described *geometrically* rather than by linear equations. In each case we take $V = W = \mathbf{R}^n$ where n is either 2 or 3, and we choose the usual basis of unit coordinate vectors for both V and W. As usual, we determine the matrix by observing how the basis vectors are transformed. Once the matrix is known, it is a simple matter to write the linear equations relating the components.

EXAMPLE 4. *Rotation in a plane.* Let $T : \mathbf{R}^2 \to \mathbf{R}^2$ denote rotation by an angle θ about an axis through the origin perpendicular to the plane $\mathbf{R}^2$. To find the matrix of T we consider the action of T on the basis elements i and j. Rotation through an angle θ

maps the point $i = (1, 0)$ onto $T(i) = (\cos \theta, \sin \theta)$, and it maps $j = (0, 1)$ onto the point $(\cos(\theta + \frac{1}{2}\pi), \sin(\theta + \frac{1}{2}\pi)) = (-\sin \theta, \cos \theta)$. Expressing these images in terms of the basis elements we have

$$T(i) = \cos \theta \, i + \sin \theta \, j$$

$$T(j) = -\sin \theta \, i + \cos \theta \, j.$$

Remember, the components of the images $T(i)$ and $T(j)$ form the *columns* of the matrix representation of T, so T is represented by the following 2×2 matrix, called a *rotation matrix*:

$$\begin{bmatrix} \cos \theta & -\sin \theta \\ \sin \theta & \cos \theta \end{bmatrix}.$$

Once we have the matrix it is a simple matter to write the linear equations describing rotation. The point (x_1, x_2) gets rotated by an angle θ onto the point (y_1, y_2), where

$$y_1 = x_1 \cos \theta - x_2 \sin \theta,$$

$$y_2 = x_1 \sin \theta + x_2 \cos \theta.$$

EXAMPLE 5.   *Rotation in 3-space.*   Let $T : \mathbf{R}^3 \to \mathbf{R}^3$ denote rotation by an angle θ about the x_1 axis. Points on the x_1 axis are fixed, but those in the $x_2 x_3$ plane are rotated through an angle θ. Applying Example 4 to the $x_2 x_3$ plane, we find the basis elements i, j, k are mapped as follows:

$$T(i) = i = 1i + 0j + 0k$$

$$T(j) = 0i + \cos \theta \, j + \sin \theta \, k$$

$$T(k) = 0i - \sin \theta \, j + \cos \theta \, k.$$

These components form the columns of the following 3×3 matrix representation of T:

$$\begin{bmatrix} 1 & 0 & 0 \\ 0 & \cos \theta & -\sin \theta \\ 0 & \sin \theta & \cos \theta \end{bmatrix}.$$

Each point (x_1, x_2, x_3) gets mapped onto (y_1, y_2, y_3), where

$$y_1 = x_1$$

$$y_2 = x_2 \cos \theta - x_3 \sin \theta$$

$$y_3 = x_2 \sin \theta + x_3 \cos \theta.$$

The next example requires a knowledge of calculus.

EXAMPLE 6.   Let V be the linear space of all real polynomials $p(x)$ of degree ≤ 3. This space has dimension 4, and we choose the ordered basis $(1, x, x^2, x^3)$. Let D be the differentiation operator that maps each polynomial $p(x)$ in V onto its derivative $p'(x)$. We can regard D as a linear transformation of V into W, where W is the 3-dimensional space of all real polynomials of degree ≤ 2. In W we choose the ordered basis $(1, x, x^2)$. To find the matrix representation of D relative to this choice of bases, we transform (differentiate) each basis element of V and express the result as a linear combination of the basis elements of W. Thus, we find that

$$D(1) = 0 = 0 + 0x + 0x^2, \qquad D(x) = 1 = 1 + 0x + 0x^2,$$
$$D(x^2) = 2x = 0 + 2x + 0x^2, \qquad D(x^3) = 3x^2 = 0 + 0x + 3x^2.$$

The coefficients in these polynomials determine the *columns* of the matrix representation of D. Therefore the required representation is given by the following 3×4 matrix:

$$\begin{bmatrix} 0 & 1 & 0 & 0 \\ 0 & 0 & 2 & 0 \\ 0 & 0 & 0 & 3 \end{bmatrix}.$$

To emphasize that the matrix representation depends not only on the basis elements but also on their order, let's reverse the order of the basis elements in W and use the ordered basis $(x^2, x, 1)$. Then the basis elements of V are transformed into the same polynomials obtained above, but the components of these polynomials relative to the new basis $(x^2, x, 1)$ appear in reversed order. Therefore, the matrix representation of D now becomes

$$\begin{bmatrix} 0 & 0 & 0 & 3 \\ 0 & 0 & 2 & 0 \\ 0 & 1 & 0 & 0 \end{bmatrix}.$$

Now we compute a third matrix representation for D, using the ordered bases

$$(1, 1 + x, 1 + x + x^2, 1 + x + x^2 + x^3) \text{ for } V, \qquad \text{and } (1, x, x^2) \text{ for } W.$$

Now the basis elements are transformed as follows:

$$D(1) = 0, \qquad D(1 + x) = 1, \qquad D(1 + x + x^2) = 1 + 2x,$$
$$D(1 + x + x^2 + x^3) = 1 + 2x + 3x^2,$$

so the matrix representation in this case is

$$\begin{bmatrix} 0 & 1 & 1 & 1 \\ 0 & 0 & 2 & 2 \\ 0 & 0 & 0 & 3 \end{bmatrix}.$$

4.11   Construction of a matrix representation in diagonal form

Since it is possible to obtain different matrix representations of a given linear transformation by different choices of bases, it is natural to try to choose the bases so the resulting matrix will have a particularly simple form, because this will also give a simple form for the set of linear equations relating the components. The next theorem shows that we can make all the entries 0 except possibly along the diagonal starting from the upper left-hand corner of the matrix (this is called the *main diagonal*). Along the main diagonal there will be a string of ones followed by zeros, the number of ones being equal to the rank of the transformation. A matrix (t_{ik}) with all entries $t_{ik} = 0$ when $i \neq k$ is said to be a *diagonal matrix*.

THEOREM 4.14. *Let V and W be finite-dimensional linear spaces, with* $\dim V = n$ *and* $\dim W = m$. *Assume* $T \in \mathcal{L}(V, W)$, *and let* $r = \dim T(V)$ *denote the rank of T. Then there exists a basis* $(v_1, \ldots, v_n)$ *for V and a basis* $(w_1, \ldots, w_m)$ *for W such that*

$$(4.14) \qquad T(v_i) = w_i \quad for \quad i = 1, 2, \ldots, r,$$

and

$$(4.15) \qquad T(v_k) = O \quad for \quad i = r + 1, 2, \ldots, n.$$

Therefore, the matrix (t_{ik}) *of T relative to these bases has all entries zero except for the r diagonal entries*

$$t_{11} = t_{22} = \cdots = t_{rr} = 1.$$

Proof. First we construct a basis for W. Since $T(V)$ is a subspace of W with dimension r, it has a basis of r elements in W, say, $w_1, \ldots, w_r$. By Theorem 4.7, these elements form a subset of some basis for W. Therefore we can adjoin elements $w_{r+1}, \ldots, w_m$ so that

$$(4.16) \qquad (w_1, \ldots, w_r, w_{r+1}, \ldots, w_m)$$

is a basis for W.

Now we construct a basis for V. Each of the first r elements w_i in (4.16) is the image of at least one element in V. Choose one such element in V and call it v_i. Then $T(v_i) = w_i$ for $i = 1, 2, \ldots, r$ so (4.14) is satisfied. Now let k be the dimension of the null space $N(T)$. By the nullity plus rank theorem (Theorem 4.3) we have $k + r = n$. Since the null space has dimension k it has a basis of k elements in V that we designate as $v_{r+1}, \ldots, v_{r+k}$. For each of these elements, Equation (4.15) is satisfied. Therefore, to complete the proof, we must show that the ordered set

$$(4.17) \qquad (v_1, \ldots, v_r, v_{r+1}, \ldots, v_{r+k})$$

is a basis for V. Since $\dim V = n = r + k$, we need only show that these elements are independent.

Suppose that some linear combination of them is zero, say

$$(4.18) \qquad \sum_{i=1}^{r+k} c_i v_i = O.$$

Applying T and using Equations (4.14) and (4.15), we find that

$$\sum_{i=1}^{r+k} c_i T(v_i) = \sum_{i=1}^{r} c_i w_i = O.$$

But $w_1, \ldots, w_r$ are independent, and hence $c_1 = \cdots = c_r = 0$. Therefore, the first r terms in (4.18) are zero, so (4.18) reduces to

$$\sum_{i=r+1}^{r+k} c_i v_i = O.$$

But $v_{r+1}, \ldots, v_{r+k}$ are independent because they form a basis for the null space of T, hence $c_{r+1} = \cdots = c_{r+k} = 0$. Therefore all the c_i in (4.18) are zero, so the elements in (4.17) form a basis for V. This completes the proof.

EXAMPLE.   Refer to Example 6 of Section 4.10, where D is the differentiation operator that maps the space V of polynomials of degree ≤ 3 into the space W of polynomials of degree ≤ 2. In this example, the range of T is W, so T has rank 3. Applying the method used to prove Theorem 4.14, we choose any basis for W, for example, the basis $(1, x, x^2)$. A set of polynomials that D maps onto these elements is $(x, \frac{1}{2}x^2, \frac{1}{3}x^3)$. We extend this set to get a basis for V by adjoining the constant polynomial 1, which is a basis for the null space of D. Therefore, if we use the basis $(x, \frac{1}{2}x^2, \frac{1}{3}x^3, 1)$ for V and the basis $(1, x, x^2)$ for W, the corresponding matrix representation of D has the diagonal form

$$\begin{bmatrix} 1 & 0 & 0 & 0 \\ 0 & 1 & 0 & 0 \\ 0 & 0 & 1 & 0 \end{bmatrix}.$$

4.12   Exercises

In all exercises involving $\mathbf{R}^n$ the usual basis of unit coordinate vectors is to be chosen unless another basis is specifically mentioned. In exercises concerned with the matrix of a linear transformation $T : V \to W$ where $V = W$, we take the same basis in both V and W unless another choice is indicated.

1. Determine the matrix of each of the following linear transformations of $\mathbf{R}^n$ into $\mathbf{R}^n$.
   (a) The identity transformation.
   (b) The zero transformation.
   (c) Multiplication by a fixed scalar c.
2. Determine the matrix for each of the following projections.
   (a) $T : \mathbf{R}^3 \to \mathbf{R}^2$, where $T(x_1, x_2, x_3) = (x_1, x_2)$.
   (b) $T : \mathbf{R}^3 \to \mathbf{R}^2$, where $T(x_1, x_2, x_3) = (x_2, x_3)$.
   (c) $T : \mathbf{R}^5 \to \mathbf{R}^3$, where $T(x_1, x_2, x_3, x_4, x_5) = (x_2, x_3, x_4)$.
3. A linear transformation $T : \mathbf{R}^2 \to \mathbf{R}^2$ maps the basis vectors i and j as follows:

$$T(i) = i + j, \ T(j) = 2i - j.$$

   (a) Compute $T(3i - 4j)$ and $T^2(3i - 4j)$ in terms of i and j.
   (b) Determine the matrix of T and of T^2.
   (c) Solve part (b) if the basis (i, j) is replaced by (e_1, e_2), where $e_1 = i - j, e_2 = 3i + j$.
4. A linear transformation $T : \mathbf{R}^2 \to \mathbf{R}^2$ is defined as follows: Each vector (x, y) is reflected in the y axis and then doubled in length to yield $T(x, y)$. Determine the matrix of T and of T^2.
5. Let $T : \mathbf{R}^3 \to \mathbf{R}^3$ be a linear transformation such that

$$T(k) = 2i + 3j + 5k, \qquad T(j + k) = i, \qquad T(i + j + k) = j - k.$$

   (a) Compute $T(i + 2j + 3k)$ and determine the nullity and rank of T.
   (b) Determine the matrix of T.
6. For the transformation in Exercise 5, choose both bases to be (e_1, e_2, e_3), where $e_1 = (2, 3, 5)$, $e_2 = (1, 0, 0)$, $e_3 = (0, 1, -1)$, and determine the matrix of T relative to the new bases.
7. A linear transformation $T : \mathbf{R}^3 \rightarrow \mathbf{R}^2$ maps the basis vectors as follows: $T(i) = (0, 0)$, $T(j) = (1, 1)$, $T(k) = (1, -1)$.
   (a) Compute $T(4i - j + k)$ and determine the nullity and rank of T.
   (b) Determine the matrix of T.
   (c) Use the basis (i, j, k) in $\mathbf{R}^3$ and the basis (w_1, w_2) in $\mathbf{R}^2$, where $w_1 = (1, 1)$, $w_2 = (1, 2)$. Determine the matrix of T relative to these bases.
   (d) Find bases (e_1, e_2, e_3) for $\mathbf{R}^3$ and (w_1, w_2) for $\mathbf{R}^2$ relative to which the matrix of T will be in diagonal form.
8. A linear transformation $T : \mathbf{R}^2 \rightarrow \mathbf{R}^3$ maps the basis vectors as follows: $T(i) = (1, 0, 1)$, $T(j) = (-1, 0, 1)$.
   (a) Compute $T(2i - 3j)$ and determine the nullity and rank of T.
   (b) Determine the matrix of T.
   (c) Find bases (e_1, e_2) for $\mathbf{R}^2$ and (w_1, w_2, w_3) for $\mathbf{R}^3$ relative to which the matrix of T will be in diagonal form.
9. Solve Exercise 8 if $T(i) = (1, 0, 1)$ and $T(j) = (1, 1, 1)$.
10. Let V and W be linear spaces, each of dimension 2 and each with basis (e_1, e_2). Let $T : V \rightarrow W$ be a linear transformation such that $T(e_1 + e_2) = 3e_1 + 9e_2$, $T(3e_1 + 2e_2) = 7e_1 + 23e_2$.
   (a) Compute $T(e_2 - e_1)$ and determine the nullity and rank of T.
   (b) Determine the matrix of T relative to the given basis.
   (c) Use the basis (e_1, e_2) for V and find a new basis of the form $(e_1 + ae_2, 2e_1 + be_2)$ for W, relative to which the matrix of T will be in diagonal form.

Exercises 11 through 20 require a knowledge of calculus.
In the linear space of all real-valued functions, each of the following sets is independent and spans a finite-dimensional subspace V. Use the given set as a basis for V and let $D : V \rightarrow V$ be the differentiation operator. In each case, find the matrix of D and of D^2 relative to this choice of basis.

11. $(\sin x, \cos x)$.   15. $(-\cos x, \sin x)$.
12. $(1, x, e^x)$.   16. $(\sin x, \cos x, x \sin x, x \cos x)$.
13. $(1, 1 + x, 1 + x + e^x)$.   17. $(e^x \sin x, e^x \cos x)$.
14. (e^x, xe^x).   18. $(e^{2x} \sin 3x, e^{2x} \cos 3x)$.

19. Choose the basis $(1, x, x^2, x^3)$ in the linear space V of all real polynomials of degree ≤ 3. Let D be the differentiation operator and let $T : V \rightarrow V$ be the linear transformation that maps $p(x)$ onto $xp'(x)$. Relative to the given basis, determine the matrix of each of the following:
   (a) T (c) TD (e) T^2
   (b) DT (d) $TD - DT$ (f) $T^2D^2 - D^2T^2$.
20. Refer to Exercise 19. Let W be the image of V under TD. Find bases for V and for W relative to which the matrix of TD is in diagonal form.

4.13   Linear spaces of matrices

We have seen how matrices arise in a natural way as representations of linear transformations. Matrices can also be regarded as objects existing in their own right, without necessarily being connected to linear transformations. As such, they form another class of mathematical objects on which algebraic operations can be defined. The connection with linear transformations serves as motivation for these definitions, but this connection will be ignored for the moment.

Let m and n be two positive integers, and let $I_{m,n}$ be the set of all pairs of integers (i, j) such that $1 \leq i \leq m$, $1 \leq j \leq n$. Any function A whose domain is $I_{m,n}$ is called an $m \times n$

matrix. The function value $A(i, j)$ is called the ij-*entry* or the ij-*element* of the matrix and will be denoted also by a_{ij}. It is customary to display all the function values as a rectangular array consisting of m rows and n columns, as follows:

$$\begin{bmatrix} a_{11} & a_{12} & \cdots & a_{1n} \\ a_{21} & a_{22} & \cdots & a_{2n} \\ \vdots & \vdots & & \vdots \\ a_{m1} & a_{m2} & \cdots & a_{mn} \end{bmatrix}.$$

The elements a_{ij} may be arbitrary objects of any kind. Usually they will be real or complex numbers, but sometimes it is convenient to consider matrices whose elements are other objects, for example, functions. We also denote matrices by the more compact notation

$$A = (a_{ik})_{i,k=1}^{m,n} \quad \text{or} \quad A = (a_{ik}).$$

If $m = n$, the matrix is said to be a *square matrix*. A $1 \times n$ matrix is called a *row matrix*; an $m \times 1$ matrix is called a *column matrix*.

Two functions are equal if and only if they have the same domain and take the same function value at each element in the domain. Since matrices are functions, two matrices $A = (a_{ij})$ and $B = (b_{ij})$ are equal if and only if they have the same number of rows, the same number of columns, and equal entries $a_{ij} = b_{ij}$ for each pair (i, j).

Now we assume the entries are numbers (real or complex) and we define addition of matrices and multiplication by scalars by the same method used for any real- or complex-valued function.

DEFINITION. *If $A = (a_{ij})$ and $B = (b_{ij})$ are two $m \times n$ matrices and if c is any scalar, we define matrices $A + B$ and cA as follows:*

$$A + B = (a_{ij} + b_{ij}), \qquad cA = (ca_{ij}).$$

The sum is defined only when A and B have the same size.

EXAMPLE. If

$$A = \begin{bmatrix} 1 & 2 & -3 \\ -1 & 0 & 4 \end{bmatrix} \quad \text{and} \quad B = \begin{bmatrix} 5 & 0 & 1 \\ 1 & -2 & 3 \end{bmatrix},$$

then we have

$$A + B = \begin{bmatrix} 6 & 2 & -2 \\ 0 & -2 & 7 \end{bmatrix}, \qquad 2A = \begin{bmatrix} 2 & 4 & -6 \\ -2 & 0 & 8 \end{bmatrix}, \qquad (-1)B = \begin{bmatrix} -5 & 0 & -1 \\ -1 & 2 & -3 \end{bmatrix}.$$

We define the zero matrix O to be the $m \times n$ matrix all of whose elements are 0. With these definitions, it is a straightforward exercise to verify that the collection of all $m \times n$ matrices is a linear space. We denote this linear space by $M_{m,n}$. If the entries are real numbers, the space $M_{m,n}$ is a real linear space; if the entries are complex, it is a complex linear space. It is easy to prove that this space has dimension mn. In fact, a basis for $M_{m,n}$ consists of the mn matrices having one entry equal to 1 and all others equal to 0. For example, the six matrices

$$\begin{bmatrix} 1 & 0 & 0 \\ 0 & 0 & 0 \end{bmatrix}, \quad \begin{bmatrix} 0 & 1 & 0 \\ 0 & 0 & 0 \end{bmatrix}, \quad \begin{bmatrix} 0 & 0 & 1 \\ 0 & 0 & 0 \end{bmatrix},$$

$$\begin{bmatrix} 0 & 0 & 0 \\ 1 & 0 & 0 \end{bmatrix}, \quad \begin{bmatrix} 0 & 0 & 0 \\ 0 & 1 & 0 \end{bmatrix}, \quad \begin{bmatrix} 0 & 0 & 0 \\ 0 & 0 & 1 \end{bmatrix},$$

form a basis for the set of all 2×3 matrices.

4.14 Isomorphism between linear transformations and matrices

We return now to the connection between matrices and linear transformations. Let V and W be finite-dimensional linear spaces with dim $V = n$ and dim $W = m$. Choose a basis $(v_1, \ldots, v_n)$ for V and a basis $(w_1, \ldots, w_m)$ for W. In this discussion, these bases are kept fixed. Let $\mathcal{L}(V, W)$ denote the linear space of all linear transformations of V into W. If $T \in \mathcal{L}(V, W)$ let $m(T)$ denote the matrix of T relative to the two given bases. We recall that $m(T)$ is defined as follows:

Express the image of each basis element v_k as a linear combination of the basis elements in W:

(4.19) $$T(v_k) = \sum_{i=1}^{m} t_{ik} w_i, \qquad \text{for} \qquad k = 1, 2, \ldots, n.$$

Then the scalar multipliers t_{ik} are the ik-entries of $m(T)$. Thus we have

(4.20) $$m(T) = (t_{ik})_{i,k=1}^{m,n}.$$

Equation (4.20) defines a new function m whose domain is $\mathcal{L}(V, W)$ and whose values are matrices in $M_{m,n}$. Since every $m \times n$ matrix is the matrix $m(T)$ for some T in $\mathcal{L}(V, W)$, the range of m is $M_{m,n}$. The next theorem tells us that the transformation $m : \mathcal{L}(V, W) \rightarrow M_{m,n}$ is linear and one-to-one on $\mathcal{L}(V, W)$.

THEOREM 4.15. ISOMORPHISM THEOREM. *For all S and T in $\mathcal{L}(V, W)$ and all scalars c, we have*

$$m(S + T) = m(S) + m(T) \qquad and \qquad m(cT) = cm(T).$$

Moreover,

$$m(S) = m(T) \qquad implies \qquad S = T,$$

so m is one-to-one on $\mathcal{L}(V, W)$.

Proof. The matrix $m(T)$ is formed from the multipliers t_{ik} in (4.19). Similarly, the matrix $m(S)$ is formed from the multipliers s_{ik} in the equations

(4.21) $$S(v_k) = \sum_{i=1}^{m} s_{ik} w_i, \qquad \text{for} \qquad k = 1, 2, \ldots, n.$$

But we have

$$(S + T)(v_k) = \sum_{i=1}^{m}(s_{ik} + t_{ik})w_i \qquad \text{and} \qquad (cT)(v_k) = \sum_{i=1}^{m}(ct_{ik})w_i,$$

so we see that $m(S + T) = (s_{ik} + t_{ik}) = m(S) + m(T)$, and $m(cT) = (ct_{ik}) = c(t_{ik}) = cm(T)$. This proves that m is linear.

To prove that m is one-to-one, suppose that $m(S) = m(T)$, where $S = (s_{ik})$ and $T = (t_{ik})$. Equations (4.19) and (4.21) show that $S(v_k) = T(v_k)$ for each basis element v_k, so $S(x) = T(x)$ for all x in V, and hence $S = T$.

Note. The function m is called an *isomorphism.* For a given choice of bases, m establishes a one-to-one correspondence between the set of all linear transformations $\mathcal{L}(V, W)$ and the set of $m \times n$ matrices $M_{m,n}$. Addition and multiplication by scalars are preserved under this correspondence. The linear spaces $\mathcal{L}(V, W)$ and $M_{m,n}$ are said to be *isomorphic.* Incidentally, Theorem 4.11 shows that the domain of a one-to-one linear transformation has the same dimension as its range. Therefore, $\dim \mathcal{L}(V, W) = \dim M_{m,n} = mn$.

If $V = W$ and if we choose the same basis in both V and W, then the matrix $m(I)$ that corresponds to the identity transformation $I : V \rightarrow V$ is an $n \times n$ diagonal matrix with each diagonal entry equal to 1 and all others equal to 0. This is called the *identity* or *unit matrix* and is denoted by I (or by I_n if we wish to emphasize that it is an $n \times n$ matrix).

4.15 Multiplication of matrices

Some linear transformations can be multiplied by means of composition. Now we shall define multiplication of matrices in such a way that the product of two matrices corresponds to the composition of the linear transformations they represent.

We recall that if $T : U \rightarrow V$ and $S : V \rightarrow W$ are linear transformations, their composition $ST : U \rightarrow W$ is the linear transformation given by

$$(ST)(x) = S[T(x)] \qquad \text{for all } x \text{ in } U.$$

Suppose that U, V, and W are finite-dimensional, say,

$$\dim U = n, \qquad \dim V = p, \qquad \dim W = m.$$

Choose bases for U, V, W. Relative to these bases, matrix $m(S)$ is an $m \times p$ matrix, matrix $m(T)$ is a $p \times n$ matrix, and matrix $m(ST)$ is an $m \times n$ matrix. The following definition of matrix multiplication will enable us to deduce the relation $m(ST) = m(S)m(T)$ and thereby extend the isomorphism property to products.

DEFINITION. PRODUCT OF MATRICES. *Let A be any $m \times p$ matrix, and let B be any $p \times n$ matrix, say*

$$A = (a_{ij})_{i,j=1}^{m,p} \qquad \text{and} \qquad B = (b_{ij})_{i,j=1}^{p,n}.$$

Then the product AB (in that order) is defined to be the m × n matrix C = (c_{ij}) whose ij-entry is given by the sum

(4.22)
$$c_{ij} = \sum_{k=1}^{p} a_{ik}b_{kj}.$$

Note. The product AB is not defined unless the number of columns of the left factor A is equal to the number of rows of the right factor B.

Let's use the notation A_i to denote the ith row of A, and B^j to denote the jth column of B, and think of these as p-dimensional vectors. Then the sum in (4.22) is simply the dot product $A_i \cdot B^j$. In other words, the ij-entry of AB is the dot product of the ith row of A with the jth column of B:

$$AB = (A_i \cdot B^j)_{i,j=1}^{m,n}.$$

Thus, matrix multiplication can be regarded as a generalization of the dot product.

EXAMPLE 1. Let $A = \begin{bmatrix} 3 & 1 & 2 \\ -1 & 1 & 0 \end{bmatrix}$ and $B = \begin{bmatrix} 4 & 6 \\ 5 & -1 \\ 0 & 2 \end{bmatrix}$. Since A is 2×3 and B is

3×2, the product AB is a 2×2 matrix. The rows of A, regarded as vectors in $\mathbf{R}^3$, are $A_1 = (3, 1, 2)$ and $A_2 = (-1, 1, 0)$. The columns of B, regarded as vectors in $\mathbf{R}^3$, are $B^1 = (4, 5, 0)$ and $B^2 = (6, -1, 2)$. Therefore the product AB is given by

$$AB = \begin{bmatrix} A_1 \cdot B^1 & A_1 \cdot B^2 \\ A_2 \cdot B^1 & A_2 \cdot B^2 \end{bmatrix} = \begin{bmatrix} 17 & 21 \\ 1 & -7 \end{bmatrix}.$$

EXAMPLE 2. Let $A = \begin{bmatrix} 2 & 1 & -3 \\ 1 & 2 & 4 \end{bmatrix}$ and $B = \begin{bmatrix} -2 \\ 1 \\ 2 \end{bmatrix}$. Here A is 2×3 and B is 3×1,

so AB is the 2×1 matrix given by

$$AB = \begin{bmatrix} A_1 \cdot B^1 \\ A_2 \cdot B^1 \end{bmatrix} = \begin{bmatrix} -9 \\ 8 \end{bmatrix}.$$

EXAMPLE 3. If A and B are square matrices of the same size, then both AB and BA are defined. For example, if

$$A = \begin{bmatrix} 1 & 2 \\ -1 & 1 \end{bmatrix} \quad \text{and} \quad B = \begin{bmatrix} 3 & 4 \\ 5 & 2 \end{bmatrix},$$

we find that

$$AB = \begin{bmatrix} 13 & 8 \\ 2 & -2 \end{bmatrix} \quad \text{and} \quad BA = \begin{bmatrix} -1 & 10 \\ 3 & 12 \end{bmatrix}.$$

This example shows that in general $AB \neq BA$. If $AB = BA$, we say that A and B *commute.*

EXAMPLE 4. If I_p is the $p \times p$ identity matrix, then $I_p A = A$ for every $p \times n$ matrix A, and $BI_p = B$ for every $m \times p$ matrix B. For example,

$$
\begin{bmatrix} 1 & 0 & 0 \\ 0 & 1 & 0 \\ 0 & 0 & 1 \end{bmatrix} \begin{bmatrix} 2 \\ 3 \\ 4 \end{bmatrix} = \begin{bmatrix} 2 \\ 3 \\ 4 \end{bmatrix}, \qquad \begin{bmatrix} 1 & 2 & 3 \\ 4 & 5 & 6 \end{bmatrix} \begin{bmatrix} 1 & 0 & 0 \\ 0 & 1 & 0 \\ 0 & 0 & 1 \end{bmatrix} = \begin{bmatrix} 1 & 2 & 3 \\ 4 & 5 & 6 \end{bmatrix}.
$$

Now we prove that the matrix of a composition ST is the product of the matrices $m(S)$ and $m(T)$.

THEOREM 4.16. *Let $T : U \to V$ and $S : V \to W$ be linear transformations, where U, V, W are finite-dimensional linear spaces. Then, for a fixed choice of bases, the matrices of S, T, and ST are related by the equation*

$$
m(ST) = m(S)m(T).
$$

Proof. Assume $\dim U = n$, $\dim V = p$, $\dim W = m$. Choose three bases for the linear spaces U, V, W, say $(u_1, \ldots, u_n)$ for U, $(v_1, \ldots, v_p)$ for V, and $(w_1, \ldots, w_m)$ for W. Relative to these bases we have

$$
m(S) = (s_{ij})_{i,j=1}^{m,p} \qquad \text{where} \qquad S(v_k) = \sum_{i=1}^{m} s_{ik} w_i, \qquad \text{for } k = 1, 2, \ldots, p,
$$

and

$$
m(T) = (t_{ij})_{i,j=1}^{p,n} \qquad \text{where} \qquad T(u_j) = \sum_{k=1}^{p} t_{kj} v_k, \qquad \text{for } j = 1, 2, \ldots, n.
$$

Therefore we have

$$
ST(u_j) = S[T(u_j)] = \sum_{k=1}^{p} t_{kj} S(v_k) = \sum_{k=1}^{p} t_{kj} \sum_{i=1}^{m} s_{ik} w_i = \sum_{i=1}^{m} \left(\sum_{k=1}^{p} s_{ik} t_{kj} \right) w_i,
$$

so we find that

$$
m(ST) = \left(\sum_{k=1}^{p} s_{ik} t_{kj} \right)_{i,j=1}^{m,n} = m(S)m(T).
$$

We have already noted that matrix multiplication is not always commutative. The next theorem shows that it does satisfy the associative and distributive laws.

THEOREM 4.17. ASSOCIATIVE AND DISTRIBUTIVE LAWS FOR MATRIX MULTIPLICATION. *Given matrices A, B, C.*

(a) *If the products $A(BC)$ and $(AB)C$ are meaningful, we have*

$$
A(BC) = (AB)C \qquad \text{(associative law)}.
$$

(b) *Assume A and B are of the same size. If AC and BC are meaningful, we have*

$$(A + B)C = AC + BC \qquad (right \ distributive \ law),$$

whereas, if CA and CB are meaningful, we have

$$C(A + B) = CA + CB \qquad (left \ distributive \ law).$$

Proof. These properties can be deduced directly from the definition of matrix multiplication, but we prefer to proceed differently, using the connection between linear transformations and matrices. Introduce finite-dimensional linear spaces U, V, W, X and three linear transformations $T : U \to V$, $S : V \to W$, $R : W \to X$ such that, for a fixed choice of bases, we have

$$A = m(R), \qquad B = m(S), \qquad C = m(T).$$

By Theorem 4.16 we have $m(RS) = AB$ and $m(ST) = BC$. From the associative law for composition, we find that $R(ST) = (RS)T$. Applying Theorem 4.16 once more to this equation, we obtain $m(R)m(ST) = m(RS)m(T)$, or $A(BC) = (AB)C$, which proves (a). The proof of (b) can be given by a similar type of argument.

DEFINITION. POWERS OF MATRICES. *If A is a square matrix, integral powers of A are defined inductively as follows:*

$$A^0 = I, \qquad A^n = AA^{n-1} \qquad for \qquad n \ge 1.$$

The reader can verify that integral powers of the same matrix A commute with one another, $A^n A^m = A^m A^n$, and satisfy the law of exponents $A^n A^m = A^{n+m}$.

4.16 Exercises

1. Given matrices $A = \begin{bmatrix} 1 & -4 & 2 \\ -1 & 4 & -2 \end{bmatrix}$, $B = \begin{bmatrix} 1 & 2 \\ -1 & 3 \\ 5 & -2 \end{bmatrix}$, $C = \begin{bmatrix} 2 & 2 \\ 1 & -1 \\ 1 & -3 \end{bmatrix}$. Compute $B + C$, AB, BA, AC, CA, and $A(2B - 3C)$.

2. Let $A = \begin{bmatrix} 0 & 1 \\ 0 & 2 \end{bmatrix}$. Find all 2×2 matrices B such that (a) $AB = O$; (b) $BA = O$.

3. In each case find a, b, c, d to satisfy the given equations.

(a) $\begin{bmatrix} 0 & 0 & 1 & 0 \\ 1 & 0 & 0 & 0 \\ 0 & 1 & 0 & 0 \\ 0 & 0 & 0 & 1 \end{bmatrix} \begin{bmatrix} a \\ b \\ c \\ d \end{bmatrix} = \begin{bmatrix} 1 \\ 9 \\ 6 \\ 5 \end{bmatrix}.$

(b) $\begin{bmatrix} a & b & c & d \\ 1 & 4 & 9 & 2 \end{bmatrix} \begin{bmatrix} 1 & 0 & 2 & 0 \\ 0 & 0 & 1 & 1 \\ 0 & 1 & 0 & 0 \\ 0 & 0 & 1 & 0 \end{bmatrix} = \begin{bmatrix} 1 & 0 & 6 & 6 \\ 1 & 9 & 8 & 4 \end{bmatrix}.$

4. Calculate $AB - BA$ for the following matrices A and B.

(a) $A = \begin{bmatrix} 1 & 2 & 2 \\ 2 & 1 & 2 \\ 1 & 2 & 3 \end{bmatrix}$, $B = \begin{bmatrix} 4 & 1 & 1 \\ -4 & 2 & 0 \\ 1 & 2 & 1 \end{bmatrix}.$

(b) $A = \begin{bmatrix} 2 & 0 & 0 \\ 1 & 1 & 2 \\ -1 & 2 & 1 \end{bmatrix}, B = \begin{bmatrix} 3 & 1 & -2 \\ 3 & -2 & 4 \\ -3 & 5 & 11 \end{bmatrix}.$

5. If A is a square matrix, prove that $A^n A^m = A^m A^n$ for all integers $m \geq 0, n \geq 0$.

6. Let $A = \begin{bmatrix} 1 & 1 \\ 0 & 1 \end{bmatrix}$. Verify that $A^2 = \begin{bmatrix} 1 & 2 \\ 0 & 1 \end{bmatrix}$, and compute A^n.

7. Let $A = \begin{bmatrix} \cos\theta & -\sin\theta \\ \sin\theta & \cos\theta \end{bmatrix}$. Verify that $A^2 = \begin{bmatrix} \cos 2\theta & -\sin 2\theta \\ \sin 2\theta & \cos 2\theta \end{bmatrix}$, and compute A^n.

8. Let $A = \begin{bmatrix} 1 & 1 & 1 \\ 0 & 1 & 1 \\ 0 & 0 & 1 \end{bmatrix}$. Verify that $A^2 = \begin{bmatrix} 1 & 2 & 3 \\ 0 & 1 & 2 \\ 0 & 0 & 1 \end{bmatrix}$. Compute A^3 and A^4. Guess a general formula for A^n and prove it by induction.

9. Let $A = \begin{bmatrix} 1 & 0 \\ -1 & 1 \end{bmatrix}$. Prove that $A^2 = 2A - I$ and compute A^{100}.

10. Find all 2×2 matrices A such that $A^2 = O$.

11. (a) Prove that a 2×2 matrix A commutes with every 2×2 matrix if and only if A commutes with each of the four matrices

$$\begin{bmatrix} 1 & 0 \\ 0 & 0 \end{bmatrix}, \quad \begin{bmatrix} 0 & 1 \\ 0 & 0 \end{bmatrix}, \quad \begin{bmatrix} 0 & 0 \\ 1 & 0 \end{bmatrix}, \quad \begin{bmatrix} 0 & 0 \\ 0 & 1 \end{bmatrix}.$$

    (b) Find all such matrices A.

12. The equation $A^2 = I$ is satisfied by each of the 2×2 matrices

$$\begin{bmatrix} 1 & 0 \\ 0 & 1 \end{bmatrix}, \quad \begin{bmatrix} 1 & 0 \\ c & -1 \end{bmatrix}, \quad \begin{bmatrix} 1 & b \\ 0 & -1 \end{bmatrix},$$

where b and c are arbitrary real numbers. Find all 2×2 matrices such that $A^2 = I$.

13. Find 2×2 matrices C and D such that $AC = B$ and $DA = B$ if

$$A = \begin{bmatrix} 2 & -1 \\ -2 & 3 \end{bmatrix} \quad \text{and} \quad B = \begin{bmatrix} 7 & 6 \\ 9 & 8 \end{bmatrix}.$$

14. (a) Verify that the algebraic identities

$$(A + B)^2 = A^2 + 2AB + B^2 \quad \text{and} \quad (A + B)(A - B) = A^2 - B^2$$

    do not hold for the 2×2 matrices $A = \begin{bmatrix} 1 & -1 \\ 0 & 2 \end{bmatrix}$ and $B = \begin{bmatrix} 1 & 0 \\ 1 & 2 \end{bmatrix}$.

    (b) Amend the right-hand members of these identities to obtain formulas valid for all square matrices A and B.

    (c) For which matrices A and B are the identities valid as stated in part (a)?

15. Prove each of the following statements about $n \times n$ matrices, or exhibit a counter example.

    (a) If $(A - B)^2 = (A + B)^2$, then $A^2B = BA^2$.

    (b) If $A^2 = I$, then either $A = I$ or $A = -I$.

4.17 Applications to systems of linear equations

Let $A = (a_{ij})$ be a given $m \times n$ matrix of numbers, and let $c_1, \ldots, c_m$ be m further numbers. A set of m equations of the form

(4.23)
$$\sum_{k=1}^{n} a_{ik}x_k = c_i \quad \text{for} \quad i = 1, 2, \ldots, m,$$

is called a system of m linear equations in n unknowns. Here $x_1, \ldots, x_n$ are regarded as unknown. A *solution* of the system is any n-tuple $(x_1, \ldots, x_n)$ for which all the equations are satisfied. The matrix $A = (a_{ij})$ is called the *coefficient-matrix* of the system.

Linear systems can be cast in the language of linear transformations. Choose the usual bases of unit coordinate vectors in $\mathbf{R}^n$ and $\mathbf{R}^m$. The coefficient-matrix A determines a linear transformation $T : \mathbf{R}^n \to \mathbf{R}^m$ that maps an arbitrary vector $x = (x_1, \ldots, x_n)$ in $\mathbf{R}^n$ onto the vector $(y_1, \ldots, y_m)$ in $\mathbf{R}^m$ given by the m linear equations

$$y_i = \sum_{k=1}^{n} a_{ik} x_k \qquad \text{for} \qquad i = 1, 2, \ldots, m.$$

The system in (4.23) can be written more simply as

$$T(x) = c,$$

where $c = (c_1, \ldots, c_m)$ is the vector in $\mathbf{R}^m$ whose components are the numbers appearing in (4.23). The system has a solution if and only if c is in the range of T. If exactly one x in $\mathbf{R}^n$ maps onto c, the system has exactly one solution. If more than one x maps onto c, the system has more than one solution.

EXAMPLE 1. *A system with no solution.* The system $x + y = 1$, $x + y = 2$, consisting of two equations in two unknowns, obviously has no solution because the sum of two numbers cannot be both 1 and 2.

EXAMPLE 2. *A system with exactly one solution.* The system $x + y = 1$, $x - y = 0$, consisting of two equations in two unknowns, has exactly one solution: $(x, y) = \left(\frac{1}{2}, \frac{1}{2}\right)$.

EXAMPLE 3. *A system with more than one solution.* The system $x + y = 1$, consisting of one equation in two unknowns, has infinitely many solutions. Any two numbers whose sum is 1 gives a solution.

With each linear system (4.23), we can associate another system

$$(4.24) \qquad \sum_{k=1}^{n} a_{ik} x_k = 0 \qquad \text{for} \qquad i = 1, 2, \ldots, m,$$

obtained by replacing each c_i in (4.23) by 0. This is called the *homogeneous system* corresponding to (4.23). If $c \neq O$, system (4.23) is called a *nonhomogeneous system*. A vector x in $\mathbf{R}^n$ will satisfy the homogeneous system if and only if

$$T(x) = O,$$

where T is the linear transformation determined by the coefficient matrix. The homogeneous system always has one solution, namely $x = O$, but it may have others. The set of solutions of the homogeneous system is the null space of T. The next theorem describes the relation between solutions of the homogeneous system and those of the nonhomogeneous system.

THEOREM 4.18. *Assume the nonhomogenous system (4.23) has a solution, say b.*

(a) *If a vector x is a solution of the nonhomogeneous system (4.23), then the vector $v = x - b$ is a solution of the corresponding homogeneous system (4.24).*

(b) *If a vector v is a solution of the homogeneous system (4.24), then the vector $x = v + b$ is a solution of the nonhomogeneous system (4.23).*

Proof. Let $T : \mathbf{R}^n \to \mathbf{R}^m$ be the linear transformation determined by the coefficient matrix, as described above. Since b is a solution of the nonhomogeneous system we have $T(b) = c$. Let x and v be two vectors in $\mathbf{R}^n$ such that $v = x - b$. Then by linearity of T we have

$$T(v) = T(x - b) = T(x) - T(b) = T(x) - c.$$

Therefore $T(x) = c$ if and only if $T(v) = O$. This proves both (a) and (b).

The foregoing theorem shows that the problem of finding all solutions of a nonhomogeneous system splits naturally into two parts: (1) Finding all solutions v of the homogeneous system (the null space of T), and (2) finding one particular solution b of the nonhomogeneous system. By adding b to each vector in the null space of T we obtain all solutions $x = v + b$ of the nonhomogeneous system.

Let k denote the dimension of $N(T)$, the null space of T. If we can find k *independent* solutions $v_1, \ldots, v_k$ of the homogeneous system, they will form a basis for $N(T)$, and we can obtain every v in $N(T)$ by forming all possible linear combinations

$$v = t_1 v_1 + \cdots + t_k v_k,$$

where $t_1, \ldots, t_k$ are arbitrary scalars. This linear combination is called the *general solution of the homogeneous system.* If b is one particular solution of the nonhomogeneous system, then all solutions x are given by

$$x = b + t_1 v_1 + \cdots + t_k v_k.$$

This linear combination is called the *general solution of the nonhomogeneous system.* Theorem 4.18 can now be restated as follows.

THEOREM 4.19. *Let $T : \mathbf{R}^n \to \mathbf{R}^m$ be the linear transformation such that $T(x) = y$, where $x = (x_1, \ldots, x_n)$, $y = (y_1, \ldots, y_m)$, and*

$$y_i = \sum_{k=1}^{n} a_{ik} x_k \qquad for \qquad i = 1, 2, \ldots, m.$$

Let k denote the nullity of T. If $v_1, \ldots, v_k$ are k independent solutions of the homogeneous system $T(x) = O$, and if b is one particular solution of the nonhomogeneous system $T(x) = c$, then the general solution of the nonhomogeneous system is

$$x = b + t_1 v_1 + \cdots + t_k v_k,$$

where $t_1, \ldots, t_k$ are arbitrary scalars.

This theorem does not tell us how to decide if a nonhomogeneous system has a particular solution b, nor does it tell us how to determine the solutions $v_1, \ldots, v_k$ of the homogeneous system. It does tell us what to expect when the nonhomogeneous system has a solution. The following example, although very simple, illustrates the theorem.

EXAMPLE. The system consisting of one equation in two unknowns, $x + y = 2$, has for its associated homogeneous system the equation $x + y = 0$. Therefore, the null space consists of all vectors in $\mathbf{R}^2$ of the form $(t, -t)$, where t is arbitrary. Since $(t, -t) = t(1, -1)$, this is a one-dimensional subspace of $\mathbf{R}^2$ with basis vector $(1, -1)$. A particular solution of the nonhomogeneous system is $(0, 2)$. Therefore, the general solution of the nonhomogeneous system is given by

$$(x, y) = (0, 2) + t(1, -1), \qquad \text{or} \qquad x = t, \quad y = 2 - t,$$

where t is arbitrary.

4.18 Computation techniques. The Gauss-Jordan method

We turn now to the problem of actually computing the solutions of a nonhomogeneous linear system. Although many methods have been developed for attacking this problem, all of them require considerable computation if the system is large.

We shall discuss a widely-used method, known as the *Gauss-Jordan elimination method*, which is relatively simple and can be easily programmed on a computer. In fact, dozens of computer programs have been developed to adapt this method to specialized large-scale systems.

The Gauss-Jordan method consists of three basic types of operations that are performed on the individual equations of a linear system:

(1) *Interchanging two equations.*
(2) *Multiplying all the terms of an equation by a nonzero scalar.*
(3) *Adding to one equation a scalar multiple of another.*

Each time we perform one of these operations on the system we obtain a new system having exactly the same solutions. Two such systems are called *equivalent*. By performing these operations over and over again in a systematic fashion we finally arrive at an equivalent system that can be solved by inspection.

We shall illustrate the method with some specific numerical examples. It will then be clear how the method is to be applied in general.

EXAMPLE 1. *A system with a unique solution.* Consider the system

$$
\begin{aligned}
2x - 5y + 4z &= -3 \\
x - 2y + z &= 5 \\
x - 4y + 6z &= 10.
\end{aligned}
$$

This particular system has a unique solution, $x = 124$, $y = 75$, $z = 31$, which we shall obtain by the Gauss-Jordan elimination process. To save labor, we do not bother to copy the letters x, y, z and the equals sign over and over again, but work instead with the *augmented matrix*

(4.25)
$$\begin{bmatrix} 2 & -5 & 4 & | & -3 \\ 1 & -2 & 1 & | & 5 \\ 1 & -4 & 6 & | & 10 \end{bmatrix}$$

obtained by adjoining the right-hand members of the system to the coefficient matrix. The three basic types of operations mentioned above are performed on the rows of the augmented matrix and are called *row operations*. At any stage of the process we can put the letters x, y, z back again and insert equals signs along the vertical bar to obtain equations. Our ultimate goal is to arrive at the augmented matrix

(4.26)
$$\begin{bmatrix} 1 & 0 & 0 & | & 124 \\ 0 & 1 & 0 & | & 75 \\ 0 & 0 & 1 & | & 31 \end{bmatrix}$$

after a succession of row operations. The corresponding system of equations is then equal to

$$x = 124, \qquad y = 75, \qquad z = 31,$$

which gives the desired solution.

The first step in the process is to obtain a 1 in the upper left-hand corner of the matrix. We can do this by interchanging the first row of the given matrix (4.25) with either the second or third row. Or, we can multiply the first row by $\frac{1}{2}$. We choose to interchange the first two rows, giving us

$$\begin{bmatrix} 1 & -2 & 1 & | & 5 \\ 2 & -5 & 4 & | & -3 \\ 1 & -4 & 6 & | & 10 \end{bmatrix}.$$

The next step is to make all the remaining entries in the first column equal to zero, leaving the first row intact. To do this we multiply the first row by -2 and add the result to the second row. Then we multiply the first row by -1 and add the result to the third row. After these two operations, we obtain

(4.27)
$$\begin{bmatrix} 1 & -2 & 1 & | & 5 \\ 0 & -1 & 2 & | & -13 \\ 0 & -2 & 5 & | & 5 \end{bmatrix}.$$

Now we repeat the process on the smaller matrix $\begin{bmatrix} -1 & 2 & | & -13 \\ -2 & 5 & | & 5 \end{bmatrix}$ that appears adjacent to the two zeros. We can obtain a 1 in *its* upper left-hand corner by multiplying the second row of (4.27) by -1. This gives us the matrix

$$\begin{bmatrix} 1 & -2 & 1 & | & 5 \\ 0 & 1 & -2 & | & 13 \\ 0 & -2 & 5 & | & 5 \end{bmatrix}.$$

Multiplying the second row by 2 and adding the result to the third, we get

(4.28)
$$\begin{bmatrix} 1 & -2 & 1 & | & 5 \\ 0 & 1 & -2 & | & 13 \\ 0 & 0 & 1 & | & 31 \end{bmatrix}.$$

At this stage the corresponding system of equations is given by

$$\begin{aligned} x - 2y + z &= 5 \\ y - 2z &= 13 \\ z &= 31. \end{aligned}$$

The last equation gives us the value of z. To find x and y we work our way up through the equations in succession, to find

$$z = 31, \qquad y = 13 + 2z = 13 + 62 = 75, \qquad x = 5 + 2y - z = 5 + 150 - 31 = 124.$$

Or, we can continue the Gauss-Jordan process and make all the entries zero above the diagonal elements in the second and third columns. Multiplying the second row of (4.28) by 2 and adding the result to the first row we obtain

$$\begin{bmatrix} 1 & 0 & -3 & | & 31 \\ 0 & 1 & -2 & | & 13 \\ 0 & 0 & 1 & | & 31 \end{bmatrix}.$$

Finally, we multiply the third row by 3 and add the result to the first row, and then multiply the third row by 2 and add the result to the second row to get the matrix in (4.26).

EXAMPLE 2. *A system with more than one solution.* Consider the following system of three equations in five unknowns:

(4.29)
$$\begin{aligned} 2x - 5y + 4z + u - v &= -3 \\ x - 2y + z - u + v &= 5 \\ x - 4y + 6z + 2u - v &= 10. \end{aligned}$$

The corresponding augmented matrix is

$$\begin{bmatrix} 2 & -5 & 4 & 1 & -1 & | & -3 \\ 1 & -2 & 1 & -1 & 1 & | & 5 \\ 1 & -4 & 6 & 2 & -1 & | & 10 \end{bmatrix}.$$

The coefficients of x, y, z and the right-hand members are the same as those in Example 1. If we perform the same row operations used in Example 1, we finally arrive at the augmented matrix

$$\begin{bmatrix} 1 & 0 & 0 & -16 & 19 & | & 124 \\ 0 & 1 & 0 & -9 & 11 & | & 75 \\ 0 & 0 & 1 & -3 & 4 & | & 31 \end{bmatrix}.$$

The corresponding system of equations can be solved for x, y, and z in terms of u and v, giving us

$$
\begin{aligned}
x &= 124 + 16u - 19v \\
y &= 75 + 9u - 11v \\
z &= 31 + 3u - 4v.
\end{aligned}
$$

The solution vector (x, y, z, u, v) in $\mathbf{R}^5$ can be expressed in terms of u and v as follows:

$$(x, y, z, u, v) = (124 + 16u - 19v, 75 + 9u - 11v, 31 + 3u - 4v, u, v).$$

By separating the parts involving u and v we can rewrite the solution vector in the form

$$(x, y, z, u, v) = (124, 75, 31, 0, 0) + u(16, 9, 3, 1, 0) + v(-19, -11, -4, 0, 1).$$

This equation gives the general solution of the system in terms of two parameters u and v. The vector $(124, 75, 31, 0, 0)$ is a particular solution of the nonhomogeneous system (4.29). The two vectors $(16, 9, 3, 1, 0)$ and $(-19, -11, -4, 0, 1)$ are solutions of the corresponding homogeneous system. Because they are independent, they form a basis for the space of all solutions of the homogeneous equation.

EXAMPLE 3. *A system with no solution.* Consider the system

(4.30)
$$
\begin{aligned}
2x - 5y + 4z &= -3 \\
x - 2y + z &= 5 \\
x - 4y + 5z &= 10.
\end{aligned}
$$

This system is almost identical to that of Example 1, except that the coefficient of z in the third equation has been changed from 6 to 5. The corresponding augmented matrix is

$$
\begin{bmatrix}
2 & -5 & 4 & -3 \\
1 & -2 & 1 & 5 \\
1 & -4 & 5 & 10
\end{bmatrix}.
$$

Applying the same row operations used in Example 1 to transform (4.25) to (4.28), we arrive at the augmented matrix

(4.31)
$$
\begin{bmatrix}
1 & -2 & 1 & 5 \\
0 & 1 & -2 & 13 \\
0 & 0 & 0 & 31
\end{bmatrix}.
$$

When the bottom row is expressed as an equation, it states that $0 = 31$, which is impossible. Therefore the original system has no solution because the two systems (4.30) and (4.31) are equivalent.

In each of the foregoing examples, the number of equations did not exceed the number of unknowns. If there are more equations than unknowns, the Gauss-Jordan process is still applicable. For example, suppose we consider the system of Example 1, which has the solution $x = 124$, $y = 75$, $z = 31$. If we adjoin a new equation to this system that is also satisfied by the same triple, for example, the equation $2x - 3y + z = 54$, then the elimination process leads to the augmented matrix

$$\begin{bmatrix} 1 & 0 & 0 & | & 124 \\ 0 & 1 & 0 & | & 75 \\ 0 & 0 & 1 & | & 31 \\ 0 & 0 & 0 & | & 0 \end{bmatrix}$$

with a row of zeros along the bottom. But if we adjoin a new equation that is not satisfied by the triple $(124, 75, 31)$, for example the equation $x + y + z = 1$, then the elimination process leads to an augmented matrix of the form

$$\begin{bmatrix} 1 & 0 & 0 & | & 124 \\ 0 & 1 & 0 & | & 75 \\ 0 & 0 & 1 & | & 31 \\ 0 & 0 & 0 & | & a \end{bmatrix}$$

where $a \neq 0$. The last row now gives a contradictory equation $0 = a$, which shows that the system has no solution.

4.19 Inverses of square matrices

DEFINITION. *NONSINGULAR MATRIX. Let $A = (a_{ij})$ be a square $n \times n$ matrix. If there is another $n \times n$ matrix B such that $BA = I$, where I is the $n \times n$ identity matrix, then A is called nonsingular, and B is called a left inverse of A.*

THEOREM 4.20. (a) *If an $n \times n$ matrix A has a left inverse B, then B is also a right inverse, $AB = I$, and there is only one left inverse.*

(b) *If an $n \times n$ matrix A has a right inverse B, then B is also a left inverse, so A is nonsingular.*

Proof of (a). Choose the usual basis of unit coordinate vectors in $\mathbf{R}^n$ and let $T : \mathbf{R}^n \to \mathbf{R}^n$ be the linear transformation with matrix $m(T) = A$. First we prove that T is invertible by showing that $T(x) = O$ implies $x = O$.

Given x such that $T(x) = O$, let X be the $n \times 1$ column matrix formed from the components of x. Since $T(x) = O$, the matrix product AX is an $n \times 1$ column matrix consisting of zeros, so $B(AX)$ is also a column matrix of zeros for any $n \times n$ matrix B. Now if B is a left inverse of A we have

$$B(AX) = (BA)X = IX = X,$$

so every component of x is 0. Therefore T is invertible and has a unique left inverse, which is also a right inverse. The equation $TT^{-1} = I$ implies $m(T)m(T^{-1}) = I$ or $Am(T^{-1}) = I$. Multiplying on the left by B we find $m(T^{-1}) = B$, and the equation $m(T)m(T^{-1}) = I$ shows that $AB = I$, so B is also a right inverse. Finally, if C is any left inverse, then $BA = CA = I$, and by multiplying the equation $CA = I$ on the right by B we get $C = B$, so there is only one left inverse.

Proof of (b). If $AB = I$, then B has a left inverse A. Hence, we can apply part (a) to matrix B, and deduce that A is also a right inverse of B. Therefore $BA = I$, which shows that B is a left inverse of A, so A is nonsingular.

Theorem 4.20 shows that a square matrix is nonsingular if it has either a left inverse or a right inverse, in which case both inverses are equal. (By contrast, a general function can have more than one right inverse, and need not have a left inverse.) If A is nonsingular with left (or right) inverse B, then $BA = AB = I$, and we call B *the inverse of* A and denote it by A^{-1}. Since $A^{-1}A = AA^{-1} = I$, the inverse A^{-1} is also nonsingular and *its* inverse is A. We note that a linear transformation T is invertible if and only if its matrix $m(T)$ is invertible.

If the inverse of a nonsingular $n \times n$ matrix A is known, it can be used to solve a system of n linear equations in n unknowns having A as coefficient-matrix. For example, the system

$$\sum_{k=1}^{n} a_{ik}x_k = c_i, \qquad i = 1, 2, \ldots, n,$$

can be written more simply as a matrix equation

$$AX = C,$$

where $A = (a_{ij})$ is the coefficient-matrix, and X and C are column matrices,

$$X = \begin{bmatrix} x_1 \\ x_2 \\ \vdots \\ x_n \end{bmatrix}, \qquad C = \begin{bmatrix} c_1 \\ c_2 \\ \vdots \\ c_n \end{bmatrix}.$$

If A is nonsingular there is unique solution of the system given by $X = A^{-1}C$. This extends the familiar result of elementary algebra, which states that a linear equation $ax = c$ has the unique solution $x = a^{-1}c$ if $a \neq 0$.

The formula $X = A^{-1}C$ for solving a linear system is deceptively simple. It is useful only if we know the inverse matrix A^{-1}. The problem of actually determining the entries of A^{-1} can be formidable if n is large. In fact, it is equivalent to solving n separate nonhomogeneous linear systems. Let $A = (a_{ij})$ be nonsingular, and let $A^{-1} = (b_{ij})$ be its inverse. The entries of A and A^{-1} are related by the n^2 equations

(4.32) $$\sum_{k=1}^{n} a_{ik}b_{kj} = c_{ij}, \qquad i, j = 1, 2, \ldots, n,$$

where c_{ij} is the ij entry of the identity matrix: $c_{ij} = 1$ if $i = j$, and $c_{ij} = 0$ if $i \neq j$. For each fixed choice of j we can regard (4.32) as a nonhomogeneous system of n linear equations in n unknowns $b_{1j}, b_{2j}, \ldots, b_{nj}$ with coefficient-matrix A. Since A is nonsingular, each of these systems has a unique solution, the jth column of B. All these systems have the same coefficient-matrix A; they differ only in their right members. For example, if A is a 3×3 matrix, there are 9 equations in (4.32) which can be expressed as 3 separate linear systems having the following augmented matrices:

$$\begin{bmatrix} a_{11} & a_{12} & a_{13} & 1 \\ a_{21} & a_{22} & a_{23} & 0 \\ a_{31} & a_{32} & a_{33} & 0 \end{bmatrix}, \quad \begin{bmatrix} a_{11} & a_{12} & a_{13} & 0 \\ a_{21} & a_{22} & a_{23} & 1 \\ a_{31} & a_{32} & a_{33} & 0 \end{bmatrix}, \quad \begin{bmatrix} a_{11} & a_{12} & a_{13} & 0 \\ a_{21} & a_{22} & a_{23} & 0 \\ a_{31} & a_{32} & a_{33} & 1 \end{bmatrix}.$$

If we apply the Gauss-Jordan process, we arrive at the respective augmented matrices

$$\begin{bmatrix} 1 & 0 & 0 & b_{11} \\ 0 & 1 & 0 & b_{21} \\ 0 & 0 & 1 & b_{31} \end{bmatrix}, \quad \begin{bmatrix} 1 & 0 & 0 & b_{12} \\ 0 & 1 & 0 & b_{22} \\ 0 & 0 & 1 & b_{32} \end{bmatrix}, \quad \begin{bmatrix} 1 & 0 & 0 & b_{13} \\ 0 & 1 & 0 & b_{23} \\ 0 & 0 & 1 & b_{33} \end{bmatrix}.$$

In actual practice, we exploit the fact that all three systems have the same coefficient-matrix and we solve all three systems simultaneously by working with the enlarged augmented matrix

$$\begin{bmatrix} a_{11} & a_{12} & a_{13} & 1 & 0 & 0 \\ a_{21} & a_{22} & a_{23} & 0 & 1 & 0 \\ a_{31} & a_{32} & a_{33} & 0 & 0 & 1 \end{bmatrix},$$

which we also write more compactly as $[A \mid I]$. The Gauss-Jordan elimination process then leads to the augmented matrix

$$\begin{bmatrix} 1 & 0 & 0 & b_{11} & b_{12} & b_{13} \\ 0 & 1 & 0 & b_{21} & b_{22} & b_{23} \\ 0 & 0 & 1 & b_{31} & b_{32} & b_{33} \end{bmatrix},$$

also written as $[I \mid B]$. The matrix on the left of the vertical bar is the 3×3 identity matrix, and, as will be shown in the next theorem, the matrix on the right of the bar is the inverse of A.

Incidentally, in order to apply the Gauss-Jordan process, it is not necessary to know in advance that A is nonsingular. If A is *singular* (not nonsingular), we can still apply the Gauss-Jordan method, but somewhere in the process one of the diagonal elements on the left will become zero, and it will not be possible to transform A to the identity matrix.

The next theorem explains why the Gauss-Jordan process produces the inverse of a nonsingular matrix.

THEOREM 4.21.   *Let A be an n × n matrix and let I be the n × n identity matrix. If the Gauss-Jordan process reduces the augmented matrix*

(4.33) $[A \mid I]$

to one of the form

(4.34) $[I \mid B],$

then A is nonsingular and B = A⁻¹.

Proof.   There are three types of elementary row operations used in the Gauss-Jordan process:
   (1) *Interchanging two rows.*
   (2) *Multiplying all the terms of one row by a nonzero scalar.*
   (3) *Adding to one row a scalar multiple of another.*

We define an *elementary matrix* to be a matrix obtained from the identity matrix I by performing one of the elementary row operations. For example, let E denote the elementary matrix obtained from the identity by interchanging the first two rows. In the 3×3 case E looks like this:

$$E = \begin{bmatrix} 0 & 1 & 0 \\ 1 & 0 & 0 \\ 0 & 0 & 1 \end{bmatrix}.$$

Now take any 3×3 matrix $A = (a_{ij})$ and multiply it on the left by E. We find

$$EA = \begin{bmatrix} 0 & 1 & 0 \\ 1 & 0 & 0 \\ 0 & 0 & 1 \end{bmatrix} \begin{bmatrix} a_{11} & a_{12} & a_{13} \\ a_{21} & a_{22} & a_{23} \\ a_{31} & a_{32} & a_{33} \end{bmatrix} = \begin{bmatrix} a_{21} & a_{22} & a_{23} \\ a_{11} & a_{12} & a_{13} \\ a_{31} & a_{32} & a_{33} \end{bmatrix}.$$

In other words, the product EA is equal to matrix A with its first two rows interchanged. This property holds in general. If matrix A is multiplied on the left by the elementary matrix E obtained by interchanging two rows of the identity, then the product EA is equal to the matrix obtained by interchanging the same two rows of A.

The same is true for the other two types of elementary row operations. For example, if the second row of the 3×3 identity matrix is multiplied by a nonzero scalar c we get the elementary matrix

$$E = \begin{bmatrix} 1 & 0 & 0 \\ 0 & c & 0 \\ 0 & 0 & 1 \end{bmatrix}.$$

If we multiply any 3×3 matrix $A = (a_{ij})$ on the left by this elementary matrix E we find

$$EA = \begin{bmatrix} 1 & 0 & 0 \\ 0 & c & 0 \\ 0 & 0 & 1 \end{bmatrix} \begin{bmatrix} a_{11} & a_{12} & a_{13} \\ a_{21} & a_{22} & a_{23} \\ a_{31} & a_{32} & a_{33} \end{bmatrix} = \begin{bmatrix} a_{11} & a_{12} & a_{13} \\ ca_{21} & ca_{22} & ca_{23} \\ a_{31} & a_{32} & a_{33} \end{bmatrix},$$

which is the matrix obtained by multiplying the second row of A by c.

Similarly, if the third row of the 3×3 identity is multiplied by c and added to the first row we get the elementary matrix

$$E = \begin{bmatrix} 1 & 0 & c \\ 0 & 1 & 0 \\ 0 & 0 & 1 \end{bmatrix}.$$

If we multiply any 3×3 matrix on the left by this elementary matrix we find

$$EA = \begin{bmatrix} 1 & 0 & c \\ 0 & 1 & 0 \\ 0 & 0 & 1 \end{bmatrix} \begin{bmatrix} a_{11} & a_{12} & a_{13} \\ a_{21} & a_{22} & a_{23} \\ a_{31} & a_{32} & a_{33} \end{bmatrix} = \begin{bmatrix} a_{11} + ca_{31} & a_{12} + ca_{32} & a_{13} + ca_{33} \\ a_{21} & a_{22} & a_{23} \\ a_{31} & a_{32} & a_{33} \end{bmatrix},$$

which is the matrix obtained by multiplying the third row of A by c and adding it to its first row.

These examples illustrate a general property that is easily verified:

> *If a square matrix A is multiplied on the left by an elementary matrix E, the product EA is equal to the matrix obtained from A by performing the same elementary row operation used to produce E from I.*

When we perform a succession of k elementary row operations on an augmented matrix $[A \mid I]$ we get a corresponding set of k augmented matrices of the form

$$[E_1 A \mid E_1 I], \qquad [E_2 E_1 A \mid E_2 E_1 I], \qquad \ldots, \qquad [E_k \cdots E_1 A \mid E_k \cdots E_1 I],$$

where $E_1, \ldots, E_k$ are the corresponding elementary matrices. If the product $E_k \cdots E_1 A$ is equal to the identity matrix I, this implies that A is nonsingular and the product $E_k \cdots E_1$ is equal to A^{-1}.

4.20 Exercises

Apply the Gauss-Jordan elimination process to each of the following systems. If a solution exists, determine the general solution.

1. $x + y + 3z = 5$
   $2x - y + 4z = 11$
   $-y + z = 3.$

2. $3x + 2y + z = 1$
   $5x + 3y + 3z = 2$
   $x + y - z = 1.$

3. $3x + 2y + z = 1$
   $5x + 3y + 3z = 2$
   $7x + 4y + 5z = 3.$

4. $3x + 2y + z = 1$
   $5x + 3y + 3z = 2$
   $7x + 4y + 5z = 3$
   $x + y - z = 0.$

5. $3x - 2y + 5z + u = 1$
   $x + y - 3z + 2u = 2$
   $6x + y - 4z + 3u = 7.$

6. $x + y - 3z + u = 5$
   $2x - y + z - 2u = 2$
   $7x + y - 7z + 3u = 3.$

7. $x + y + 2z + 3u + 4v = 0$
   $2x + 2y + 7z + 11u + 14v = 0$
   $3x + 3y + 6z + 10u + 15v = 0.$

8. $x - 2y + z + 2u = -2$
   $2x + 3y - z - 5u = 9$
   $4x - y + z - u = 5$
   $5x - 3y + 2z + u = 3.$

9. Prove that the system $x + y + 2z = 2$, $2x - y + 3z = 2$, $5x - y + az = 6$, has a unique solution if $a \neq 8$. Find all solutions when $a = 8$.

10. (a) Determine all solutions of the system

$$5x + 2y - 6z + 2u = -1$$
$$x - y + z - u = -2.$$

   (b) Determine all solutions of the system

$$5x + 2y - 6z + 2u = -1$$
$$x - y + z - u = -2.$$
$$x + y + z = 6.$$

11. This exercise tells how to determine all nonsingular 2×2 matrices. Prove that

$$\begin{bmatrix} a & b \\ c & d \end{bmatrix} \begin{bmatrix} d & -b \\ -c & a \end{bmatrix} = (ad - bc)I.$$

Deduce that $\begin{bmatrix} a & b \\ c & d \end{bmatrix}$ is nonsingular if and only if $ad - bc \neq 0$, in which case its inverse is

$$\frac{1}{(ad - bc)} \begin{bmatrix} d & -b \\ -c & a \end{bmatrix}.$$

Determine the inverse of each of the matrices in Exercises 12 through 16.

12. $\begin{bmatrix} 2 & 3 & 4 \\ 2 & 1 & 1 \\ -1 & 1 & 2 \end{bmatrix}$.

15. $\begin{bmatrix} 1 & 2 & 3 & 4 \\ 0 & 1 & 2 & 3 \\ 0 & 0 & 1 & 2 \\ 0 & 0 & 0 & 1 \end{bmatrix}$.

13. $\begin{bmatrix} 1 & 2 & 2 \\ 2 & -1 & 1 \\ 1 & 3 & 2 \end{bmatrix}$.

16. $\begin{bmatrix} 0 & 1 & 0 & 0 & 0 & 0 \\ 2 & 0 & 2 & 0 & 0 & 0 \\ 0 & 3 & 0 & 1 & 0 & 0 \\ 0 & 0 & 1 & 0 & 2 & 0 \\ 0 & 0 & 0 & 3 & 0 & 1 \\ 0 & 0 & 0 & 0 & 2 & 0 \end{bmatrix}$.

14. $\begin{bmatrix} 1 & -2 & 1 \\ -2 & 5 & -4 \\ 1 & -4 & 6 \end{bmatrix}$.

4.21 Miscellaneous exercises on matrices

1. If a square matrix has a row of zeros or a column of zeros, prove that it is singular.
2. Prove of each of the following statements about $n \times n$ matrices, or exhibit a counter example.
   (a) If $AB + BA = O$, then $A^2B^3 = B^3A^2$.
   (b) If A and B are nonsingular, then $A + B$ is nonsingular.
   (c) If A and B are nonsingular, then AB is nonsingular.
   (d) If the product AB is nonsingular, then both A and B are nonsingular.
   (e) If A, B, and $A + B$ are nonsingular, then $A - B$ is nonsingular.
   (f) If $A^2 = O$, then $A = O$.
   (g) If $A^2 = O$, then $A + I$ is nonsingular.
   (h) If $A^3 = O$, then $A - I$ is nonsingular.
   (i) If $(A + 2I)^2 = O$, then A is nonsingular.
   (j) If $AB = BA$, then $AB^k = B^kA$ for every integer $k \geq 1$.
3. If $A = \begin{bmatrix} 1 & 2 \\ 5 & 4 \end{bmatrix}$, find a nonsingular matrix P such that $P^{-1}AP = \begin{bmatrix} 6 & 0 \\ 0 & -1 \end{bmatrix}$.

   Hint: Try a matrix P with undetermined entries such that $AP = P \begin{bmatrix} 6 & 0 \\ 0 & -1 \end{bmatrix}$.

4. The matrix $A = \begin{bmatrix} a & i \\ i & b \end{bmatrix}$, where $i^2 = -1$, $a = \frac{1}{2}(1 + \sqrt{5})$, and $b = \frac{1}{2}(1 - \sqrt{5})$, has the property that $A^2 = A$. Describe all 2×2 matrices A with complex entries such that $A^2 = A$.
5. If $A^2 = A$, prove that $(A + I)^k = I + (2^k - 1)A$.
6. The special theory of relativity makes use of a set of equations of the form $x' = a(x - vt)$, $y' = y$, $z' = z$, $t' = a(t - vx/c^2)$. Here v represents the velocity of a moving object, c the speed of light, and $a = c/\sqrt{c^2 - v^2}$, where $|v| < c$. The linear transformation that maps the two-dimensional vector (x, t) onto (x', t') is called a *Lorentz transformation*, in honor of the Dutch physicist Hendrik A. Lorentz (1853–1928). Its matrix relative to the usual bases is denoted by $L(v)$ and is given by

$$L(v) = a \begin{bmatrix} 1 & -v \\ -vc^{-2} & 1 \end{bmatrix}.$$

   Note that $L(v)$ is nonsingular and that $L(0) = I$. Prove that $L(v)L(u) = L(w)$, where $w = (u + v)c^2/(uv + c^2)$. In other words, the product of two Lorentz transformations is another Lorentz transformation.
7. When we interchange the rows and columns of a rectangular matrix A, the new matrix so obtained is called the *transpose* of A and is denoted by A^t (read: A transpose). Thus, if

$$A = \begin{bmatrix} 1 & 2 & 3 \\ 4 & 5 & 6 \end{bmatrix}, \quad \text{then} \quad A^t = \begin{bmatrix} 1 & 4 \\ 2 & 5 \\ 3 & 6 \end{bmatrix}.$$

Prove that transposes have the following properties:
(a) $(A^t)^t = A$.
(b) $(A + B)^t = A^t + B^t$.
(c) $(cA)^t = cA^t$.
(d) $(AB)^t = B^t A^t$ (note reversal of order).
(e) $(A^t)^{-1} = (A^{-1})^t$ if A is nonsingular.

8. A square matrix A is said to be orthogonal if $AA^t = I$.

   (a) Verify that the 2×2 matrix $\begin{bmatrix} \cos \theta & -\sin \theta \\ \sin \theta & \cos \theta \end{bmatrix}$ is orthogonal for each real θ.

   (b) If A is any real orthogonal $n \times n$ matrix, prove that its rows, considered as vectors in $\mathbf{R}^n$, form an orthonormal set.

9. Prove each of the following statements about $n \times n$ matrices, or exhibit a counter example.
   (a) If A and B are orthogonal, then $A + B$ is orthogonal.
   (b) If A and B are orthogonal, then AB is orthogonal.
   (c) If A and AB are orthogonal, then B is orthogonal.

10. *Hadamard matrices*, named for the French mathematician Jacques Hadamard (1865–1963), are those $n \times n$ matrices with the following three properties:

   I. Each entry is 1 or -1.
   II. Each row, considered as a vector in $\mathbf{R}^n$, has length $\sqrt{n}$.
   III. The dot product of any two distinct rows is 0.

   Hadamard matrices arise in certain problems in geometry and the theory of numbers, and they have also been applied to the construction of optimum code words in space communication.

   In spite of their apparent simplicity, they present many unsolved problems. The main unsolved problem at this time is to determine all n for which an $n \times n$ Hadamard matrix exists. This exercise outlines a partial solution.

   (a) Determine all 2×2 Hadamard matrices. (There are exactly 8.)
   (b) This part of the exercise outlines a simple proof of the following theorem:

If A is an $n \times n$ Hadamard matrix, where $n > 2$, then n is a multiple of 4.

The proof is based on two very simple lemmas concerning vectors in $\mathbf{R}^n$. Prove each of these lemmas and apply them to the rows of a Hadamard matrix to prove the theorem.

   LEMMA 4.22. *If X, Y, Z are orthogonal vectors in $\mathbf{R}^n$, then we have*

$$(X + Y) \cdot (X + Z) = \|X\|^2.$$

   LEMMA 4.23. *Write $X = (x_1, \ldots, x_n)$, $Y = (y_1, \ldots, y_n)$, $Z = (z_1, \ldots, z_n)$. If each component x_i, y_i, z_i, is either 1 or -1, then the product $(x_i + y_i)(x_i + z_i)$ is either 0 or 4.*

5

DETERMINANTS

5.1 Introduction

In many applications of linear algebra to geometry and analysis the concept of a determinant plays a useful role. This chapter studies the basic properties of determinants and some of their applications.

Determinants of order two and three were introduced in Chapter 2 as a convenient notation for expressing certain formulas in compact form. We recall that a determinant of order two was defined by the formula

$$(5.1) \qquad \begin{vmatrix} a_{11} & a_{12} \\ a_{21} & a_{22} \end{vmatrix} = a_{11}a_{22} - a_{12}a_{21}.$$

Despite similarity in notation, the determinant $\begin{vmatrix} a_{11} & a_{12} \\ a_{21} & a_{22} \end{vmatrix}$ (written with vertical bars) is conceptually distinct from the matrix $\begin{bmatrix} a_{11} & a_{12} \\ a_{21} & a_{22} \end{bmatrix}$ (written with square brackets).

The determinant is a *number* assigned to the matrix according to Formula (5.1). To emphasize this connection we also write

$$\begin{vmatrix} a_{11} & a_{12} \\ a_{21} & a_{22} \end{vmatrix} = \det \begin{bmatrix} a_{11} & a_{12} \\ a_{21} & a_{22} \end{bmatrix}.$$

In Chapter 2 determinants of order three were defined in terms of those of order two by the formula

$$(5.2) \quad \det \begin{bmatrix} a_{11} & a_{12} & a_{13} \\ a_{21} & a_{22} & a_{23} \\ a_{31} & a_{32} & a_{33} \end{bmatrix} = a_{11} \begin{vmatrix} a_{22} & a_{23} \\ a_{32} & a_{33} \end{vmatrix} - a_{12} \begin{vmatrix} a_{21} & a_{23} \\ a_{31} & a_{33} \end{vmatrix} + a_{13} \begin{vmatrix} a_{21} & a_{22} \\ a_{31} & a_{32} \end{vmatrix}.$$

This chapter discusses the general case, the determinant of a square matrix of order n for any $n \geq 1$. For a 1×1 matrix with entry a_{11} we define the determinant to be that entry: $\det [a_{11}] = a_{11}$.

Our point of view is to treat the determinant as a function that assigns to each square matrix A a number called the determinant of A and denoted by $\det A$. One way to define this function is by an explicit formula generalizing (5.1) and (5.2). This formula is a sum containing $n!$ products of entries of A. For large n the formula is unwieldy and is rarely used in practice. It seems preferable to study determinants from another point of view that emphasizes their essential properties. These properties, which are important in the applications, will be taken as *axioms* for a determinant function.

Initially, our program will consist of three parts: (1) To motivate the choice of axioms. (2) To deduce further properties of determinants from the axioms. (3) To prove that for each n there is one and only one function that satisfies these axioms.

5.2 Motivation for the choice of axioms for a determinant function

In Chapter 2 we proved that the scalar triple product of three vectors A_1, A_2, A_3 in 3-space can be expressed as the determinant of a matrix whose rows are the given vectors. Thus we have

$$A_1 \times A_2 \cdot A_3 = \det \begin{bmatrix} a_{11} & a_{12} & a_{13} \\ a_{21} & a_{22} & a_{23} \\ a_{31} & a_{32} & a_{33} \end{bmatrix},$$

where $A_1 = (a_{11}, a_{12}, a_{13})$, $A_2 = (a_{21}, a_{22}, a_{23})$, and $A_3 = (a_{31}, a_{32}, a_{33})$.

If the rows are linearly independent the scalar triple product is nonzero; the absolute value of the product is equal to the volume of the parallelepiped determined by the three vectors A_1, A_2, A_3. (An example is shown in Figure 5.1.) If the rows are dependent the scalar triple product is zero. In this case the vectors A_1, A_2, A_3 are coplanar and the parallelepiped degenerates to a plane figure of zero volume.

Some of the properties of the scalar triple product will serve as motivation for the choice of axioms for a determinant function in the higher-dimensional case. To state these properties in a form suitable for generalization, we regard the scalar triple product as a function of the row vectors A_1, A_2, A_3, and we denote this function by d:

$$d(A_1, A_2, A_3) = A_1 \times A_2 \cdot A_3.$$

We focus our attention on the following properties:

(a) *If a row is multiplied by a scalar t, then the scalar triple product is multiplied by t.*
(b) *If one row is replaced by itself plus another row, the scalar triple product is unchanged.*
(c) *The scalar triple product of the three unit coordinate vectors $i \times j \cdot k$ is equal to 1.*

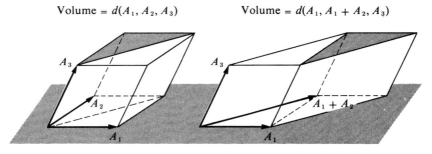

Volume $= d(A_1, A_2, A_3)$ Volume $= d(A_1, A_1 + A_2, A_3)$

FIGURE 5.1   Geometric interpretation of the property $d(A_1, A_2 + A_1, A_3) = d(A_1, A_2, A_3)$. The two parallelepipeds have equal volumes.

All these properties follow easily from properties of dot and cross products. For example, if the first row is multiplied by t we find $d(tA_1, A_2, A_3) = td(A_1, A_2, A_3)$ for every scalar t. If the second row is replaced by itself plus the first row, property (b) states that

$$(5.3) \qquad d(A_1, A_2 + A_1, A_3) = d(A_1, A_2, A_3).$$

This property is illustrated in Figure 5.1 which shows a parallelepiped determined by A_1, A_2, A_3 and another parallelepiped determined by A_1, $A_2 + A_1$, A_3. Eq. (5.3) states that these two parallelepipeds have equal volumes. This is evident geometrically because the parallelepipeds have equal altitudes and bases of equal area.

5.3 A set of axioms for a determinant function

The properties of the scalar triple product mentioned in the foregoing section can be suitably generalized and used as axioms for determinants of order n. If $A = (a_{ij})$ is an $n \times n$ matrix with real or complex entries, we denote its rows by $A_1, \ldots, A_n$. Thus, the ith row of A is a vector in n-space:

$$A_i = (a_{i1}, a_{i2}, \ldots, a_{in}).$$

We regard the determinant as a function of A or as a function of its n rows $A_1, \ldots, A_n$, and we denote its values by $\det A$ or by $d(A_1, \ldots, A_n)$.

AXIOMATIC DEFINITION OF A DETERMINANT FUNCTION. *A scalar-valued function d of an $n \times n$ matrix A is called a determinant function of order n if it satisfies the following axioms:*

AXIOM 1. HOMOGENEITY IN EACH ROW. *If matrix B is obtained from matrix A by multiplying one row of A by a scalar factor t, then* $\det B = t \det A$.

AXIOM 2. INVARIANCE UNDER ROW ADDITION. *If matrix B is obtained from matrix A by adding one row of A to another row of A, then* $\det B = \det A$.

AXIOM 3. THE DETERMINANT OF THE IDENTITY MATRIX IS EQUAL TO 1.

$$d(I_1, \ldots, I_n) = 1, \qquad \textit{where } I_k \textit{ is the kth unit coordinate vector.}$$

It is a remarkable fact that for a given n there is one and only one function d that satisfies these three axioms. A proof of this fact will be given later (Theorems 5.6 and 5.20). But first we deduce further properties of determinants that follow from the axioms.

THEOREM 5.1. *If some row of matrix A is the zero vector, then* $\det A = 0$.

Proof. Suppose row $A_k = O$, and let B be the matrix obtained from A by multiplying row A_k by the scalar -1. Then $B = A$, so $\det B = \det A$. On the other hand, by Axiom 1 we have $\det B = -\det A$, so $\det A = 0$.

THEOREM 5.2. *If matrix B is obtained from matrix A by adding a scalar multiple of one row of A to another row of A, then* $\det B = \det A$.

Proof. Suppose that B is obtained by adding a scalar t times row A_k to row A_j, where $j \neq k$. In other words, $B_j = A_j + tA_k$ and $B_i = A_i$ for all $i \neq j$. If $t = 0$, then $B = A$ and $\det B = \det A$.

Now assume that $t \neq 0$. Let C be the matrix whose kth row is tA_k and whose ith row is A_i for all $i \neq k$. By Axiom 1, $\det C = t \det A$. Let D be the matrix obtained from C by adding its kth row, tA_k, to its jth row A_j. In other words, $D_k = tA_k$, $D_j = A_j + tA_k$, and $D_i = A_i$ for $i \neq j, i \neq k$. By Axiom 2, $\det D = \det C$. But if we multiply the kth row of D by $1/t$, we get matrix B. Therefore, by Axiom 1, we have $\det B = (1/t) \det C = \det A$.

THEOREM 5.3.   (a) *If matrix B is obtained from matrix A by interchanging two rows A_i and A_j with $i \neq j$, then $\det B = -\det A$.*

(b) *If two rows of a matrix are equal, then its determinant is zero.*

Proof. (a) We introduce three new matrices C, D, and E that have the same determinant as A. We describe only their ith and jth rows as indicated here:

$$C_i = A_i + A_j, \qquad C_j = A_j.$$
$$D_i = C_i, \qquad D_j = C_j + (-1)C_i,$$
$$E_i = D_i + D_j, \qquad E_j = D_j.$$

All other rows of C, D, E are the same as those of A. By Axiom 2 we have $\det C = \det A$. By Theorem 5.2 we have $\det D = \det C$, and by Axiom 2 we have $\det E = \det D$. So C, D, E all have determinant equal to $\det A$. Expressing the ith and jth rows of D and E in terms of the rows of A, we find

$$D_i = A_i + A_j, \qquad D_j = A_j - (A_i + A_j) = -A_i$$

and

$$E_i = A_j, \qquad E_j = -A_i.$$

Therefore matrix B can be obtained from E by multiplying row E_j by (-1), so

$$\det B = -\det E = -\det A.$$

This proves part (a). To prove part (b) interchange the two equal rows and use part (a).

- It should be noted that Axiom 3 has not been used in proving Theorems 5.1, 5.2 or 5.3.

5.4   The determinant of a diagonal matrix

Now we have enough information to calculate determinants, assuming throughout that a determinant function exists. We begin with a special case called a *diagonal matrix*.

DEFINITION.   DIAGONAL MATRIX.   *A square matrix (a_{ij}) is called a diagonal matrix if each a_{ij} with $i \neq j$ (off the main diagonal) is zero.*

In other words a diagonal matrix has the form

$$A = \begin{bmatrix} a_{11} & 0 & \cdots & 0 \\ 0 & a_{22} & \cdots & 0 \\ \vdots & \vdots & \ddots & \vdots \\ 0 & 0 & \cdots & a_{nn} \end{bmatrix}.$$

We use the notation $A = \text{diag}(a_{11}, a_{22}, \ldots, a_{nn})$ to denote this diagonal matrix. The next theorem tells us that the determinant of a diagonal matrix is equal to the product of its diagonal entries.

THEOREM 5.4.   *If* $A = \text{diag}(a_{11}, a_{22}, \ldots, a_{nn})$ *then*

$$(5.4) \qquad \det A = a_{11} a_{22} \cdots a_{nn}.$$

Proof.   Matrix A can be obtained from the $n \times n$ identity matrix I by multiplying the rows of I successively by the diagonal elements $a_{11}, a_{22}, \ldots, a_{nn}$. Applying Axiom 1 repeatedly we find

$$(5.5) \qquad \det A = a_{11} a_{22} \cdots a_{nn} \det I.$$

Note that (5.5) was deduced from Axiom 1 alone. When we use Axiom 3 in (5.5) we get (5.4).

5.5   The determinant of an upper triangular matrix

The Gauss-Jordan elimination process for solving systems of linear equations is also one of the best methods for computing determinants. We recall that the method performs three types of operations on the rows of a matrix:
   (1) *Interchanging two rows.*
   (2) *Multiplying all the elements in a row by a nonzero scalar.*
   (3) *Adding to one row a scalar multiple of another.*
By Theorem 5.3, operation (1) produces a sign change in the determinant of the matrix. By Axiom 1, operation (2) multiplies the determinant by the nonzero scalar. And by Theorem 5.2, operation (3) leaves the determinant unchanged. We show next that we can calculate the determinant of any upper triangular matrix by transforming it to a diagonal matrix, using only row operations of type (3).

DEFINITION.   UPPER TRIANGULAR MATRIX.   *A square matrix of the form*

$$(5.6) \qquad U = \begin{bmatrix} u_{11} & u_{12} & \cdots & u_{1n} \\ 0 & u_{22} & \cdots & u_{2n} \\ \vdots & \vdots & \ddots & \vdots \\ 0 & 0 & \cdots & u_{nn} \end{bmatrix}$$

is called an upper triangular matrix. All entries below the main diagonal are zero.

THEOREM 5.5. *The determinant of an upper triangular matrix is equal to the product of its diagonal entries. In other words, if U is given by (5.6) we have*

$$\det U = u_{11}u_{22}\cdots u_{nn}. \tag{5.7}$$

Proof. If $u_{nn} = 0$, the last row of U is the zero vector so $\det U = 0$ by Theorem 5.1, and Eq. (5.7) holds trivially because both members are zero. Now assume that $u_{nn} \neq 0$, and apply operation (3) of the Gauss-Jordan process. Multiply the last row of U by the scalar $-u_{1n}/u_{nn}$ and add it to the first row. We get a new matrix with u_{1n} replaced by 0 and with the remaining entries unchanged. By repeated use of operation (3) all entries lying above u_{nn} in the last column can be replaced by 0 without changing any other entries or the determinant. This produces an upper triangular matrix U' containing zeros in the first $n-1$ entries of the last column:

$$U' = \begin{bmatrix} u_{11} & u_{12} & \cdots & 0 \\ 0 & u_{22} & \cdots & 0 \\ \vdots & \vdots & \ddots & \vdots \\ 0 & 0 & \cdots & u_{nn} \end{bmatrix},$$

with $\det U' = \det U$. If the next-to-last diagonal entry $u_{n-1,n-1}$ of U' is 0, then the next-to-last row of U' is equal to zero so $\det U' = 0$ and Eq. (5.7) holds trivially, with both members being zero. Otherwise we use Gauss-Jordan operation (3) repeatedly. If no diagonal entry u_{kk} is zero we eventually arrive at a diagonal matrix whose determinant is the product of the diagonal entries. And if some diagonal entry $u_{kk} = 0$ the process will produce a matrix with a row of zeros, in which case Eq. (5.7) holds trivially, with both members being zero.

5.6 Computation of any determinant by the Gauss-Jordan process

By repeated use of all three types of row operations we can transform any square matrix A to an upper triangular matrix U whose determinant we know how to compute. Each time operation (1) is performed the determinant changes sign. Each time (2) is performed with a nonzero scalar c the determinant is multiplied by c. And each time (3) is performed the determinant is unchanged. Therefore, if operation (1) is performed p times and if $c_1, \ldots, c_q$ are the nonzero scalar multipliers used in connection with operation (2), we find $\det U = (-1)^p c_1 \cdots c_q \det A$, and hence

$$\det A = (-1)^p (c_1 \cdots c_q)^{-1} \det U.$$

Note that this formula is a consequence of the first two axioms alone. Its proof does not depend on Axiom 3, which was used only to deduce (5.4) from (5.5). In other words, using only Axioms 1 and 2 we have shown that for every $n \times n$ matrix A there is a scalar $c(A)$ (depending on A) such that

$$\det A = c(A) \det I. \tag{5.8}$$

5.7 Uniqueness of the determinant function

Using Eq. (5.8), which is a consequence of Axioms 1 and 2 alone, we can prove that there is at most one determinant function of order n.

THEOREM 5.6. UNIQUENESS THEOREM FOR DETERMINANTS. *Let d be a function satisfying all three axioms for a determinant function of order n. If f is another function satisfying Axioms* 1 *and* 2, *then for every n × n matrix A we have*

$$(5.9) \qquad\qquad f(A) = d(A)f(I).$$

Consequently, if f also satisfies Axiom 3 *then* $f(A) = d(A)$.

Proof. Let $g(A) = f(A) - d(A)f(I)$. Note that g satisfies both Axioms 1 and 2, but not Axiom 3. In fact, $g(I) = f(I) - d(I)f(I) = 0$ because $d(I) = 1$. Since g satisfies Axioms 1 and 2 we can apply Eq. (5.8) to g because the proof of (5.8) depended only on Axioms 1 and 2. Therefore there is a scalar $c(A)$ such that $g(A) = c(A)g(I)$. But $g(I) = 0$ so $g(A) = 0$ for every A, hence $f(A) = d(A)f(I)$.

5.8 Exercises

In this set of exercises you may assume existence of a determinant function. Determinants of order 3 may be computed by using Equation (5.2).

1. Compute each of the following determinants:

(a) $\begin{vmatrix} 2 & 1 & 1 \\ 1 & 4 & -4 \\ 1 & 0 & 2 \end{vmatrix}$. (b) $\begin{vmatrix} 3 & 0 & 8 \\ 5 & 0 & 7 \\ -1 & 4 & 2 \end{vmatrix}$. (c) $\begin{vmatrix} a & 1 & 0 \\ 2 & a & 2 \\ 0 & 1 & a \end{vmatrix}$.

2. If $\det \begin{bmatrix} x & y & z \\ 3 & 0 & 2 \\ 1 & 1 & 1 \end{bmatrix} = 1$, compute the determinant of each of the following matrices:

(a) $\begin{bmatrix} 2x & 2y & 2z \\ \frac{3}{2} & 0 & 1 \\ 1 & 1 & 1 \end{bmatrix}$. (b) $\begin{bmatrix} x & y & z \\ 3x+3 & 3y & 3z+2 \\ x+1 & y+1 & z+1 \end{bmatrix}$.

(c) $\begin{bmatrix} x-1 & y-1 & z-1 \\ 4 & 1 & 3 \\ 1 & 1 & 1 \end{bmatrix}$.

3. (a) Prove that $\begin{vmatrix} 1 & 1 & 1 \\ a & b & c \\ a^2 & b^2 & c^2 \end{vmatrix} = (b-a)(c-a)(c-b)$.

(b) Find corresponding formulas for the determinants $\begin{vmatrix} 1 & 1 & 1 \\ a & b & c \\ a^3 & b^3 & c^3 \end{vmatrix}$ and $\begin{vmatrix} 1 & 1 & 1 \\ a^2 & b^2 & c^2 \\ a^3 & b^3 & c^3 \end{vmatrix}$.

4. If A is an $n \times n$ matrix, prove that $\det(cA) = c^n \det A$ for every scalar c.

5. Compute the determinant of each of the following matrices by transforming each of them to an upper triangular matrix.

(a) $\begin{bmatrix} 1 & -1 & 1 & 1 \\ 1 & -1 & -1 & -1 \\ 1 & 1 & -1 & -1 \\ 1 & 1 & 1 & -1 \end{bmatrix}$. (b) $\begin{bmatrix} 1 & 1 & 1 & 1 \\ a & b & c & d \\ a^2 & b^2 & c^2 & d^2 \\ a^3 & b^3 & c^3 & d^3 \end{bmatrix}$.

(c) $\begin{bmatrix} 1 & 1 & 1 & 1 \\ a & b & c & d \\ a^2 & b^2 & c^2 & d^2 \\ a^4 & b^4 & c^4 & d^4 \end{bmatrix}$. (d) $\begin{bmatrix} a & 1 & 0 & 0 & 0 \\ 4 & a & 2 & 0 & 0 \\ 0 & 3 & a & 3 & 0 \\ 0 & 0 & 2 & a & 4 \\ 0 & 0 & 0 & 1 & a \end{bmatrix}$.

$$(e) \quad \begin{bmatrix} 1 & 1 & 1 & 1 & 1 & 1 \\ 1 & 1 & 1 & -1 & -1 & -1 \\ 1 & 1 & -1 & -1 & 1 & 1 \\ 1 & -1 & -1 & 1 & -1 & 1 \\ 1 & -1 & 1 & -1 & 1 & 1 \\ 1 & -1 & -1 & 1 & 1 & -1 \end{bmatrix}.$$

6. Let U and V be two $n \times n$ upper triangular matrices.
   (a) Prove that each of $U + V$ and UV is an upper triangular matrix.
   (b) Prove that $\det(UV) = (\det U)(\det V)$.
   (c) If $\det U \neq 0$, prove that there is an upper triangular matrix U^{-1} such that $UU^{-1} = I$, and deduce that $\det(U^{-1}) = 1/\det U$.

7. Calculate $\det A$, $\det(A^{-1})$, and A^{-1} for the following upper triangular matrix:

$$A = \begin{bmatrix} 2 & 3 & 4 & 5 \\ 0 & 2 & 3 & 4 \\ 0 & 0 & 2 & 3 \\ 0 & 0 & 0 & 2 \end{bmatrix}.$$

8. A lower triangular matrix $A = (a_{ij})$ is a square matrix with all entries above the main diagonal equal to 0; that is, $a_{ij} = 0$ whenever $i < j$. Prove that the determinant of such a matrix is equal to the product of its diagonal entries: $\det A = a_{11}a_{22} \cdots a_{nn}$.

Exercises 9, 10 and 11 require a knowledge of differential calculus.

9. Let f_1, f_2, g_1, g_2 be four functions differentiable on an interval (a, b). Define $F(x)$ as follows:

$$F(x) = \begin{vmatrix} f_1(x) & f_2(x) \\ g_1(x) & g_2(x) \end{vmatrix}$$

for each x in (a, b). Prove that

$$F'(x) = \begin{vmatrix} f_1'(x) & f_2'(x) \\ g_1(x) & g_2(x) \end{vmatrix} + \begin{vmatrix} f_1(x) & f_2(x) \\ g_1'(x) & g_2'(x) \end{vmatrix}.$$

10. State and prove a generalization of Exercise 9 for the determinant

$$F(x) = \begin{vmatrix} f_1(x) & f_2(x) & f_3(x) \\ g_1(x) & g_2(x) & g_3(x) \\ h_1(x) & h_2(x) & h_3(x) \end{vmatrix}.$$

11. (a) If $F(x) = \begin{vmatrix} f_1(x) & f_2(x) \\ f_1'(x) & f_2'(x) \end{vmatrix}$, prove that $F'(x) = \begin{vmatrix} f_1(x) & f_2(x) \\ f_1''(x) & f_2''(x) \end{vmatrix}$.
    (b) State and prove a corresponding result for 3×3 determinants. You may use Equation (5.2).

5.9 Multilinearity of determinants

As an application of the uniqueness theorem we deduce an important property of determinants known as *multilinearity*. We begin with the following special case.

THEOREM 5.7. ADDITIVITY IN EACH ROW. *Given an $n \times n$ matrix $A = (A_1, \ldots, A_n)$ and given any n-dimensional vector V, let B be the matrix obtained from A by replacing one row, say A_k, by V, and let C be the matrix obtained by replacing row A_k by $A_k + V$. Then*

$$(5.10) \qquad \det C = \det A + \det B.$$

This property, called *additivity in the kth row*, can be expressed as follows:

$$d(A_1, \ldots, A_k + V, \ldots, A_n) = d(A_1, \ldots, A_k, \ldots, A_n) + d(A_1, \ldots, V, \ldots, A_n).$$

Proof. Let $f(A) = \det C - \det B$. We will show that f satisfies all three axioms for a determinant function. Then by the uniqueness theorem we must have $f(A) = \det A$, which implies (5.10).

Each of $\det C$ and $\det B$, when regarded as a function of the n rows of A, satifies Axioms 1 and 2, so the same is true for their difference $f(A)$. It remains to verify that f satisfies Axiom 3: $f(I) = 1$, where I is the identity matrix. Taking $A = I$ in the definition of f we get

(5.11) $$f(I) = \det(I_1, \ldots, I_k + V, \ldots, I_n) - \det(I_1, \ldots, V, \ldots, I_n).$$

Write V as an n-tuple, say, $V = (v_1, \ldots, v_n)$. If $k = 1$, both C and B are upper triangular matrices with determinants $1 + v_1$ and v_1, respectively, so

$$f(I) = \det C - \det B = (1 + v_1) - v_1 = 1.$$

If $k = 2$, the second rows of matrices C and B are given by

$$C_2 = (v_1, 1 + v_2, v_3, \ldots, v_n) \qquad \text{and} \qquad B_2 = (v_1, v_2, v_3, \ldots, v_n),$$

while their remaining rows are identical to those of I. By applying a row operation of type (3) to each of matrices C and B on the right of Eq. (5.11) we can transform them to upper triangular matrices without changing their determinants. In fact, if we multiply the first row of each of these matrices by $-v_1$ and add it to the second row the new second rows become, respectively,

$$(0, 1 + v_2, v_3, \ldots, v_n) \qquad \text{and} \qquad (0, v_2, v_3, \ldots, v_n).$$

Now we have upper triangular matrices with respective determinants $(1 + v_2)$ and v_2 so the difference of the determinants, $f(I)$, is equal to 1.

The same type of argument works for a general $k \geq 2$. By a sequence of row operations of type (3) we transform each of the matrices on the right of (5.11) to upper triangular form with respective determinants $1 + v_k$ and v_k whose difference is 1. Therefore $f(I) = 1$, and this proves Eq. (5.10).

If instead of one vector V we use a sum of two vectors $V + U$ and apply Theorem 5.7 twice we find

$$d(A_1, \ldots, A_k + V + U, \ldots, A_n) = d(A_1, \ldots, A_k, \ldots, A_n) + d(A_1, \ldots, V, \ldots, A_n)$$
$$+ d(A_1, \ldots, U, \ldots, A_n).$$

More generally, we can use any number of vectors and multiply each of these vectors by a scalar (using homogeneity) to obtain a property known as *linearity in each row*. Linearity combines both homogeneity and additivity. For example, linearity in the first row can be expressed as follows:

$$d\left(\sum_{i=1}^{p} t_i U_i, A_2, \ldots, A_n\right) = \sum_{i=1}^{p} t_i\, d(U_i, A_2, \ldots, A_n),$$

for any scalars $t_1, \ldots, t_p$ and any vectors $U_1, \ldots, U_p$ in n-space. Because this property holds for *every* row it is often described by saying that the determinant is a *multilinear function* of its rows.

5.10   Applications of multilinearity

Our first application is the following extension of Theorem 5.3(b).

THEOREM 5.8.   *If the rows of a matrix A are dependent, then* $d(A) = 0$.

Proof.   Suppose scalars $c_1, \ldots, c_n$ exist, not all zero, such that $\sum_{i=1}^{n} c_i A_i = O$. Then any row with $c_k \neq 0$ can be expressed as a linear combination of the other rows. For simplicity, suppose that the first row A_1 is a linear combination of the others, say $A_1 = \sum_{i=2}^{n} t_i A_i$. By linearity in the first row we have

$$d(A_1, A_2, \ldots, A_n) = d\left(\sum_{i=2}^{n} t_i A_i, A_2, \ldots, A_n\right) = \sum_{i=2}^{n} t_i\, d(A_i, A_2, \ldots, A_n).$$

But, by Theorem 5.3(b), each term of the last sum is zero because row A_i is equal to at least one of rows $A_2, \ldots, A_n$. Hence the whole sum is zero.

If row A_k is a linear combination of the other rows we argue the same way, using linearity in the kth row. This proves that the determinant of a matrix vanishes if its rows are dependent.

The next application establishes existence of a determinant function for matrices of orders 2 and 3 and justifies the formulas (5.1) and (5.2) introduced earlier.

THEOREM 5.9.   (a) *For matrices of order* 2 *there is one and only one determinant function:*

$$\det \begin{bmatrix} a_{11} & a_{12} \\ a_{21} & a_{22} \end{bmatrix} = \begin{vmatrix} a_{11} & a_{12} \\ a_{21} & a_{22} \end{vmatrix} = a_{11}a_{22} - a_{12}a_{21}.$$

(b) *For matrices of order* 3 *there is one and only one determinant function:*

$$(5.12) \quad \det \begin{bmatrix} a_{11} & a_{12} & a_{13} \\ a_{21} & a_{22} & a_{23} \\ a_{31} & a_{32} & a_{33} \end{bmatrix} = a_{11} \begin{vmatrix} a_{22} & a_{23} \\ a_{32} & a_{33} \end{vmatrix} - a_{12} \begin{vmatrix} a_{21} & a_{23} \\ a_{31} & a_{33} \end{vmatrix} + a_{13} \begin{vmatrix} a_{21} & a_{22} \\ a_{31} & a_{32} \end{vmatrix}.$$

Proof.   The easiest way to prove this theorem is to verify that each formula satisfies the axioms for a determinant function and then invoke the uniqueness theorem. However, to better understand the origin of these formulas, we choose to derive them directly, using the multilinearity property.

Start with any 2×2 matrix $A = \begin{bmatrix} a_{11} & a_{12} \\ a_{21} & a_{22} \end{bmatrix}$ and write its row vectors as linear combinations of the unit coordinate vectors $i = (1, 0)$ and $j = (0, 1)$:

$$A_1 = (a_{11}, a_{12}) = a_{11}i + a_{12}j, \qquad A_2 = (a_{21}, a_{22}) = a_{21}i + a_{22}j.$$

Using linearity in the first row we have

$$d(A_1, A_2) = d(a_{11}i + a_{12}j, A_2) = a_{11}d(i, A_2) + a_{12}d(j, A_2).$$

Now we use linearity in the second row to obtain

$$d(i, A_2) = d(i, a_{21}i + a_{22}j) = a_{21}d(i, i) + a_{22}d(i, j) = a_{22}d(i, j),$$

since $d(i, i) = 0$ because its two rows are equal. Similarly we find

$$d(j, A_2) = d(j, a_{21}i + a_{22}j) = a_{21}d(j, i) + a_{22}d(j, j) = -a_{21}d(i, j),$$

the sign change occuring when we interchange i and j in $d(j, i)$. Hence we obtain

$$d(A_1, A_2) = (a_{11}a_{22} - a_{12}a_{21})d(i, j).$$

But $d(i, j) = 1$ by Axiom 3, so $d(A_1, A_2) = a_{11}a_{22} - a_{12}a_{21}$ as asserted.

By a similar argument we can derive Eq. (5.12) for determinants of order 3. Details are left as an exercise for the reader.

5.11 The product formula for determinants

In this section we use the uniqueness theorem to prove that the determinant of a product of two square matrices is equal to the product of their determinants,

$$\det(AB) = (\det A)(\det B),$$

assuming that a determinant function exists.

We recall that the product AB of two matrices $A = (a_{ij})$ and $B = (b_{ij})$ is the matrix $C = (c_{ij})$ whose ij entry is given by the formula

(5.13)
$$c_{ij} = \sum_{k=1}^{n} a_{ik}b_{kj}.$$

The product is defined only if the number of columns of the left-hand factor A is equal to the number of rows of the right-hand factor B. This is always the case if both A are B are square matrices of the same size.

The proof of the product formula will make use of a simple relation that holds between the rows of AB and the rows of A. We state this as a lemma. As usual, A_i denotes the ith row of matrix A.

LEMMA 5.10.   *If A is an m $\times$ n matrix and B is an n $\times$ p matrix, then we have*

(5.14) $(AB)_i = A_i B.$

That is, the ith row of the product AB is equal to the product of row matrix A_i with B.

Proof.   Let B^j denote the *j*th column of *B* and let $C = AB$. Then the sum in (5.13) can be regarded as the dot product of the *i*th row of *A* with the *j*th column of *B*,

$$c_{ij} = A_i \cdot B^j.$$

Therefore the *i*th row C_i is a row matrix whose entries are dot products:

$$C_i = [A_i \cdot B^1, A_i \cdot B^2, \ldots, A_i \cdot B^p].$$

But this is also the result of matrix multiplication of row matrix A_i with matrix *B*, since

$$A_i B = [a_{i1}, \ldots, a_{in}] \begin{bmatrix} b_{11} & b_{12} & \cdots & b_{1p} \\ b_{21} & b_{22} & \cdots & b_{2p} \\ \vdots & \vdots & \vdots & \vdots \\ b_{n1} & b_{n2} & \cdots & b_{np} \end{bmatrix} = [A_i \cdot B^1, A_i \cdot B^2, \ldots, A_i \cdot B^p].$$

Therefore $C_i = A_i B$, which proves the lemma.

THEOREM 5.11.   PRODUCT FORMULA FOR DETERMINANTS.   *For any two square matrices A and B we have*

(5.15) $\det(AB) = (\det A)(\det B).$

Proof.   By the Lemma we have $(AB)_i = A_i B$, hence

$$\det(AB) = d(A_1 B, \ldots, A_n B).$$

Keep *B* fixed and let $f(A) = \det(AB)$. Expressing this relation as a function of the rows of *A* we have

$$f(A_1, \ldots, A_n) = d(A_1 B, \ldots, A_n B).$$

It is a simple matter to verify that *f* satisfies the first two axioms for a determinant function. Therefore, by the uniqueness theorem we have $f(A) = d(A)f(I)$. Taking $A = I$ in the definition of *f* we find $f(I) = d(B)$ and hence $f(A) = d(A)d(B)$, which proves Eq. (5.15).

5.12   The determinant of the inverse of a nonsingular matrix

We recall that a square matrix *A* is called *nonsingular* if it has a left inverse *B* such that $BA = I$. If a left inverse exists it is unique and is also a right inverse, $AB = I$. We denote the inverse by A^{-1}, so $A^{-1}A = AA^{-1} = I$. The relation between $\det A$ and $\det(A^{-1})$ is as natural as one could expect.

THEOREM 5.12. *If matrix A is nonsingular, then* $\det A \neq 0$ *and we have*

(5.16)
$$\det(A^{-1}) = \frac{1}{\det A}.$$

Proof. The product formula implies $(\det A)\det(A^{-1}) = \det(AA^{-1}) = \det I = 1$. Hence $\det A \neq 0$ and (5.16) holds.

Theorem 5.12 shows that nonvanishing of $\det A$ is a necessary condition for A to be nonsingular. Later (Theorem 5.16) we will prove that this condition is also sufficient. That is, if $\det A \neq 0$ then the inverse matrix A^{-1} exists.

5.13 Determinants and independence of vectors

A simple criterion for testing independence of vectors can be deduced from Theorem 5.12.

THEOREM 5.13. *A set of n vectors* $A_1, A_2, \ldots, A_n$ *in n-space is independent if, and only if, the determinant* $d(A_1, A_2, \ldots, A_n)$ *is nonzero.*

Proof. In Theorem 5.8 we proved that dependence implies $d(A_1, A_2, \ldots, A_n) = 0$. To prove the converse we assume that $A_1, A_2, \ldots, A_n$ are independent and prove that $d(A_1, A_2, \ldots, A_n) \neq 0$.

Let V denote the linear space of n-tuples of scalars. Since $A_1, A_2, \ldots, A_n$ are n independent elements in an n-dimensional space they form a basis for V. By Theorem 4.12 there is a linear transformation $T : V \to V$ that maps these n vectors onto the unit coordinate vectors:

$$T(A_k) = I_k \qquad \text{for } k = 1, 2, \ldots, n.$$

Therefore there is an $n \times n$ matrix B such that

$$A_k B = I_k \qquad \text{for } k = 1, 2, \ldots, n.$$

But, by Lemma 5.10 we have $A_k B = (AB)_k$, where A is the matrix with rows $A_1, A_2, \ldots, A_n$. Hence $AB = I$, so A is nonsingular and $\det A \neq 0$.

5.14 The determinant of a block-diagonal matrix

A square matrix C of the form

$$C = \begin{bmatrix} A & O \\ O & B \end{bmatrix},$$

where A and B are square matrices and each O denotes a matrix of zeros, is called a *block-diagonal matrix* with two diagonal blocks A and B. An example is the 5×5 matrix

$$C = \begin{bmatrix} 1 & 0 & 0 & 0 & 0 \\ 0 & 1 & 0 & 0 & 0 \\ 0 & 0 & 1 & 2 & 3 \\ 0 & 0 & 4 & 5 & 6 \\ 0 & 0 & 7 & 8 & 9 \end{bmatrix}.$$

The diagonal blocks in this case are

$$A = \begin{bmatrix} 1 & 0 \\ 0 & 1 \end{bmatrix} \qquad \text{and} \qquad B = \begin{bmatrix} 1 & 2 & 3 \\ 4 & 5 & 6 \\ 7 & 8 & 9 \end{bmatrix}.$$

The next theorem shows that the determinant of a block-diagonal matrix is equal to the product of the determinants of its diagonal blocks.

THEOREM 5.14.   *For any two square matrices A and B we have*

(5.17) $$\det \begin{bmatrix} A & O \\ O & B \end{bmatrix} = (\det A)(\det B).$$

Proof.   Assume A is $n \times n$ and B is $m \times m$. We note that the given block-diagonal matrix can be expressed as a product of the form

$$\begin{bmatrix} A & O \\ O & B \end{bmatrix} = \begin{bmatrix} A & O \\ O & I_m \end{bmatrix} \begin{bmatrix} I_n & O \\ O & B \end{bmatrix},$$

where I_n and I_m are identity matrices of orders n and m, respectively. Therefore, by the product formula for determinants we have

(5.18) $$\det \begin{bmatrix} A & O \\ O & B \end{bmatrix} = \det \begin{bmatrix} A & O \\ O & I_m \end{bmatrix} \det \begin{bmatrix} I_n & O \\ O & B \end{bmatrix}.$$

Now we regard the determinant $\det \begin{bmatrix} A & O \\ O & I_m \end{bmatrix}$ as a function of the n rows of A. This is possible because of the block of zeros in the upper right-hand corner. It is easily verified that this function satifies all three axioms for a determinant function of order n. Therefore, by the uniqueness theorem, we must have

$$\det \begin{bmatrix} A & O \\ O & I_m \end{bmatrix} = \det A.$$

A similar argument shows that $\det \begin{bmatrix} I_n & O \\ O & B \end{bmatrix} = \det B$. Hence (5.18) implies (5.17).

5.15   Exercises

1. Prove each of the following statements about $n \times n$ matrices, or exhibit a counter example.
   (a) $\det(A + B) = \det A + \det B$.
   (b) $\det\{(A + B)^2\} = \{\det(A + B)\}^2$.
   (c) $\det\{(A + B)^2\} = \det(A^2 + 2AB + B^2)$.
   (d) $\det\{(A + B)^2\} = \det(A^2 + B^2)$.
2. (a) Extend Theorem 5.14 to block-diagonal matrices with three diagonal blocks:

$$\det \begin{bmatrix} A & O & O \\ O & B & O \\ O & O & C \end{bmatrix} = (\det A)(\det B)(\det C).$$

(b) State and prove a generalization for any number of diagonal blocks.

(c) If A, B, C are nonsingular matrices, prove that the block-diagonal matrix $\begin{bmatrix} A & O & O \\ O & B & O \\ O & O & C \end{bmatrix}$ is

nonsingular and that its inverse is $\begin{bmatrix} A^{-1} & O & O \\ O & B^{-1} & O \\ O & O & C^{-1} \end{bmatrix}$.

3. If $A = \begin{bmatrix} 1 & 0 & 0 & 0 \\ 0 & 1 & 0 & 0 \\ a & b & c & d \\ e & f & g & h \end{bmatrix}$, and $B = \begin{bmatrix} a & b & c & d \\ e & f & g & h \\ 0 & 0 & 1 & 0 \\ 0 & 0 & 0 & 1 \end{bmatrix}$, prove that $\det A = \begin{vmatrix} c & d \\ g & h \end{vmatrix}$ and

$\det B = \begin{vmatrix} a & b \\ e & f \end{vmatrix}$.

4. State and prove a generalization of Exercise 3 for $n \times n$ matrices.

5. If $A = \begin{bmatrix} a & b & 0 & 0 \\ c & d & 0 & 0 \\ e & f & g & h \\ x & y & z & w \end{bmatrix}$, prove that $\det A = \det \begin{bmatrix} a & b \\ c & d \end{bmatrix} \det \begin{bmatrix} g & h \\ z & w \end{bmatrix}$.

6. State and prove a genereralization of Exercise 5 for $n \times n$ matrices of the form

$$A = \begin{bmatrix} B & O \\ C & D \end{bmatrix},$$

where B, C, D denote square matrices and O denotes a matrix of zeros.

7. Use Theorem 5.13 to determine whether the following sets of vectors are dependent or independent.

(a) $A_1 = (1, -1, 0)$, $A_2 = (0, 1, -1)$, $A_3 = (2, 3, -1)$.

(b) $A_1 = (1, -1, 2, 1)$, $A_2 = (-1, 2, -1, 0)$, $A_3 = (3, -1, 1, 0)$, $A_4 = (1, 0, 0, 1)$.

(c) $A_1 = (1, 0, 0, 0, 1)$, $A_2 = (1, 1, 0, 0, 0)$, $A_3 = (1, 0, 1, 0, 1)$, $A_4 = (1, 1, 0, 1, 1)$, $A_5 = (0, 1, 0, 1, 0)$.

5.16 Expansion formulas by cofactors

In Section 5.10 we evaluated determinants of orders 2 and 3 by expressing each row of the matrix as a linear combination of the unit coordinate vectors. More generally, every row of an $n \times n$ matrix A can be expressed as a linear combination of the n unit coordinate vectors $I_1, I_2, \ldots, I_n$. For example, the first row $A_1 = (a_{11}, a_{12}, \ldots, a_{1n})$ can be written as follows:

$$A_1 = \sum_{j=1}^{n} a_{1j} I_j.$$

Since determinants are linear in the first row we have

(5.19)
$$d(A_1, A_2, \ldots, A_n) = d\left(\sum_{j=1}^{n} a_{1j} I_j, A_2, \ldots, A_n \right)$$

$$= \sum_{j=1}^{n} a_{1j} d(I_j, A_2, \ldots, A_n).$$

Therefore to compute $\det A$ it suffices to compute $d(I_j, A_2, \ldots, A_n)$ for each unit coordinate vector I_j. We shall use the notation A'_{1j} to denote the matrix obtained from A by replacing the first row A_1 by the unit vector I_j. For example, if $n = 3$ there are three such matrices:

$$A'_{11} = \begin{bmatrix} 1 & 0 & 0 \\ a_{21} & a_{22} & a_{23} \\ a_{31} & a_{32} & a_{33} \end{bmatrix}, \qquad A'_{12} = \begin{bmatrix} 0 & 1 & 0 \\ a_{21} & a_{22} & a_{23} \\ a_{31} & a_{32} & a_{33} \end{bmatrix}, \qquad A'_{13} = \begin{bmatrix} 0 & 0 & 1 \\ a_{21} & a_{22} & a_{23} \\ a_{31} & a_{32} & a_{33} \end{bmatrix}.$$

Note that $\det A'_{1j} = d(I_j, A_2, \ldots, A_n)$. Equation (5.19) can now be written in the form

(5.20)
$$\det A = \sum_{j=1}^{n} a_{1j} \det A'_{1j}.$$

This is called an *expansion formula*; it expresses the determinant of A as a linear combination of the elements in its first row. The number $\det A'_{1j}$ that multiplies a_{1j} is called the *cofactor* of entry a_{1j} and is denoted by the symbol cof a_{1j}. Therefore Equation (5.20) can be written as follows:

(5.21)
$$\det A = \sum_{j=1}^{n} a_{1j} \operatorname{cof} a_{1j}.$$

This equation is called *expansion by first row cofactors*.

More generally, we define the cofactor of any element a_{kj} as follows:

DEFINITION.   COFACTOR OF AN ENTRY.   *Let A'_{kj} denote the matrix obtained from A by replacing its kth row A_k by the unit coordinate vector I_j. Then the scalar $\det A'_{kj}$ is called the cofactor of entry a_{kj} and is denoted by* cof a_{kj}. *Thus we have*

(5.22)
$$\operatorname{cof} a_{kj} = \det A'_{kj}.$$

The argument used to derive Equation (5.21) can be applied to the kth row of A instead of the first row, and the result is an expansion formula in terms of elements of the kth row:

(5.23)
$$\det A = \sum_{j=1}^{n} a_{kj} \det A'_{kj}.$$

But $\det A'_{kj} = \operatorname{cof} a_{kj}$ so we can express this result as follows:

THEOREM 5.15.   EXPANSION BY kTH ROW COFACTORS.   *For any $n \times n$ matrix A and any $k = 1, 2, \ldots, n$ we have*

(5.24)
$$\det A = \sum_{j=1}^{n} a_{kj} \operatorname{cof} a_{kj}.$$

5.17   The cofactor matrix

Theorem 5.12 tells us that if A is nonsingular then $\det A \neq 0$. The next theorem proves the converse. That is, if $\det A \neq 0$ then A^{-1} exists. Moreover it gives an explicit formula for expressing A^{-1} in terms of a matrix formed from the cofactors of the entries of A. This is called the cofactor matrix of A, and is defined as follows:

DEFINITION.   COFACTOR MATRIX.   *Given an $n \times n$ matrix $A = (a_{ij})$, the $n \times n$ matrix whose ij-entry is $\operatorname{cof} a_{ij}$ is denoted by $\operatorname{cof} A$ and is called the cofactor matrix of A. Thus, we have*

$$\operatorname{cof} A = \left(\operatorname{cof} a_{ij}\right)_{i,j=1}^{n}.$$

The next theorem shows that the product of A with the transpose of its cofactor matrix is a scalar times the identity matrix I. (The transpose of any matrix A is obtained by interchanging its rows and columns and is denoted by A^t. Properties of transposes are stated in Exercise 7 of Section 4.21.)

THEOREM 5.16.   *For any $n \times n$ matrix A with $n \geq 2$ we have*

$$(5.25) \qquad\qquad A(\operatorname{cof} A)^t = (\det A)I.$$

In particular, if $\det A \neq 0$ the inverse of A exists and is given by

$$(5.26) \qquad\qquad A^{-1} = \frac{1}{\det A}(\operatorname{cof} A)^t.$$

Proof.   Using Theorem 5.15 we expand $\det A$ by its kth row cofactors:

$$(5.27) \qquad\qquad \det A = \sum_{j=1}^{n} a_{kj} \operatorname{cof} a_{kj}.$$

Keep k fixed and apply this relation to a new matrix B whose ith row is equal to the kth row of A for some $i \neq k$, and whose remaining rows are the same as those of A. Then $\det B = 0$ because the ith and kth rows of B are equal. Expanding $\det B$ by its ith row cofactors we have

$$(5.28) \qquad\qquad \det B = \sum_{j=1}^{n} b_{ij} \operatorname{cof} b_{ij} = 0.$$

But since the ith row of B is equal to the kth row of A we have

$$b_{ij} = a_{kj} \qquad \text{and} \qquad \operatorname{cof} b_{ij} = \operatorname{cof} a_{ij} \qquad \text{for every } j.$$

Hence (5.28) states that

$$(5.29) \qquad\qquad \sum_{j=1}^{n} a_{kj} \operatorname{cof} a_{ij} = 0 \qquad \text{if } k \neq i.$$

Because $k \neq i$ in (5.29), this equation is often called *expansion by alien cofactors*. Equations (5.27) and (5.29) together can be written as follows:

$$(5.30) \qquad \sum_{j=1}^{n} a_{kj} \operatorname{cof} a_{ij} = \begin{cases} \det A & \text{if } i = k \\ 0 & \text{if } i \neq k. \end{cases}$$

But the sum appearing on the left of (5.30) is the k, i entry of the product $A(\operatorname{cof} A)^t$. Therefore (5.30) implies (5.25).

As a direct corollary of Theorems 5.12 and 5.16 we have the following necessary and sufficient condition for a square matrix to be nonsingular.

THEOREM 5.17. *A square matrix A is nonsingular if and only if* $\det A \neq 0$.

5.18 Cramer's rule

Theorem 5.16 can be used to give explicit formulas for the solutions of a system of linear equations with a nonsingular coefficient matrix. The formulas are called *Cramer's rule*, in honor of the Swiss mathematician Gabriel Cramer (1704–1752). The case $n = 3$ was discussed earlier in Section 2.14.

THEOREM 5.18. CRAMER'S RULE. *A system of n linear equations in n unknowns* $x_1, \ldots, x_n$, *say*

$$\sum_{j=1}^{n} a_{ij}x_j = b_i \qquad (i = 1, 2, \ldots, n),$$

with a nonsingular coefficient-matrix $A = (a_{ij})$, *has a unique solution given by the formulas*

$$(5.31) \qquad x_j = \frac{1}{\det A} \sum_{k=1}^{n} b_k \operatorname{cof} a_{kj}, \qquad \text{for } j = 1, 2, \ldots, n.$$

Proof. The system can be written as a matrix equation, $AX = B$, where X and B are column matrices, $X = [x_1, \ldots, x_n]^t$ and $B = [b_1, \ldots, b_n]^t$. Since A is nonsingular there is a unique solution given by

$$(5.32) \qquad X = A^{-1}B = \frac{1}{\det A}(\operatorname{cof} A)^t B.$$

The formulas in (5.31) follow by equating components in (5.32).

Note that the sum in Eq. (5.31) is the determinant of a matrix C_j obtained from A by replacing the jth column of A by column matrix B. Therefore the formula for each x_j in (5.31) can be expressed as the quotient of two determinants:

$$x_j = \frac{\det C_j}{\det A}.$$

5.19   Expansion formulas by minors

The kth row cofactor expansion of the determinant of a matrix A of order n states that

$$\det A = \sum_{j=1}^{n} a_{kj} \operatorname{cof} a_{kj},$$

where $\operatorname{cof} a_{kj}$ is the determinant of the matrix A'_{kj} obtained from A by replacing its kth row by the unit coordinate vector I_j. This section shows that $\det A'_{kj}$ is, except for a plus or minus sign, equal to the determinant of a matrix of order $n - 1$. These smaller matrices are called *minors*.

DEFINITION.   MINOR OF AN ENTRY.   *Given a square matrix A of order $n \geq 2$, the square matrix of order $n - 1$ obtained by deleting the kth row and the jth column of A is called the k, j minor of A and is denoted by A_{kj}.*

EXAMPLE.   A matrix $A = (a_{ij})$ of order 3 has nine minors, one for each entry. The minors of the first row entries are

$$A_{11} = \begin{bmatrix} a_{22} & a_{23} \\ a_{32} & a_{33} \end{bmatrix}, \qquad A_{12} = \begin{bmatrix} a_{21} & a_{23} \\ a_{31} & a_{33} \end{bmatrix}, \qquad A_{13} = \begin{bmatrix} a_{21} & a_{22} \\ a_{31} & a_{32} \end{bmatrix}.$$

Equation (5.2) expresses the determinant of a 3×3 matrix as a linear combination of determinants of the first row minors. The formula can be written as follows:

$$\det A = a_{11} \det A_{11} - a_{12} \det A_{12} + a_{13} \det A_{13}.$$

The next theorem extends this formula to the $n \times n$ case.

THEOREM 5.19.   EXPANSION BY kTH ROW MINORS.   *For any $n \times n$ matrix A with $n \geq 2$, the cofactor of a_{kj} is related to the minor A_{kj} by the formula*

$$(5.33) \qquad \det A'_{kj} = (-1)^{k+j} \det A_{kj}.$$

Therefore the expansion of $\det A$ *by kth row minors is given by*

$$(5.34) \qquad \det A = \sum_{j=1}^{n} (-1)^{k+j} a_{kj} \det A_{kj}.$$

Proof.   We illustrate the idea of the proof by considering first the special case $k = j = 1$. The matrix A'_{11} has the form

$$A'_{11} = \begin{bmatrix} 1 & 0 & 0 & \cdots & 0 \\ a_{21} & a_{22} & a_{23} & \cdots & a_{2n} \\ a_{31} & a_{32} & a_{33} & \cdots & a_{3n} \\ \vdots & \vdots & \vdots & \vdots & \vdots \\ a_{n1} & a_{n2} & a_{n3} & \cdots & a_{nn} \end{bmatrix}.$$

By applying elementary row operations of type (3) we can make every entry below the 1 in the first column equal to zero, leaving all the remaining entries intact. For example, if we multiply the first row of A_{11}' by $-a_{21}$ and add it to the second row, the new second row becomes $(0, a_{22}, a_{23}, \ldots, a_{2n})$. By a succession of such elementary row operations we obtain a new matrix which is identical to A_{11}' except that its first column is $[1, 0, 0, \ldots, 0]^t$. This new matrix is a block diagonal matrix so, by Theorem 5.14, its determinant is equal to $\det A_{11}$, where A_{11} is the 1, 1 minor of A:

$$A_{11} = \begin{bmatrix} a_{22} & a_{23} & \cdots & a_{2n} \\ a_{32} & a_{33} & \cdots & a_{3n} \\ \vdots & \vdots & \vdots & \vdots \\ a_{n2} & a_{n3} & \cdots & a_{nn} \end{bmatrix}.$$

Therefore $\det A_{11}' = \det A_{11}$, which proves (5.33) for $k = j = 1$.

We consider next the special case $k = 1$, j arbitrary, and prove that

(5.35) $\det A_{1j}' = (-1)^{j-1} \det A_{1j}.$

Once we prove (5.35) the more general formula (5.33) follows at once because matrix A_{kj}' can be transformed to a matrix of the form B_{1j}' by $k - 1$ successive interchanges of adjacent rows. The determinant changes sign at each interchange so we have

(5.36) $\det A_{kj}' = (-1)^{k-1} \det B_{1j}',$

where B is an $n \times n$ matrix whose first row is I_j and whose 1, j minor B_{1j} is A_{kj}. By (5.35) we have

$$\det B_{1j}' = (-1)^{j-1} \det B_{1j} = (-1)^{j-1} \det A_{kj},$$

so (5.36) gives us

$$\det A_{kj}' = (-1)^{k-1}(-1)^{j-1} \det A_{kj} = (-1)^{k+j} \det A_{kj}.$$

Therefore if we prove (5.35) we also prove (5.33).

We turn now to the proof of (5.35). The matrix A_{1j}' has the form

$$A_{1j}' = \begin{bmatrix} 0 & \cdots & 1 & \cdots & 0 \\ a_{21} & \cdots & a_{2j} & \cdots & a_{2n} \\ a_{31} & \cdots & a_{3j} & \cdots & a_{3n} \\ \vdots & \vdots & \vdots & \vdots & \vdots \\ a_{n1} & \cdots & a_{nj} & \cdots & a_{nn} \end{bmatrix}$$

By elementary row operations of type (3) we introduce a column of zeros below the 1 and transform this to a matrix we denote by A_{1j}^0:

$$A_{1j}^0 = \begin{bmatrix} 0 & \cdots & 1 & \cdots & 0 \\ a_{21} & \cdots & 0 & \cdots & a_{2n} \\ a_{31} & \cdots & 0 & \cdots & a_{3n} \\ \vdots & \vdots & \vdots & \vdots & \vdots \\ a_{n1} & \cdots & 0 & \cdots & a_{nn} \end{bmatrix}.$$

As before, the determinant is unchanged so $\det A^0_{1j} = \det A'_{1j}$. The $1, j$ minor A_{1j} has the form

$$A_{1j} = \begin{bmatrix} a_{21} & \cdots & a_{2j-1} & a_{2j+1} & \cdots & a_{2n} \\ a_{31} & \cdots & a_{3j-1} & a_{3j+1} & \cdots & a_{3n} \\ \vdots & \vdots & \vdots & \vdots & \vdots & \vdots \\ a_{n1} & \cdots & a_{nj-1} & a_{nj+1} & \cdots & a_{nn} \end{bmatrix}.$$

Now we regard $\det A^0_{1j}$ as a function of the $n - 1$ rows of A_{1j}, say $\det A^0_{1j} = f(A_{1j})$. The function f satisfies the first two axioms for a determinant function of order $n - 1$. Therefore, by the uniqueness theorem we can write

$$f(A_{1j}) = f(J) \det A_{1j},$$

where J is the identity matrix of order $n - 1$. Therefore, to prove (5.35) we must show that $f(J) = (-1)^{j-1}$. Let C denote the following $n \times n$ matrix:

$$\begin{bmatrix} 0 & \cdots & 0 & 1 & 0 & \cdots & 0 \\ 1 & \cdots & 0 & 0 & 0 & \cdots & 0 \\ \vdots & \ddots & \vdots & \vdots & \vdots & & \vdots \\ 0 & \cdots & 1 & 0 & 0 & \cdots & 0 \\ 0 & \cdots & 0 & 0 & 1 & \cdots & 0 \\ 0 & & \vdots & \vdots & \vdots & \ddots & \vdots \\ 0 & \cdots & 0 & 0 & 0 & \cdots & 1 \end{bmatrix} \leftarrow j\text{th row}$$

$$\uparrow$$
$$j\text{th column}$$

All the entries along the sloping dotted lines are equal to 1. The remaining entries not shown are equal to 0. Because of the way f was defined we have $f(J) = \det C$. By interchanging the first row of C successively with rows $2, 3, \ldots, j$ we arrive at the $n \times n$ identity matrix after $j - 1$ interchanges. The determinant changes sign at each interchange, so $\det C = (-1)^{j-1}$. Hence $f(J) = (-1)^{j-1}$, which proves (5.35) and hence (5.33).

5.20 Exercises

1. Determine the cofactor matrix of each of the following matrices:

(a) $\begin{bmatrix} 1 & 2 \\ 3 & 4 \end{bmatrix}$. (b) $\begin{bmatrix} 2 & -1 & 3 \\ 0 & 1 & 1 \\ -1 & -2 & 0 \end{bmatrix}$. (c) $\begin{bmatrix} 3 & 1 & 2 & 4 \\ 2 & 0 & 5 & 1 \\ 1 & -1 & -2 & 6 \\ -2 & 3 & 2 & 3 \end{bmatrix}$.

2. Determine the inverse of each of the nonsingular matrices in Exercise 1.

3. Find all values of the scalar λ for which the matrix $\lambda I - A$ is singular, if A is equal to

(a) $\begin{bmatrix} 1 & 0 \\ 0 & 1 \end{bmatrix}$. (b) $\begin{bmatrix} 1 & 0 & 2 \\ 0 & -1 & -2 \\ 2 & -2 & 0 \end{bmatrix}$. (c) $\begin{bmatrix} 11 & -2 & 8 \\ 19 & -3 & 14 \\ -8 & 2 & -5 \end{bmatrix}$.

4. If A is an $n \times n$ matrix with $n \geq 2$, prove each of the following properties of its cofactor matrix:

(a) $\operatorname{cof}(A^t) = (\operatorname{cof} A)^t$.

(b) $(\operatorname{cof} A)^t A = (\det A)I$.

(c) $A(\operatorname{cof} A)^t = (\operatorname{cof} A)^t A$. ($A$ commutes with the transpose of its cofactor matrix).

5. Use Cramer's rule to solve each of the following systems:
   (a) $x + 2y + 3z = 8$, $2x - y + 4z = 7$, $-y + z = 1$.
   (b) $x + 2y + 2z = 0$, $3x - y - z = 3$, $2x + 5y + 3z = 4$.

5.21   Existence of the determinant function

In this section we use induction on the size of a matrix to prove that a determinant function of order n exists for every n. We have already established existence for $n = 1, 2, 3$.

Assuming a determinant function exists of order $n - 1$, a logical candidate for a determinant function of order n would be one of the expansion formulas by minors, for example, the expansion by first-row minors. However, it is easier to verify the axioms if we use a different but analogous formula in terms of first-*column* minors.

THEOREM 5.20.   *Assume determinants of order $n - 1$ exist. For any $n \times n$ matrix $A = (a_{ij})$, let f be the function defined by the formula*

$$(5.37) \qquad f(A_1, \ldots, A_n) = \sum_{j=1}^{n} (-1)^{j+1} a_{j1} \det A_{j1}.$$

Then f satisfies all three axioms for a determinant function of order n. Therefore, by induction, determinants of order n exist for every n.

Proof.   Because A_{j1} is the $j, 1$ minor of A, it consists of $n - 1$ truncated rows of A (all rows of A except row A_j, with the first entry deleted in each row). By the induction hypothesis, $\det A_{j1}$ satisfies all the properties of a determinant function of order $n - 1$. Now we verify that the function f defined by Eq. (5.37) satisfies all three axioms for a determinant of order n.

Consider first Axiom 1, the homogeneity property, and write (5.37) in the form

$$f(A_1, \ldots, A_n) = \sum_{j=1}^{n} f_j(A_1, \ldots, A_n)$$

where

$$f_j(A_1, \ldots, A_n) = (-1)^{j+1} a_{j1} \det A_{j1}.$$

We will show that each f_j satisfies the homogeneity property, so the same will be true for f.

Let's examine the effect of multiplying the first row of A by a scalar t. The minor A_{11} is not affected because it does not involve row A_1. But the coefficient a_{11} gets multiplied by t so we have

$$f_j(tA_1, \ldots, A_n) = t f_j(A_1, \ldots, A_n).$$

Therefore, each f_j is homogeneous in the first row.

If the kth row of A is multiplied by t, where $k > 1$, the minor A_{k1} is not affected, but a_{k1} is multiplied by t, so f_k is homogeneous in the kth row. If $j \neq k$, the coefficient a_{j1} is not affected but exactly one row of A_{j1} gets multiplied by t. Hence every f_j is homogeneous in the kth row, so f satisfies Axiom 1.

Now we show that f satisfies Axiom 2, invariance under row addition. In this case, exactly two of the terms f_j in the sum are affected and the others are unchanged. But the *sum* of the two affected terms is unchanged, so the entire sum is unchanged. To illustrate, let B denote the matrix obtained from A by adding row A_1 to row A_k, where $k \neq 1$. Then

$$B_k = A_1 + A_k \qquad \text{and} \qquad B_i = A_i \qquad \text{for all } i \neq k.$$

Hence

$$b_{kj} = a_{1j} + a_{kj} \qquad \text{and} \qquad b_{ij} = a_{ij} \qquad \text{for all } i \neq k \text{ and all } j.$$

In the sum (5.37) defining $f(B_1, \ldots, B_n)$ the only two terms affected are those with $j = 1$ and $j = k$. For example, suppose $k = 2$, so that

$$B_2 = A_1 + A_2 \qquad \text{and} \qquad b_{2j} = a_{1j} + a_{2j}.$$

In the sum (5.37) defining $f(B_1, \ldots, B_n)$ the term with $j = 1$ is

$$b_{11} \det B_{11} = b_{11} d(B_2, B_3, \ldots, B_n) = a_{11} d(A_1 + A_2, A_3, \ldots, A_n)$$
$$= a_{11} d(A_1, A_3, \ldots, A_n) + a_{11} d(A_2, A_3, \ldots, A_n).$$

The term with $j = 2$ is

$$-b_{21} \det B_{21} = -b_{21} d(B_1, B_3, \ldots, B_n) = -(a_{11} + a_{21}) d(A_1, A_3, \ldots, A_n)$$
$$= -a_{11} d(A_1, A_3, \ldots, A_n) - a_{21} d(A_1, A_3, \ldots, A_n).$$

Adding this to the term with $j = 1$ and taking into account cancellation we find

$$b_{11} \det B_{11} - b_{21} \det B_{21} = a_{11} d(A_2, A_3, \ldots, A_n) - a_{21} d(A_1, A_3, \ldots, A_n)$$
$$= a_{11} \det A_{11} - a_{21} \det A_{21}.$$

The remaining terms are unchanged, so $f(B_1, \ldots, B_n) = f(A_1, \ldots, A_n)$ in this case. Similarly, if $k = 3$ we have

$$B_3 = A_1 + A_3 \qquad \text{and} \qquad b_{3j} = a_{1j} + a_{3j}.$$

In the sum (5.37) defining $f(B_1, \ldots, B_n)$ the term with $j = 1$ is

$$b_{11} \det B_{11} = b_{11} d(B_2, B_3, \ldots, B_n) = a_{11} d(A_2, A_1 + A_3, A_4, \ldots, A_n)$$
$$= a_{11} d(A_2, A_1, A_4, \ldots, A_n) + a_{11} d(A_2, A_3, A_4, \ldots, A_n).$$

The term with $j = 3$ is

$$b_{31} \det B_{31} = b_{31} d(B_1, B_2, B_4, \ldots, B_n) = (a_{11} + a_{31}) d(A_1, A_2, A_4, \ldots, A_n)$$
$$= a_{11} d(A_1, A_2, A_4, \ldots, A_n) + a_{31} d(A_1, A_2, A_4, \ldots, A_n)$$
$$= -a_{11} d(A_2, A_1, A_4, \ldots, A_n) + a_{31} d(A_1, A_2, A_4, \ldots, A_n).$$

Adding this to the term with $j = 1$ and taking into account cancellation we find

$$b_{11} \det B_{11} + b_{31} \det B_{31} = a_{11} d(A_2, A_3, A_4, \ldots, A_n) + a_{31} d(A_1, A_2, A_4, \ldots, A_n)$$

$$= a_{11} \det A_{11} + a_{31} \det A_{31}.$$

The remaining terms are unchanged, so $f(B_1, \ldots, B_n) = f(A_1, \ldots, A_n)$ in this case as well. The same kind of argument works in the general case and shows that f satisfies Axiom 2.

Finally, we verify that f satisfies Axiom 3. When $A = I$ we have $a_{11} = 0$ and $a_{j1} = 0$ for $j > 1$. Also, the minor A_{11} is the identity matrix of order $n - 1$, so each term in (5.37) is zero except the first, which is equal to 1. Hence $f(I_1, \ldots, I_n) = 1$, so f satisfies Axiom 3.

In the foregoing proof we could just as well have used a function f defined in terms of the kth-column minors A_{jk} instead of the first-column minors A_{j1}. In fact, if we let

$$(5.38) \qquad\qquad f(A_1, \ldots, A_n) = \sum_{j=1}^{n} (-1)^{j+k} a_{jk} \det A_{jk},$$

exactly the same type of proof shows that this f satisfies all three axioms for a determinant function. By the uniqueness theorem the expansion by kth column minors is equal to det A.

The expansion formulas in (5.38) not only establish the existence of determinant functions but also reveal a new aspect of the theory of determinants—a connection between row properties and column properties. This connection is discussed further in the next theorem.

THEOREM 5.21.   *A matrix and its transpose have equal determinants.*

Proof.   The proof is by induction on n, the size of the matrix. For $n = 1$ the result is trivial, and for $n = 2$ the result is easily verified by use of Eq. (5.1). Assume, then, that the theorem is true for matrices of order $n - 1$. Let $A = (a_{ij})$ and let $B = A^t = (b_{ij})$ be the transpose of A. Expand det A by its first-column minors and expand det B by its first-row minors to get

$$\det A = \sum_{j=1}^{n} (-1)^{j+1} a_{j1} \det A_{j1}, \qquad \det B = \sum_{j=1}^{n} (-1)^{j+1} b_{1j} \det B_{1j}.$$

But from the definition of transpose we have $b_{1j} = a_{j1}$ and $B_{1j} = (A_{j1})^t$. Since we are assuming the theorem is true for matrices of order $n - 1$ we have $\det B_{1j} = \det A_{j1}$. Hence the foregoing sums are equal term by term, so det $A = \det B$.

5.22 Miscellaneous exercises on determinants

1. (a) Explain why each of the following is a Cartesian equation for a straight line in the xy plane passing through two distinct points (x_1, y_1) and (x_2, y_2).

$$\det \begin{bmatrix} x - x_1 & y - y_1 \\ x_2 - x_1 & y_2 - y_1 \end{bmatrix} = 0; \qquad \det \begin{bmatrix} x & y & 1 \\ x_1 & y_1 & 1 \\ x_2 & y_2 & 1 \end{bmatrix} = 0.$$

   (b) State and prove corresponding relations for a plane in $\mathbf{R}^3$ passing through three distinct points.

(c) State and prove corresponding relations for a circle in the xy plane passing through three noncollinear points.

2. This exercise gives a useful method for constructing nonsingular matrices with integer entries whose inverses also have integer entries. Let $A = \begin{bmatrix} 1 & x & y \\ 0 & 1 & x \\ 0 & 0 & 1 \end{bmatrix}$.

(a) Calculate the cofactor matrix of A.
(b) Let $B = AA'$, where A' is the transpose of A. Prove that B is nonsingular.
(c) If the entries x and y are integers, prove that both B and B^{-1} have integer entries.

3. This exercise calculates the determinant of a special $n \times n$ matrix, called the *Vandermonde*, which is defined as follows:

$$V_n(a_1, a_2, \ldots, a_n) = \det \begin{bmatrix} 1 & a_1 & a_1^2 & \cdots & a_1^{n-1} \\ 1 & a_2 & a_2^2 & \cdots & a_2^{n-1} \\ \vdots & \vdots & \vdots & & \vdots \\ 1 & a_n & a_n^2 & \cdots & a_n^{n-1} \end{bmatrix}.$$

The numbers $a_1, a_2, \ldots, a_n$ can be real or complex. If any two of the a_i are equal the determinant is zero because it will have two equal rows. So, we assume that the a_i are distinct. Replace a_1 by x and let $P(x) = V_n(x, a_2, \ldots, a_n)$, so the first row becomes $1, x, x^2, \ldots, x^{n-1}$.

(a) Show that $P(x)$ is a polynomial in x of degree $\leq n - 1$. [*Hint:* Expand along first row minors.]
(b) Show that $P(x)$ has $n - 1$ distinct roots $a_2, \ldots, a_n$ and therefore has the factorization

$$P(x) = k \prod_{i=2}^{n} (x - a_i),$$

where the constant factor k is the coefficient of x^{n-1}.

(c) Show that $k = (-1)^{n-1} V_{n-1}(a_2, \ldots, a_n)$.
(d) Use parts (b) and (c) to deduce the recursion formula

$$V_n(a_1, a_2, \ldots, a_n) = \prod_{i=2}^{n} (a_i - a_1) V_{n-1}(a_2, \ldots, a_n).$$

(e) Use part (d) to deduce that

$$V_n(a_1, a_2, \ldots, a_n) = \prod_{\substack{i,j=1 \\ i<j}}^{n} (a_j - a_i).$$

If the a_i are distinct, this shows that the determinant is nonzero, so the Vandermonde matrix is nonsingular. This property is used in many applications.

Exercises 4 through 7 require a knowledge of differential calculus.

4. Given n^2 functions f_{ij}, each differentiable on an interval (a, b), define $F(x) = \det[f_{ij}(x)]$ for each x in (a, b). Prove that the derivative $F'(x)$ is the sum of n determinants,

$$F'(x) = \sum_{i=1}^{n} \det A_i(x),$$

where $A_i(x)$ is the matrix obtained by differentiating the functions in the ith row of $[f_{ij}(x)]$.

5. If $F(x) = \det \begin{bmatrix} 1 & x & x^2 & x^3 \\ 0 & 1 & 2x & 3x \\ 0 & 0 & 2 & 3 \\ 1 & e^x & e^{2x} & e^{3x} \end{bmatrix}$, calculate the derivative $F'(x)$.

6. If $F(x) = \det \begin{bmatrix} x & x^2/2 & x^3/3 & x^4/4 & x^5/5 \\ 1 & 1 & 1 & 1 & 1 \\ 1 & 2 & 4 & 8 & 16 \\ 1 & 3 & 9 & 27 & 81 \\ 1 & 4 & 16 & 64 & 256 \end{bmatrix}$, show that $F'(x) = c(x^2 - 3x + 2)(x^2 + ax + b)$

for some choice of constants a, b, c.

7. An $n \times n$ matrix of functions of the form $W(x) = [u_j^{(i-1)}(x)]$, in which each row after the first is obtained by differentiating the entries of the previous row, is called a *Wronskian matrix*, in honor of the Polish mathematician J. M. H. Wronski. (We shall encounter this matrix in Chapter 8.) Prove that the derivative of $\det W(x)$ is the determinant of the matrix obtained by differentiating each entry in the last row of $W(x)$. [*Hint:* Use Exercise 4.]

6

EIGENVALUES AND EIGENVECTORS

6.1 Linear transformations with diagonal matrix representations

Let $T : V \to V$ be a linear transformation acting on a finite-dimensional linear space V. Properties of T that are independent of any coordinate system (basis) for V are called *intrinsic properties* of T. They are shared by all matrix representations of T. If a basis can be chosen so that the resulting matrix has a particularly simple form it may be possible to detect some of the intrinsic properties directly from the matrix representation.

Diagonal matrices are especially simple. Therefore we might ask whether every linear transformation has a diagonal matrix representation. In Chapter 4 we treated the problem of finding a diagonal matrix representation of a linear transformation $T : V \to W$, where $\dim V = n$ and $\dim W = m$. Theorem 4.14 showed that there always exists a basis $(v_1, \ldots, v_n)$ for V and a basis $(w_1, \ldots, w_n)$ for W such that the matrix of T relative to this choice of bases is a diagonal matrix. In particular, if $V = W$ the matrix will be a square diagonal matrix. The new feature now is that we want to use the *same* basis for both V and W. With this restriction it is not always possible to find a diagonal matrix representation for T. We turn, then, to the central problem of this chapter: Given a finite-dimensional linear space V, is it possible to find a basis for V relative to which the linear transformation $T : V \to V$ has a diagonal matrix representation?

Notation. If $A = (a_{ij})$ is a diagonal matrix, we write $A = \text{diag}(a_{11}, a_{22}, \ldots, a_{nn})$.

The first theorem exploits the fact that a linear transformation is completely determined by its action on the basis elements.

THEOREM 6.1. *Given a linear transformation $T : V \to V$, where $\dim V = n$. Then T has a diagonal matrix representation relative to a basis $(v_1, \ldots, v_n)$ if and only if there exist scalars $\lambda_1, \ldots, \lambda_n$ such that*

$$(6.1) \qquad T(v_k) = \lambda_k v_k \quad for \quad k = 1, 2, \ldots, n,$$

in which case the matrix representation is $\text{diag}(\lambda_1, \ldots, \lambda_n)$.

Proof. If T has a diagonal matrix representation, say $A = \text{diag}(\lambda_1, \ldots, \lambda_n)$ relative to a basis $(v_1, \ldots, v_n)$, then the action of T on the basis elements is given by (6.1). Conversely, if scalars $\lambda_1, \ldots, \lambda_n$ exist such that each basis element v_k is mapped onto $\lambda_k v_k$, then $\text{diag}(\lambda_1, \ldots, \lambda_n)$ is a diagonal matrix representation relative to the basis $(v_1, \ldots, v_n)$.

Thus, the problem of finding a diagonal matrix representation is equivalent to finding n independent elements $(v_1, \ldots, v_n)$ in V and a set of scalars $\lambda_1, \ldots, \lambda_n$ such that Eq. (6.1) holds for each basis element v_k.

Elements v_k and scalars λ_k satisfying Equation (6.1) are called *eigenvectors* and *eigenvalues* of T, respectively. In the next section we study eigenvectors and eigenvalues[*] in a more general setting.

6.2 Eigenvalues and eigenvectors of a linear transformation

In this discussion, V denotes a linear space and S denotes a subspace of V. The spaces S and V are not required to be finite-dimensional.

DEFINITION. EIGENVALUE AND EIGENVECTOR. *Let $T : S \to V$ be a linear transformation of S into V. A scalar λ is called an eigenvalue of T if there is a nonzero element x in S such that*

$$(6.2) \qquad\qquad T(x) = \lambda x.$$

The corresponding x is called an eigenvector of T belonging to λ. The scalar λ is called an eigenvalue corresponding to x.

There is exactly one eigenvalue corresponding to a given eigenvector x. If fact, if $T(x) = \lambda x$ and $T(x) = \mu x$ for some $x \ne O$, then $\lambda x = \mu x$, so $\lambda = \mu$.

If $V = S = \mathbf{R}^n$, the equation $T(x) = \lambda x$ has a simple geometric meaning when $\lambda \ne 0$. It states that each eigenvector x is mapped onto a vector parallel to x.

Note. Although Equation (6.2) always holds for $x = O$ and any scalar λ, the definition specifically excludes O as an eigenvector. One reason for this prejudice against O is to have exactly one eigenvalue λ associated with a given eigenvector x.

The following examples illustrate the meaning of these concepts.

EXAMPLE 1. *Multiplication by a fixed scalar.* Let $T : S \to V$ be the linear transformation defined by the equation $T(x) = cx$ for each x in S, where c is a fixed scalar. In this example every nonzero element of S is an eigenvector belonging to the scalar c.

EXAMPLE 2. *The eigenspace $E(\lambda)$ consisting of all x such that $T(x) = \lambda x$.* Let $T : S \to V$ be a linear transformation having an eigenvalue λ. Let $E(\lambda)$ be the set of all elements x in S such that $T(x) = \lambda x$. This set contains the zero element O and all eigenvectors belonging to λ. It is easy to show that $E(\lambda)$ is a subspace of S, because if x and y are in $E(\lambda)$ we have

$$T(ax + by) = aT(x) + bT(y) = a\lambda x + b\lambda y = \lambda(ax + by)$$

for all scalars a and b. Hence $(ax + by) \in E(\lambda)$ so $E(\lambda)$ is a subspace. The space $E(\lambda)$ is called the *eigenspace* corresponding to λ. It may be finite- or infinite-dimensional. If $E(\lambda)$ is finite-dimensional then its dimension is at least 1 because $E(\lambda)$ contains at least one nonzero element x corresponding to λ.

---

[*]The words *eigenvector* and *eigenvalue* are partial translations of the German words *Eigenvektor* and *Eigenwert*, respectively. Some authors use the terms *characteristic vector*, or *proper vector* as synonyms for eigenvector. Eigenvalues are also called *characteristic values*, *proper values*, or *latent roots*.

EXAMPLE 3.   *Existence of zero eigenvalues.*   If an eigenvector exists it cannot be zero, by definition. However, the zero scalar can be an eigenvalue. In fact, if 0 is an eigenvalue for x then $T(x) = 0x = O$, so x is in the null space of T. Conversely, if the null space of T contains any nonzero elements, then each of these is an eigenvector with eigenvalue 0. For a general eigenvalue λ, zero or not, $E(\lambda)$ is the null space of the operator $T - \lambda I$.

EXAMPLE 4.   *Reflection in the xy plane.*   Let $S = V = \mathbf{R}^3$, and let T be a reflection in the xy plane. That is, let T act on the basis vectors i, j, k as follows: $T(i) = i$, $T(j) = j$, and $T(k) = -k$. Every nonzero vector in the xy plane is an eigenvector with eigenvalue 1. The remaining eigenvectors are those of the form ck, where $c \neq 0$; each of them has eigenvalue -1.

EXAMPLE 5.   *Rotation of the plane through a fixed angle α.*   This example is of special interest because it shows that the existence of eigenvectors may depend on the underlying field of scalars. The plane can be regarded as a linear space in two different ways: (1) As a 2-dimensional *real* linear space, $V = \mathbf{R}^2$, with two basis elements $i = (1, 0)$ and $j = (0, 1)$, and with real numbers as scalars; or (2) as a 1-dimensional *complex* linear space, $V = \mathbf{C}$, with one basis element 1, and with complex numbers as scalars.

Consider the second interpretation first. Every nonzero element z of $\mathbf{C}$ can be expressed in polar form, $z = re^{i\theta}$, where $r = |z|$ and $\theta = \arg(z)$. If T rotates z through an angle α, then

$$T(z) = re^{i(\theta + \alpha)} = e^{i\alpha}z.$$

Thus, each nonzero z is an eigenvector with corresponding eigenvalue $\lambda = e^{i\alpha} = \cos \alpha + i \sin \alpha$. Note that $e^{i\alpha}$ is not real unless α is an integer multiple of π. If α is an even multiple of π, then T is the identity mapping, $T(z) = z$. And if α is an odd multiple of π, then T reflects each point through the origin, so that $T(z) = -z$.

Now consider the plane as a *real* linear space $\mathbf{R}^2$. Since the scalars in this case are real numbers, the rotation T has real eigenvalues only if the rotation angle α is an integer multiple of π. In other words, if α is not an integer multiple of π then T has no real eigenvalues and hence no eigenvectors. Thus, the existence of eigenvectors and eigenvalues may depend on the choice of scalars for V.

EXAMPLE 6.   *The subspace spanned by an eigenvector.*   Let $T : S \to V$ be a linear transformation having an eigenvalue λ. Let x be an eigenvector belonging to λ, and let $L(x)$ be the subspace spanned by x. That is, $L(x)$ is the set of all scalar multiples of x. It is easy to show that T maps $L(x)$ into itself. In fact, if $y = cx$ we have

$$T(y) = T(cx) = cT(x) = c(\lambda x) = \lambda(cx) = \lambda y.$$

If $c \neq 0$ then $y \neq O$ so every nonzero element y of $L(x)$ is also an eigenvector belong to λ.

A subspace U of S is called *invariant* under T if T maps each element of U onto an element of U. Example 6 shows that the subspace spanned by an eigenvector is invariant under T.

The next two examples require a knowledge of calculus.

EXAMPLE 7.   *The differentiation operator.*   Let V be the linear space of all real functions f having derivatives of every order on a given open interval. Let D be the linear transformation that maps each f onto its derivative, $D(f) = f'$. The eigenvectors of D are those nonzero functions f satisfying an equation of the form

$$f' = \lambda f$$

for some real λ. This is a first-order linear differential equation. All its solutions are given by the formula

$$f(x) = ce^{\lambda x},$$

where c is an arbitrary real constant. Therefore the eigenvectors of D are all exponential functions $f(x) = ce^{\lambda x}$ with $c \neq 0$. The eigenvalue corresponding to $f(x) = ce^{\lambda x}$ is λ. In examples like this one, where V is a function space, the eigenvectors are called *eigenfunctions*.

EXAMPLE 8.   *The integration operator.*   Let V be the linear space of all real functions continuous on a finite interval $[a, b]$. If $f \in V$ define $g = T(f)$ to be that function given by

$$g(x) = \int_a^x f(t)\,dt \qquad \text{if} \qquad a \leq x \leq b.$$

The eigenfunctions of T (if any exist) are those nonzero f satisfying an equation of the form

(6.3) $$\int_a^x f(t)\,dt = \lambda f(x)$$

for some real λ. If an eigenfunction exists, we can differentiate this equation to obtain the relation $f(x) = \lambda f'(x)$, from which we find $f(x) = ce^{x/\lambda}$, provided $\lambda \neq 0$. In other words, the only candidates for eigenfunctions are those exponential functions of the form $f(x) = ce^{x/\lambda}$ with $c \neq 0$ and $\lambda \neq 0$. However, if we put $x = a$ in Equation (6.3) we obtain

$$0 = \lambda f(a) = \lambda ce^{a/\lambda}.$$

But the exponential $e^{a/\lambda}$ is never zero, so the equation $T(f) = \lambda f$ cannot be satisfied with a nonzero f. Therefore T has no eigenfunctions and no eigenvalues.

6.3   Linear independence of eigenvectors corresponding to distinct eigenvalues

The next theorem describes a fundamental property of eigenvectors. As in the previous section, S denotes a subspace of a linear space V.

THEOREM 6.2.   *Let $u_1, \dots, u_k$ be eigenvectors of a linear transformation $T : S \to V$, and assume that the corresponding eigenvalues $\lambda_1, \dots, \lambda_k$ are distinct. Then the eigenvectors $u_1, \dots, u_k$ are independent.*

Proof.   The proof is by induction on k, the number of eigenvectors. The result holds trivially if $k = 1$. Assume, then, that it has been proved for every set of $k - 1$ eigenvectors. Let $u_1, \dots, u_k$ be k eigenvectors belonging to distinct eigenvalues $\lambda_1, \dots, \lambda_k$ and assume scalars c_i exist such that

(6.4)
$$\sum_{i=1}^{k} c_i u_i = O.$$

Applying T to both members of (6.4) and using the fact that $T(u_i) = \lambda_i u_i$ we find

(6.5)
$$\sum_{i=1}^{k} c_i \lambda_i u_i = O.$$

Multiplying (6.4) by λ_k and subtracting from (6.5) we obtain the equation

$$\sum_{i=1}^{k-1} c_i(\lambda_i - \lambda_k)u_i = O.$$

Because $u_1, \ldots, u_{k-1}$ are independent we must have $c_i(\lambda_i - \lambda_k) = 0$ for each $i = 1, 2, \ldots, k - 1$. But the eigenvalues are distinct, so $(\lambda_i - \lambda_k) \neq 0$, hence $c_i = 0$ for each $i = 1, 2, \ldots, k - 1$. Using these values in (6.4) we see that c_k is also 0, so the eigenvectors $u_1, \ldots, u_k$ are independent.

Note that Theorem 6.2 would not be true if the zero element were allowed to be an eigenvector. This is another reason for excluding O from being an eigenvector.

Theorem 6.2 has important consequences for the finite-dimensional case.

THEOREM 6.3. *If* dim $V = n$, *every linear transformation* $T : V \to V$ *has at most* n *distinct eigenvalues. If T has exactly n distinct eigenvalues, then the corresponding eigenvectors form a basis for V, and the matrix of T relative to this basis is a diagonal matrix with the eigenvalues as diagonal entries.*

Proof. If there were $n + 1$ distinct eigenvalues then, by Theorem 6.2, V would contain $n + 1$ independent elements, contradicting the fact that dim $V = n$. The second assertion follows from Theorems 6.1 and 6.2.

Note. Theorem 6.3 tells us that the existence of n distinct eigenvalues is a *sufficient* condition for T to have a diagonal matrix representation. This condition is not necessary. There exist linear transformations with fewer than n distinct eigenvalues that can be represented by diagonal matrices. The identity transformation is an example. All its eigenvalues are equal to 1, but the identity transformation is represented by the identity matrix relative to any choice of basis. A *necessary and sufficient* condition for T to have a diagonal matrix representation is that described in Theorem 6.1.

6.4 Exercises

1. (a) If T has an eigenvalue λ, prove that aT has the eigenvalue $a\lambda$.
   (b) If x is an eigenvector for both T_1 and T_2, prove that x is also an eigenvector for $aT_1 + bT_2$ for all scalars a and b. How are the eigenvalues related?
2. Assume $T : V \to V$ has an eigenvector x belonging to an eigenvalue λ. Prove that x is an eigenvector of T^2 belonging to λ^2 and, more generally, x is an eigenvector of T^n belonging to λ^n. Then use Exercise 1 to show that, for any polynomial P, x is an eigenvector of $P(T)$ belonging to $P(\lambda)$.

3. Consider the plane as the real linear space $\mathbf{R}^2$, and let T be a rotation of $\mathbf{R}^2$ through an angle of $\pi/2$ radians about the origin. Although T has no eigenvectors, prove that every nonzero vector is an eigenvector for T^2.

4. If $T : V \rightarrow V$ has the property that T^2 has a nonnegative eigenvalue λ^2, prove that at least one of λ or $-\lambda$ is an eigenvalue for T. [*Hint:* $T^2 - \lambda^2 I = (T + \lambda I)(T - \lambda I)$.]

5. Assume that a linear transformation T has two eigenvectors x and y belonging to distinct eigenvalues $\lambda \neq \mu$. If $ax + by$ is an eigenvector of T, prove that $a = 0$ or $b = 0$.

6. Let $T : S \rightarrow V$ be a linear transformation such that every nonzero element of S is an eigenvector. Prove that there exists a scalar c such that $T(x) = cx$ for all x. In other words, the only transformation with this property is a scalar times the identity. [*Hint:* Use Exercise 5.]

7. Let V be the linear space of all real polynomials $p(x)$ of degree $\leq n$. If $p \in V$ define $q = T(p)$ to mean that $q(t) = p(t + 1)$ for all real t. Prove that T has only the eigenvalue 1. What are the eigenfunctions belonging to this eigenvalue?

Exercises 8 through 11 require a knowledge of calculus.

8. Let V be the linear space of all real functions differentiable on $(0, 1)$. If $f \in V$ define $g = T(f)$ to mean that $g(t) = tf'(t)$ for all t in $(0, 1)$. Prove that every real λ is an eigenvalue for T, and determine the eigenfunctions corresponding to λ.

9. Let V be the linear space of all functions continuous on $(-\infty, +\infty)$ and such that the integral $\int_{-\infty}^{x} f(t)\,dt$ exists for all real x. If $f \in V$, let $g = T(f)$ be defined by $g(x) = \int_{-\infty}^{x} f(t)\,dt$. Prove that every $\lambda > 0$ is an eigenvalue for T and determine the eigenfunctions corresponding to λ.

10. Let V be the linear space of all functions continuous on $(-\infty, +\infty)$ and such that the integral $\int_{-\infty}^{x} tf(t)\,dt$ exists for all real x. If $f \in V$, let $g = T(f)$ be defined by $g(x) = \int_{-\infty}^{x} tf(t)\,dt$. Prove that every $\lambda < 0$ is an eigenvalue for T and determine the eigenfunctions corresponding to λ.

11. Let V be the linear space of all real convergent sequences $\{x_n\}$. Define $T : V \rightarrow V$ as follows: If $x = \{x_n\}$ is a convergent sequence with limit a, let $T(x) = \{y_n\}$, where $y_n = a - x_n$ for $n \geq 1$. Prove that T has only two eigenvalues, $\lambda = 0$ and $\lambda = -1$, and determine the eigenvectors belonging to each λ.

6.5 The finite-dimensional case

In Example 5 of Section 6.2 we showed that the existence of eigenvalues for a given linear operator may depend on the underlying field of scalars. The next theorem shows that if the scalars are complex and if the linear space being acted on is finite-dimensional, then an eigenvalue always exists.

THEOREM 6.4.   *Let V be a finite-dimensional linear space with complex scalars. Then every linear transformation $T : V \rightarrow V$ has an eigenvalue.*

Proof.   Let $n = \dim V$, and choose any fixed nonzero x in V. Consider the following $n + 1$ elements of V:

$$x, T(x), T^2(x), \ldots, T^n(x).$$

These are dependent because $\dim V = n$. Therefore there exist scalars $c_0, c_1, \ldots, c_n$, not all zero, such that

$$\sum_{k=0}^{n} c_k T^k(x) = O.$$

This equation can also be written as $F(x) = O$, where F is the linear operator

$$F = \sum_{k=0}^{n} c_k T^k.$$

Use the same scalars c_k as coefficients to form a complex-valued polynomial, say,

$$f(z) = \sum_{k=0}^{n} c_k z^k.$$

If c_m is the last nonzero coefficient among $c_0, c_1, \ldots, c_n$, this polynomial has degree m, where $m \leq n$. Write the polynomial in factored form:

$$f(z) = c_m(z - \alpha_1) \cdots (z - \alpha_m),$$

where $\alpha_1, \ldots, \alpha_m$ are the complex roots of the polynomial $f(z)$. Then the corresponding operator F has factored form

$$F = c_m(T - \alpha_1 I) \cdots (T - \alpha_m I),$$

where I is the identity operator and $c_m \neq 0$. Applying F to the nonzero element x we get

$$F(x) = c_m(T - \alpha_1 I) \cdots (T - \alpha_m I)(x) = O.$$

Let $x_1 = (T - \alpha_m I)(x)$. If $x_1 = O$ then $T(x) = \alpha_m x$ and x is an eigenvector with α_m as corresponding eigenvalue. If $x_1 \neq O$, define elements $x_2, x_3, \ldots, x_m$ successively as follows:

$$x_2 = (T - \alpha_{m-1} I)(x_1), \qquad x_3 = (T - \alpha_{m-2} I)(x_2), \qquad \ldots, x_m = (T - \alpha_1 I)(x_{m-1}).$$

Because $F(x) = O$ and $c_m \neq 0$, at least one of $x_2, x_3, \ldots, x_m$ must be zero. Let x_k be the last nonzero element in this sequence. Then $x_{k+1} = O$, hence $T(x_k) = \alpha_{m-k} x_k$ so x_k is an eigenvector with α_{m-k} as corresponding eigenvalue.

6.6 The triangularization theorem

Theorem 6.1 tells us that a linear transformation T acting on a linear space V of dimension n can be diagonalized if, and only if, T has n independent eigenvectors. Consequently, if T does not have n independent eigenvectors it cannot be diagonalized. However, if V has complex scalars, we can show that T can always be *triangularized*, even when it has fewer than n independent eigenvectors.

THEOREM 6.5.   TRIANGULARIZATION   THEOREM.   *Given any linear transformation* $T : V \to V$, *where V has complex scalars and* $\dim V = n$. *Then there is a basis for V, relative to which the matrix of T is upper triangular. Also, each diagonal entry of this matrix is an eigenvalue of T.*

Proof.   The proof is by induction on n, the dimension of V. If $n = 1$ the result is trivial because every 1×1 matrix is upper triangular and its single entry is an eigenvalue. So now we assume that the triangularization theorem is true for all operators acting on linear spaces of dimension $n - 1$ and prove that it is also true for all operators acting on spaces of dimension n.

Let λ_1 be an eigenvalue for T. (Theorem 6.4 guarantees that an eigenvalue exists because V has complex scalars.) Let u_1 be an eigenvector belonging to λ_1. Then we have

$$T(u_1) = \lambda_1 u_1 \qquad \text{and} \qquad u_1 \neq O.$$

By Theorem 3.7(a), u_1 is part of a basis for V, say the basis

$$U = (u_1, u_2, \ldots, u_n).$$

Let $A = (a_{ij})$ be the matrix of T relative to U. Then we have

$$(6.6) \qquad\qquad T(u_k) = \sum_{i=1}^{n} a_{ik} u_i \qquad \text{for} \qquad k = 1, 2, \ldots, n.$$

Now let $V^* = L(u_2, \ldots, u_n)$ be the linear space spanned by the $n - 1$ elements $u_2, \ldots, u_n$. These elements are independent so dim $V^* = n - 1$. To apply the induction hypotheses, we define a new linear operator $T^* : V^* \rightarrow V^*$ by describing its action on the basis elements of V^*:

$$(6.7) \qquad\qquad T^*(u_k) = \sum_{i=2}^{n} a_{ik} u_i \qquad \text{for} \qquad k = 2, \ldots, n.$$

Note that the sum in Eq. (6.7) is like that in Eq. (6.6), except the first term in the sum has been omitted. Only the elements $u_2, \ldots, u_n$ occur in the sum, so T^* maps V^* into V^*. Because T^* acts on a linear space of dimension $n - 1$, the induction hypothesis tells us that there is a basis for V^*, say

$$W^* = (w_2, \ldots, w_n),$$

relative to which the matrix of T^* is upper triangular. Denote this matrix by (b_{ij}). Then we have

$$T^*(w_2) = b_{22} w_2,$$

$$T^*(w_3) = b_{23} w_2 + b_{33} w_3,$$

$$\vdots$$

$$T^*(w_n) = b_{2n} w_2 + b_{3n} w_3 + \cdots + b_{nn} w_n,$$

and, by the induction hypothesis, each diagonal entry $b_{22}, b_{33}, \ldots, b_{nn}$ is an eigenvalue of T^*.

Now let $W = (u_1, w_2, \ldots, w_n)$. We will show that W is a basis for V, and that the matrix of T relative to W is upper triangular. Since W contains n elements of V, to prove that W is a basis it suffices to show that W spans V. If $x \in V$ then x is a linear combination of $u_1, u_2, \ldots, u_n$. But the elements $u_2, \ldots, u_n$ are in V^*, so each of these is a linear combination of $w_2, \ldots, w_n$. Hence x is a linear combination of $u_1, w_2, \ldots, w_n$, which means that W spans V.

To calculate the matrix of T relative to the basis W we apply T to the basis elements. For the first basis element u_1 we have

$$T(u_1) = \lambda_1 u_1$$

because of the way u_1 was chosen. To find the images of $w_2, \ldots, w_n$ under T let's compare the action of T with that of T^* on a typical element of V^*. If $x \in V^*$ we can write

$$x = \sum_{k=2}^{n} c_k u_k$$

for some choice of scalars $c_2, c_3, \ldots, c_n$, so

$$T(x) = \sum_{k=2}^{n} c_k T(u_k)$$

and

$$T^*(x) = \sum_{k=2}^{n} c_k T^*(u_k).$$

Therefore the difference is given by

$$T(x) - T^*(x) = \sum_{k=2}^{n} c_k \{T(u_k) - T^*(u_k)\}.$$

In view of Eqs. (6.6) and (6.7) we have $T(u_k) - T^*(u_k) = a_{1k}u_1$, hence,

$$T(x) - T^*(x) = \left(\sum_{k=2}^{n} c_k a_{1k} \right) u_1 = c(x)u_1,$$

where

$$c(x) = \sum_{k=2}^{n} c_k a_{1k}$$

is a scalar depending on x. In other words, we have

(6.8) $$T(x) = c(x)u_1 + T^*(x).$$

Now apply this relation repeatedly, taking $x = w_2, \ldots, w_n$, to obtain

$$T(w_2) = c(w_2)u_1 + T^*(w_2) = c(w_2)u_1 + b_{22}w_2,$$

$$T(w_3) = c(w_3)u_1 + T^*(w_3) = c(w_3)u_1 + b_{23}w_2 + b_{33}w_3,$$

$$\vdots$$

$$T(w_n) = c(w_n)u_1 + T^*(w_n) = c(w_n)u_1 + b_{2n}w_2 + b_{3n}w_3 + \cdots + b_{nn}w_n.$$

Therefore the matrix of T relative to W is upper triangular, with diagonal entries λ_1, $b_{22}, \ldots, b_{nn}$.

It remains to show that each diagonal element is an eigenvalue of T. We already know that λ_1 is an eigenvalue of T. By the induction hypothesis, the elements $b_{22}, \ldots, b_{nn}$ are eigenvalues of T^*. Hence there exist eigenvectors $v_2, v_3, \ldots, v_n$ in V^* such that $T^*(v_k) = b_{kk}v_k$ for $k = 2, 3, \ldots, n$. By Eq. (6.8) we have

$$T(v_k) = c(v_k)u_1 + T^*(v_k)$$

$$= c(v_k)u_1 + b_{kk}v_k.$$

If $c(v_k) = 0$ then $T(v_k) = b_{kk}v_k$ and v_k is an eigenvector for T belonging to the eigenvalue b_{kk}. But if $c(v_k) \neq 0$ then v_k is not an eigenvector for T, but we can find an eigenvector of the form

$$v'_k = u_1 + \lambda_k v_k$$

by choosing the scalar λ_k appropriately. For any choice of λ_k this element v'_k is nonzero because u_1 and v_k are independent. To find the choice that makes v'_k an eigenvector let's apply T to v'_k. We get

$$T(v'_k) = T(u_1) + \lambda_k T(v_k) = \lambda_1 u_1 + \lambda_k\{c(v_k)u_1 + b_{kk}v_k\}$$

$$= \{\lambda_1 + \lambda_k c(v_k)\}u_1 + b_{kk}\{v'_k - u_1\}$$

$$= b_{kk}v'_k + \{\lambda_1 + \lambda_k c(v_k) - b_{kk}\}u_1.$$

If $c(v_k) \neq 0$ there is a unique λ_k such that $\{\lambda_1 + \lambda_k c(v_k) - b_{kk}\} = 0$, namely, $\lambda_k = \frac{b_{kk} - \lambda_1}{c(v_k)}$. For this λ_k we have

$$T(v'_k) = b_{kk}v'_k,$$

so v'_k is an eigenvector belonging to the eigenvalue b_{kk}. Thus, all diagonal entries λ_1, $b_{22}, \ldots, b_{nn}$ are eigenvalues of T, and this completes the proof of the triangularization theorem.

6.7 Characteristic polynomials

We turn next to the problem of actually finding the eigenvalues of a transformation $T : V \to V$ when they exist. We seek those scalars λ such that the equation $T(x) = \lambda x$ has a solution x with $x \neq O$. The equation $T(x) = \lambda x$ can be written in the form

$$(6.9) \qquad\qquad (\lambda I - T)(x) = O,$$

where I is the identity transformation. Therefore, λ is an eigenvalue if and only if Equation (6.9) has a nonzero solution x, in which case, by Theorem 4.10, $\lambda I - T$ is not invertible.

If V is finite-dimensional we can determine the eigenvalues with the help of determinants. If A is a matrix representation for T, then $\lambda I - A$ is a matrix representation for $\lambda I - T$. Moreover, $\lambda I - T$ is invertible if and only if $\lambda I - A$ is nonsingular. Consequently, λ is an eigenvalue for T if and only if $\lambda I - A$ is singular, and, by Theorem 5.17, this occurs if and only if $\det(\lambda I - A) = 0$. In other words, every eigenvalue λ for T satisfies the equation

(6.10)
$$\det(\lambda I - A) = 0.$$

Conversely, any λ that satisfies (6.10) *and also lies in the underlying field of scalars* is an eigenvalue for T. This suggests that we study the determinant $\det(\lambda I - A)$ as a function of λ.

THEOREM 6.6. *If $A = (a_{ij})$ is an $n \times n$ matrix and if I is the $n \times n$ identity matrix, the function f defined by the equation*

$$f(\lambda) = \det(\lambda I - A)$$

is a polynomial in λ of degree n. Moreover, the term of highest degree is λ^n, and the constant term is $f(0) = \det(-A) = (-1)^n \det A$.

Proof. The statement $f(0) = \det(-A)$ follows at once from the definition of f. We prove that f is a polynomial of degree n only for $n \leq 3$. The proof for general n can be given by induction and is left as an exercise. (See Exercise 9 in Section 6.10.)

For $n = 1$ the determinant is the linear polynomial $f(\lambda) = \lambda - a_{11}$. For $n = 2$ we have

$$\det(\lambda I - A) = \begin{vmatrix} \lambda - a_{11} & -a_{12} \\ -a_{21} & \lambda - a_{22} \end{vmatrix} = (\lambda - a_{11})(\lambda - a_{22}) - a_{12}a_{21}$$

$$= \lambda^2 - (a_{11} + a_{22})\lambda + (a_{11}a_{22} - a_{12}a_{21}),$$

a quadratic polynomial in λ. For $n = 3$ we have

$$\det(\lambda I - A) = \begin{vmatrix} \lambda - a_{11} & -a_{12} & -a_{13} \\ -a_{21} & \lambda - a_{22} & -a_{23} \\ -a_{31} & -a_{32} & \lambda - a_{33} \end{vmatrix}$$

$$= (\lambda - a_{11}) \begin{vmatrix} \lambda - a_{22} & -a_{23} \\ -a_{32} & \lambda - a_{33} \end{vmatrix} + a_{12} \begin{vmatrix} -a_{21} & -a_{23} \\ -a_{31} & \lambda - a_{33} \end{vmatrix}$$

$$- a_{13} \begin{vmatrix} -a_{21} & \lambda - a_{22} \\ -a_{31} & -a_{32} \end{vmatrix}.$$

The last two terms are linear polynomials in λ. The first term is a cubic polynomial, the term of highest degree being λ^3.

DEFINITION.   *If A is an n × n matrix, the determinant*

$$f(\lambda) = \det(\lambda I - A)$$

is called the characteristic polynomial of A.

The roots of the characteristic polynomial of A are complex numbers, some of which may be real. If we let F denote either the real field $\mathbf{R}$ or the complex field $\mathbf{C}$, we have the following theorem:

THEOREM 6.7.   *Let T : V $\rightarrow$ V be a linear transformation, where* dim $V = n$ *and V has scalars in F. Let A be a matrix representation of T. Then the set of eigenvalues of T consists of those roots of the characteristic polynomial of A that lie in F.*

Proof.   The discussion preceding Theorem 6.6 shows that every eigenvalue of T satisfies the equation $\det(\lambda I - A) = 0$. Moreover, any root of the characteristic polynomial of A that lies in F is an eigenvalue of T.

Matrix A depends on the choice of basis for V, but the eigenvalues of T were defined without reference to a basis. Therefore, the *set* of roots of the characteristic polynomial of A must be independent of the choice of basis. More than this is true. In a later section we shall prove that the characteristic polynomial itself is independent of the choice of basis.

We turn now to the problem of actually calculating the eigenvalues and eigenvectors in the finite-dimensional case.

6.8   Calculation of eigenvalues and eigenvectors in the finite-dimensional case

In the finite-dimensional case the eigenvalues and eigenvectors of a linear transformation T are also called eigenvalues and eigenvectors of each matrix representation of T. Thus, the eigenvalues of a square matrix A are those roots of the characteristic polynomial $f(\lambda) = \det(\lambda I - A)$ that lie in the underlying field of scalars. The eigenvectors corresponding to an eigenvalue λ are those nonzero vectors $X = (x_1, \ldots, x_n)$, regarded as $n \times 1$ column matrices, that satisfy the matrix equation

$$AX = \lambda X, \quad \text{or} \quad (\lambda I - A)X = O.$$

This is a system of n linear equations for the components $x_1, \ldots, x_n$. Once we know λ we can obtain the eigenvectors by solving this system. We illustrate with three examples that exhibit different features.

EXAMPLE 1.   *A matrix with all its eigenvalues distinct.*   The matrix

$$A = \begin{bmatrix} 2 & 1 & 1 \\ 2 & 3 & 4 \\ -1 & -1 & -2 \end{bmatrix}$$

has the characteristic polynomial

$$\det(\lambda I - A) = \begin{vmatrix} \lambda - 2 & -1 & -1 \\ -2 & \lambda - 3 & -4 \\ 1 & 1 & \lambda + 2 \end{vmatrix} = (\lambda - 1)(\lambda + 1)(\lambda - 3),$$

so there are three distinct eigenvalues: $1, -1$, and 3. To find the eigenvectors corresponding to $\lambda = 1$ we solve the system $AX = X$, or

$$\begin{bmatrix} 2 & 1 & 1 \\ 2 & 3 & 4 \\ -1 & -1 & -2 \end{bmatrix} \begin{bmatrix} x_1 \\ x_2 \\ x_3 \end{bmatrix} = \begin{bmatrix} x_1 \\ x_2 \\ x_3 \end{bmatrix}.$$

This gives us three linear equations

$$2x_1 + x_2 + x_3 = x_1$$
$$2x_1 + 3x_2 + 4x_3 = x_2$$
$$-x_1 - x_2 - 2x_3 = x_3,$$

which can be rewritten as

$$x_1 + x_2 + x_3 = 0$$
$$2x_1 + 2x_2 + 4x_3 = 0$$
$$-x_1 - x_2 - 3x_3 = 0.$$

Add the first and third equations to get $x_3 = 0$; all three equations then reduce to $x_1 + x_2 = 0$. Thus, the eigenvectors corresponding to $\lambda = 1$ are $X = t(1, -1, 0)$, where t is any nonzero scalar.

By similar calculations we find the eigenvectors $X = t(0, 1, -1)$ corresponding to $\lambda = -1$, and $X = t(2, 3, -1)$ corresponding to $\lambda = 3$, with t any nonzero scalar. Because the eigenvalues are distinct the corresponding eigenvectors are independent. The results can be summarized in tabular form as follows. In the third column we have listed the dimension of the eigenspace $E(\lambda)$.

Eigenvalue λ	Eigenvectors	$\dim E(\lambda)$
1	$t(1, -1, 0), t \neq 0$	1
-1	$t(0, 1, -1), t \neq 0$	1
3	$t(2, 3, -1), t \neq 0$	1

EXAMPLE 2.   *A matrix with repeated eigenvalues.*   The matrix

$$A = \begin{bmatrix} 2 & -1 & 1 \\ 0 & 3 & -1 \\ 2 & 1 & 3 \end{bmatrix}$$

has the characteristic polynomial

$$\det(\lambda I - A) = \begin{vmatrix} \lambda - 2 & 1 & -1 \\ 0 & \lambda - 3 & 1 \\ -2 & -1 & \lambda - 3 \end{vmatrix} = (\lambda - 2)(\lambda - 2)(\lambda - 4).$$

The eigenvalues are 2, 2, and 4. (We list the eigenvalue 2 twice to emphasize that it is a double root of the characteristic polynomial.) To find the eigenvectors corresponding to $\lambda = 2$ we solve the system $AX = 2X$, which reduces to

$$-x_2 + x_3 = 0$$

$$x_2 - x_3 = 0$$

$$2x_1 + x_2 + x_3 = 0.$$

This has the solution $x_2 = x_3 = -x_1$ so the eigenvectors corresponding to $\lambda = 2$ are $t(-1, 1, 1)$, where $t \neq 0$. Similarly we find the eigenvectors $t(1, -1, 1)$ corresponding to the eigenvalue $\lambda = 4$. The results can be summarized as follows:

Eigenvalue λ	Eigenvectors	dim $E(\lambda)$
2, 2	$t(-1, 1, 1), t \neq 0$	1
4	$t(1, -1, 1), t \neq 0$	1

EXAMPLE 3.   *Another matrix with repeated eigenvalues.*   The matrix

$$A = \begin{bmatrix} 2 & 1 & 1 \\ 2 & 3 & 2 \\ 3 & 3 & 4 \end{bmatrix}$$

has the characteristic polynomial $(\lambda - 1)(\lambda - 1)(\lambda - 7)$. When $\lambda = 7$ the system $AX = 7X$ becomes

$$5x_1 - x_2 - x_3 = 0$$

$$-2x_1 + 4x_2 - 2x_3 = 0$$

$$-3x_1 - 3x_2 + 3x_3 = 0.$$

This has the solution $x_2 = 2x_1$, $x_3 = 3x_1$, so the eigenvectors corresponding to $\lambda = 7$ are $t(1, 2, 3)$, where $t \neq 0$. For the eigenvalue $\lambda = 1$, the system $AX = X$ consists of the equation $x_1 + x_2 + x_3 = 0$ repeated three times. To solve this equation we may take $x_1 = a$, $x_2 = b$, where a and b are arbitrary, and then take $x_3 = -a - b$. Thus every eigenvector corresponding to $\lambda = 1$ has the form

$$(a, b, -a - b) = a(1, 0, -1) + b(0, 1, -1),$$

where $a \neq 0$ or $b \neq 0$. This means that the independent vectors $(1, 0, -1)$ and $(0, 1, -1)$ form a basis for $E(1)$. Hence dim $E(\lambda) = 2$ when $\lambda = 1$. The results can be summarized as follows:

Eigenvalue λ	Eigenvectors	dim $E(\lambda)$
7	$t(1, 2, 3), t \neq 0$	1
1, 1	$a(1, 0, -1) + b(0, 1, -1), a, b$ not both 0	2

Note. In this example there are three independent eigenvectors but only two distinct eigenvalues.

6.9 The product and sum of the roots of a characteristic polynomial

Let $f(\lambda) = \det(\lambda I - A)$ be the characteristic polynomial of an $n \times n$ matrix $A = (a_{ij})$. We denote the n roots of $f(\lambda)$ by $\lambda_1, \ldots, \lambda_n$, with each root written as often as its multiplicity indicates. (Double roots are written twice, triple roots three times, etc.) Then we have the factorization

$$f(\lambda) = (\lambda - \lambda_1) \cdots (\lambda - \lambda_n).$$

We can also write the polynomial $f(\lambda)$ in decreasing powers of λ as follows,

$$f(\lambda) = \lambda^n + c_{n-1}\lambda^{n-1} + \cdots + c_1\lambda + c_0.$$

Comparing this with the factored form we find that the constant term c_0 and the coefficient of λ^{n-1} are given by the formulas

$$c_0 = (-1)^n \lambda_1 \cdots \lambda_n \quad \text{and} \quad c_{n-1} = -(\lambda_1 + \cdots + \lambda_n).$$

Since we also have $c_0 = f(0) = (-1)^n \det A$, we see that

$$\lambda_1 \cdots \lambda_n = \det A.$$

In other words:

The product of the roots of the characteristic polynomial of A is equal to the determinant of A.

In this product, repeated roots (eigenvalues) must appear as repeated factors. For example, matrix A of Example 2 in Section 6.8 has eigenvalues 2, 2, 4. The product of the eigenvalues (with multiplicities taken into account) is $2 \times 2 \times 4 = 16$, which agrees with the fact that $\det A = 16$.

The *sum* of the roots of $f(\lambda)$ is called the *trace of A*, denoted by tr A. Thus, by definition,

$$\text{tr } A = \sum_{i=1}^{n} \lambda_i,$$

where again repeated roots are written as often as their multiplicity indicates. The coefficient of λ^{n-1} is given by $c_{n-1} = -\text{tr } A$. We can also compute this coefficient from the determinant form for $f(\lambda)$ and we find that

$$c_{n-1} = -(a_{11} + \cdots + a_{nn}).$$

(A proof of this formula is requested in Exercise 12 of Section 6.10.) The two formulas for c_{n-1} show that

$$\operatorname{tr} A = \sum_{i=1}^{n} a_{ii}.$$

In other words, *the trace of a matrix is also equal to the sum of its diagonal entries.*

Since the sum of the diagonal entries of a matrix is easy to compute it can be used as a numerical check in calculations of eigenvalues. Further properties of the trace are described in the next set of exercises.

6.10   Exercises

Determine the eigenvalues and eigenvectors of each of the matrices in Exercises 1 through 3. Also, for each eigenvalue λ compute the dimension of the eigenspace $E(\lambda)$.

1. (a) $\begin{bmatrix} 1 & 0 \\ 0 & 1 \end{bmatrix}$. (b) $\begin{bmatrix} 1 & 1 \\ 0 & 1 \end{bmatrix}$. (c) $\begin{bmatrix} 1 & 0 \\ 1 & 1 \end{bmatrix}$. (d) $\begin{bmatrix} 1 & 1 \\ 1 & 1 \end{bmatrix}$.

2. $\begin{bmatrix} 1 & a \\ b & 1 \end{bmatrix}$, $a > 0, b > 0$.

3. $\begin{bmatrix} \cos \theta & -\sin \theta \\ \sin \theta & \cos \theta \end{bmatrix}$.

4. The three matrices $P_1 = \begin{bmatrix} 0 & 1 \\ 1 & 0 \end{bmatrix}$, $P_2 = \begin{bmatrix} 0 & -i \\ i & 0 \end{bmatrix}$, $P_3 = \begin{bmatrix} 1 & 0 \\ 0 & -1 \end{bmatrix}$ occur in the quantum mechanical theory of electon spin and are called *Pauli spin matrices,* in honor of the physicist Wolfgang Pauli (1900–1958). Verify that they all have eigenvalues 1 and -1. Then determine all 2×2 matrices with complex entries having the two eigenvalues 1 and -1.

5. Determine all 2×2 matrices with real entries whose eigenvalues are: (a) real and distinct, (b) real and equal, (c) complex conjugates.

6. Determine a, b, c, d, e, f, given that the vectors $(1, 1, 1), (1, 0, -1)$, and $(1, -1, 0)$ are eigenvectors of the matrix

$$\begin{bmatrix} 1 & 1 & 1 \\ a & b & c \\ d & e & f \end{bmatrix}.$$

7. Calculate the eigenvalues and eigenvectors of each of the following matrices. Also, compute the dimension of the eigenspace $E(\lambda)$ for each eigenvalue λ.

(a) $\begin{bmatrix} 1 & 0 & 0 \\ -3 & 1 & 0 \\ 4 & -7 & 1 \end{bmatrix}$. (b) $\begin{bmatrix} 2 & 1 & 3 \\ 1 & 2 & 3 \\ 3 & 3 & 20 \end{bmatrix}$. (c) $\begin{bmatrix} 5 & -6 & -6 \\ -1 & 4 & 2 \\ 3 & -6 & -4 \end{bmatrix}$.

8. Calculate the eigenvalues of each of the following five matrices.

(a) $\begin{bmatrix} 0 & 0 & 1 & 0 \\ 0 & 0 & 0 & 1 \\ 1 & 0 & 0 & 0 \\ 0 & 1 & 0 & 0 \end{bmatrix}$. (c) $\begin{bmatrix} 0 & 1 & 0 & 0 \\ 1 & 0 & 0 & 0 \\ 0 & 0 & 0 & 1 \\ 0 & 0 & 1 & 0 \end{bmatrix}$. (e) $\begin{bmatrix} 1 & 0 & 0 & 0 \\ 0 & -1 & 0 & 0 \\ 0 & 0 & 1 & 0 \\ 0 & 0 & 0 & -1 \end{bmatrix}$.

(b) $\begin{bmatrix} 1 & 0 & 0 & 0 \\ 0 & 1 & 0 & 0 \\ 0 & 0 & -1 & 0 \\ 0 & 0 & 0 & -1 \end{bmatrix}$. (d) $\begin{bmatrix} 0 & -i & 0 & 0 \\ i & 0 & 0 & 0 \\ 0 & 0 & 0 & -i \\ 0 & 0 & i & 0 \end{bmatrix}$.

These are called *Dirac matrices* in honor of Paul M. Dirac (1902–1984), the English physicist. They occur in the solution of the relativistic wave equation in quantum mechanics.

9. If A and B are $n \times n$ matrices, with B a diagonal matrix, prove (by induction) that the determinant $f(\lambda) = \det(\lambda B - A)$ is a polynomial in λ with $f(0) = (-1)^n \det A$, and with the coefficient of λ^n equal to the product of the diagonal entries of B.

10. Prove that a square matrix A and its transpose A^t have the same characteristic polynomial.

11. If A and B are $n \times n$ matrices, with A nonsingular, prove that AB and BA have the same set of eigenvalues. *Note:* It can be shown that AB and BA have the same characteristic polynomial, even if A is singular, but you are not required to prove this.

12. Let A be an $n \times n$ matrix with characteristic polynomial $f(\lambda)$. Prove (by induction) that the coefficient of λ^{n-1} in $f(\lambda)$ is $- \operatorname{tr} A$.

13. Let A and B be $n \times n$ matrices with $\det A = \det B$ and $\operatorname{tr} A = \operatorname{tr} B$. Prove that A and B have the same characteristic polynomial if $n = 2$ but that this need not be the case if $n > 2$.

14. Prove each of the following statements about the trace of a matrix.
    (a) $\operatorname{tr}(A + B) = \operatorname{tr} A + \operatorname{tr} B$.   (c) $\operatorname{tr}(AB) = \operatorname{tr}(BA)$.
    (b) $\operatorname{tr}(cA) = c \operatorname{tr} A$.   (d) $\operatorname{tr} A^t = \operatorname{tr} A$.

6.11 Matrices representing the same linear transformation. Similar matrices

In this section we show that two different matrix representations of a linear transformation have the same characteristic polynomial. To do this, we investigate more closely the relation between matrices that represent the same transformation.

Let us recall how matrix representations are defined. Suppose $T : V \to W$ is a linear mapping of an n-dimensional space V into an m-dimensional space W. Let $(v_1, \ldots, v_n)$ and $(w_1, \ldots, w_m)$ be ordered bases for V and W, respectively. The matrix representation of T relative to this choice of bases is the $m \times n$ matrix whose columns consist of the components of $T(v_1), \ldots, T(v_n)$ relative to the basis $(w_1, \ldots, w_m)$. Different matrix representations arise from different choice of the bases.

Now consider the case in which the two linear spaces V and W are equal, and assume that the same ordered basis $(v_1, \ldots, v_n)$ is used for each copy of this space. If $A = (a_{ij})$ is the matrix of T relative to this basis, we have

$$(6.11) \qquad T(v_k) = \sum_{i=1}^{n} a_{ik} v_i \qquad \text{for} \qquad k = 1, 2, \ldots, n.$$

Now choose another ordered basis $(u_1, \ldots, u_n)$ for the same space, and let $B = (b_{ij})$ be the matrix of T relative to this new basis. We will show that the two matrices A and B are related by a simple matrix equation

$$(6.12) \qquad B = C^{-1}AC,$$

where C is a nonsingular matrix.

To determine C we observe that the definition of B implies that

$$(6.13) \qquad T(u_j) = \sum_{k=1}^{n} b_{kj} u_k \qquad \text{for} \qquad j = 1, 2, \ldots, n.$$

But each element u_j is in the space spanned by $v_1, \ldots, v_n$ so it can be written as a linear combination of these elements, say

$$(6.14) \qquad u_j = \sum_{k=1}^{n} c_{kj} v_k \qquad \text{for} \qquad j = 1, 2, \ldots, n,$$

for some set of scalars c_{kj}. The $n \times n$ matrix $C = (c_{kj})$ determined by these scalars is called the *transition matrix*. It transforms the basis $v_1, \ldots, v_n$ into the basis $u_1, \ldots, u_n$. Matrix C is nonsingular because it represents a linear transformation that maps one basis onto another basis, so the inverse C^{-1} exists. We will show that Eq. (6.12) holds for this C.

Equation (6.14), which relates the two bases, can be written as a simple matrix equation,

$$(6.15) \qquad\qquad U = VC,$$

where U and V are $1 \times n$ row matrices whose entries are not scalars but are the basis elements in question:

$$U = [u_1, \ldots, u_n] \qquad \text{and} \qquad V = [v_1, \ldots, v_n].$$

The matrix product in (6.15) is equivalent to the n equations in (6.14). Multiplying each member of (6.15) on the right by C^{-1} we find

$$(6.16) \qquad\qquad V = UC^{-1}.$$

Equation. (6.11) can also be written as a matrix equation,

$$(6.17) \qquad\qquad V' = VA,$$

where $V' = [T(v_1), \ldots, T(v_n)]$. Similarly, Eq. (6.13) can be written as a matrix equation,

$$(6.18) \qquad\qquad U' = UB,$$

where $U' = [T(u_1), \ldots, T(u_n)]$. Now if we apply T to each member of Eq. (6.14) we find

$$T(u_j) = \sum_{k=1}^{n} c_{kj} T(v_k) \qquad \text{for } j = 1, 2, \ldots, n,$$

and this, too, can be written as a matrix equation,

$$(6.19) \qquad\qquad U' = V'C.$$

Using (6.17) in (6.19) we find $U' = VAC$, which, in view of (6.16), gives us $U' = UC^{-1}AC$. Comparing this with (6.18) we see that

$$UB = UC^{-1}AC.$$

But each entry in this matrix equation is a linear combination of the basis elements $u_1, \ldots, u_n$. Because these elements are independent we must have $B = C^{-1}AC$, which is Eq. (6.12). Thus, we have proved the following theorem.

THEOREM 6.8. *If two $n \times n$ matrices A and B represent the same linear transformation T, then there is a nonsingular matrix C such that*

$$B = C^{-1}AC.$$

Moreover, if A is the matrix of T relative to a basis V = $[v_1, \ldots, v_n]$ *and if B is the matrix of T relative to a basis U* = $[u_1, \ldots, u_n]$, *then for C we can take the nonsingular transition matrix relating the two bases according to the equation U* = *VC.*

The converse of Theorem 6.8 is also true.

THEOREM 6.9.   *Let A and B be two n* × *n matrices related by an equation of the form B* = $C^{-1}AC$, *where C is a nonsingular n* × *n matrix. Then A and B represent the same linear transformation.*

Proof.   Choose a basis *V* = $[v_1, \ldots, v_n]$ for the underlying linear space, and let *U* = $[u_1, \ldots, u_n]$ be the elements determined by the equations

$$(6.20) \qquad u_j = \sum_{k=1}^{n} c_{kj} v_k \qquad \text{for} \qquad j = 1, 2, \ldots, n,$$

where the scalar multipliers c_{kj} are the entries of *C*. Because *C* is nonsingular it represents an invertible linear transformation, so *U* = $[u_1, \ldots, u_n]$ is also a basis. The relation between the two bases *V* and *U* can also be expressed more simply in matrix form. First, Equation (6.20) can be written as

$$U = VC,$$

from which we find

$$V = UC^{-1}.$$

Because *B* = $C^{-1}AC$ we also have the equation

$$(6.21) \qquad UB = UC^{-1}AC = VAC,$$

which will help us show that *A* and *B* represent the same linear transformation.

Let *T* be the linear transformation having matrix representation *A* relative to basis *V*, and let *S* be the linear transformation having matrix representation *B* relative to basis *U*. Then we have

$$(6.22) \qquad T(v_k) = \sum_{i=1}^{n} a_{ik} v_i \qquad \text{for} \qquad k = 1, 2, \ldots, n,$$

and

$$(6.23) \qquad S(u_j) = \sum_{k=1}^{n} b_{kj} u_k \qquad \text{for} \qquad j = 1, 2, \ldots, n.$$

We shall prove that *S* = *T* by showing that $S(u_j) = T(u_j)$ for each basis element u_j. To do this, we write Equations (6.22) and (6.23) more simply in matrix form as follows:

$$[T(v_1), \ldots, T(v_n)] = VA \qquad \text{and} \qquad [S(u_1), \ldots, S(u_n)] = UB.$$

Applying T to Eq. (6.20) we also obtain the relation $T(u_j) = \sum_{k=1}^{n} c_{kj} T(v_k)$, which, in matrix form, states that

$$[T(u_1), \ldots, T(u_n)] = [T(v_1), \ldots, T(v_n)]C = VAC.$$

But by (6.21) we have $UB = VAC$, hence,

$$[S(u_1), \ldots, S(u_n)] = [T(u_1), \ldots, T(u_n)].$$

Equating entries, we see that $S(u_j) = T(u_j)$ for each basis element u_j. Therefore $S = T$, so the two matrices A and B represent the same linear transformation.

DEFINITION. *Two $n \times n$ matrices A and B are called similar if there is a nonsingular matrix C such that $B = C^{-1}AC$.*

Theorems 6.8 and 6.9 can be combined to give us

THEOREM 6.10. *Two $n \times n$ matrices are similar if and only if they represent the same linear transformation.*

Similar matrices share many properties. For example, *they have the same determinant* because

$$\det(C^{-1}AC) = \det(C^{-1})(\det A)(\det C) = \det A.$$

This property gives us the following theorem.

THEOREM 6.11. *Similar matrices have the same characteristic polynomial and therefore the same eigenvalues.*

Proof. If A and B are similar there is a nonsingular matrix C such that $B = C^{-1}AC$. Therefore

$$\lambda I - B = \lambda I - C^{-1}AC = \lambda C^{-1}IC - C^{-1}AC = C^{-1}(\lambda I - A)C.$$

This shows that $\lambda I - B$ and $\lambda I - A$ are similar, so they have the same determinant,

$$\det(\lambda I - B) = \det(\lambda I - A).$$

Theorems 6.10 and 6.11 together show that all matrix representations of a given linear transformation T have the same characteristic polynomial. *This polynomial is also called the characteristic polynomial of T.*

The next theorem is a combination and restatement of Theorems 6.7, 6.2, and 6.8. In this theorem, F denotes either the real field **R** or the complex field **C**.

THEOREM 6.12. *Let T be a linear transformation that maps an n-dimensional linear space into itself, and let F denote the field of scalars. Assume the characteristic polynomial of T has n distinct roots $\lambda_1, \ldots, \lambda_n$ lying in F. Then we have:*

(a) *The corresponding eigenvectors $u_1, \ldots, u_n$ form a basis for the underlying linear space.*

(b) *The matrix of T relative to the ordered basis $U = [u_1, \ldots, u_n]$ is the diagonal matrix Λ having the eigenvalues as diagonal entries:*

$$\Lambda = \text{diag}(\lambda_1, \ldots, \lambda_n).$$

(c) *If A is the matrix of T relative to another basis $V = [v_1, \ldots, v_n]$, then $\Lambda = C^{-1}AC$, where C is the nonsingular transition matrix relating the two bases by the equation*

$$U = VC.$$

Proof. By Theorem 6.7, each λ_i is an eigenvalue. Since there are n distinct roots, Theorem 6.2 tells us that the corresponding eigenvectors $u_1, \ldots, u_n$ are independent and hence form a basis. This proves (a). Because $T(u_i) = \lambda_i u_i$, the matrix of T relative to U is the diagonal matrix Λ, which proves (b). To prove (c) we use Theorem 6.8.

Note. The nonsingular transition matrix C in Theorem 6.12 is also called the *diagonalizing matrix*. If V is the basis of unit coordinate vectors $(I_1, \ldots, I_n)$, then the equation $U = VC$ in Theorem 6.12 shows that the kth column of C consists of the components of the eigenvector u_k relative to $(I_1, \ldots, I_n)$.

If the eigenvalues of A are distinct then A is similar to a diagonal matrix. If the eigenvalues are not distinct, then A still might be similar to a diagonal matrix. This will happen if and only if there are k independent eigenvectors corresponding to each eigenvalue of multiplicity k. Examples occur in the next set of exercises.

6.12 Exercises

1. Prove that the matrices $\begin{bmatrix} 1 & 1 \\ 0 & 1 \end{bmatrix}$ and $\begin{bmatrix} 1 & 0 \\ 0 & 1 \end{bmatrix}$ have the same eigenvalues but are not similar.

2. In each case find a nonsingular matrix C such that $C^{-1}AC$ is a diagonal matrix or explain why no such C exists.

   (a) $A = \begin{bmatrix} 1 & 0 \\ 1 & 3 \end{bmatrix}$.   (b) $A = \begin{bmatrix} 1 & 2 \\ 5 & 4 \end{bmatrix}$.   (c) $A = \begin{bmatrix} 2 & 1 \\ -1 & 4 \end{bmatrix}$.   (d) $A = \begin{bmatrix} 2 & 1 \\ -1 & 0 \end{bmatrix}$.

3. Three bases in the plane are given. With respect to these bases a point has components (x_1, x_2), (y_1, y_2), and (z_1, z_2), respectively. Suppose that $[y_1, y_2] = [x_1, x_2]A$, $[z_1, z_2] = [x_1, x_2]B$, and $[z_1, z_2] = [y_1, y_2]C$, where A, B, C are 2×2 matrices. Express C in terms of A and B.

4. In each case, show that the eigenvalues of A are not distinct but that A has three independent eigenvectors. Find a nonsingular matrix C such that $C^{-1}AC$ is a diagonal matrix.

   (a) $A = \begin{bmatrix} 0 & 0 & 1 \\ 0 & 1 & 0 \\ 1 & 0 & 0 \end{bmatrix}$.   (b) $A = \begin{bmatrix} 1 & -1 & -1 \\ 1 & 3 & 1 \\ -1 & -1 & 1 \end{bmatrix}$.

5. Show that none of the following matrices is similar to a diagonal matrix, but that each is similar to a triangular matrix of the form $\begin{bmatrix} \lambda & 0 \\ 1 & \lambda \end{bmatrix}$, where λ is an eigenvalue.

   (a) $\begin{bmatrix} 2 & -1 \\ 0 & 2 \end{bmatrix}$.   (b) $\begin{bmatrix} 2 & 1 \\ -1 & 4 \end{bmatrix}$.

6. Determine the eigenvalues and eigenvectors of the matrix $\begin{bmatrix} 0 & -1 & 0 \\ 0 & 0 & 1 \\ -1 & -3 & 3 \end{bmatrix}$ and thereby show that it is not similar to a diagonal matrix.

7. (a) Prove that a square matrix A is nonsingular if and only if 0 is not an eigenvalue of A.
   (b) If A is nonsingular, prove that the eigenvalues of A^{-1} are the reciprocals of the eigenvalues of A.
8. Given an $n \times n$ matrix A with real entries such that $A^2 = -I$. Prove the following:
   (a) A is nonsingular.   (c) A has no real eigenvalues.
   (b) n is even.   (d) $\det A = 1$.
9. For each of the following, give a proof if you think the statement is always true, or else explain (by a counter example or otherwise) why it is sometimes false. Unless otherwise noted, A, B, and C are square matrices with complex entries.
   (a) If A can be diagonalized, then its eigenvalues are distinct.
   (b) If all the eigenvalues of A are equal to 2, then $B^{-1}AB = 2I$ for some nonsingular B.
   (c) If all the eigenvalues of A are zero, then A is the zero matrix.
   (d) If all the eigenvalues of A are zero, then A is similar to the zero matrix.
   (e) If $B = C^{-1}AC$ for some nonsingular C, then $B^3 = C^{-1}A^3C$ for the same matrix C.
   (f) If A is nonsingular, then all its eigenvalues are nonzero.
   (g) There exist 3×3 real matrices with no real eigenvalue.
   (h) The matrix $A = \begin{bmatrix} 1 & 1 & 0 \\ 0 & 1 & 1 \\ 0 & 0 & 1 \end{bmatrix}$ is similar to a diagonal matrix.

6.13 The Cayley-Hamilton theorem

The Cayley-Hamilton theorem states that every square matrix is a root of the characteristic polynomial satisfied by its eigenvalues. This was first proved by William Rowan Hamilton in 1853 for a special class of matrices. A few years later Arthur Cayley (1821–1895) announced that the theorem is true for all matrices, but gave no proof. Many different proofs of this theorem are known. We will state the theorem in two equivalent forms, one for linear operators and another for matrices. The proof we give for operators gives insight into why the theorem is valid.

THEOREM 6.13. CAYLEY-HAMILTON THEOREM FOR LINEAR OPERATORS. *Given a linear transformation $T : V \to V$, where dim $V = n$ and the scalars are complex. Let $\lambda_1, \ldots, \lambda_n$ be the eigenvalues of T, and let*

$$(6.24) \qquad f_T(\lambda) = (\lambda - \lambda_1) \cdots (\lambda - \lambda_n)$$

be the characteristic polynomial of T. Let $g(T)$ denote the constant-coefficient operator obtained from $f_T(\lambda)$ by replacing λ by T and by replacing each λ_k by $\lambda_k I$. That is, let

$$(6.25) \qquad g(T) = (T - \lambda_1 I) \cdots (T - \lambda_n I).$$

Then $g(T)$ is the zero operator; that is, $g(T)$ annihilates every element of V:

$$g(T)(x) = O \qquad \text{for all } x \text{ in } V.$$

Note. Each eigenvalue λ_k satisfies the polynomial equation $f_T(\lambda) = 0$. The Cayley-Hamilton Theorem says that T satisfies the operator equation $f_T(T) = O$.

Proof. Because the scalars are complex, Theorem 6.5 tells us that we can choose a basis for V such that the matrix of T relative to this basis is upper triangular. Denote this upper triangular matrix by U. Its diagonal entries are the eigenvalues of T, so U has the form

$$U = \begin{bmatrix} \lambda_1 & u_{12} & \cdots & u_{1n} \\ 0 & \lambda_2 & \cdots & u_{2n} \\ 0 & 0 & \ddots & \vdots \\ 0 & 0 & \cdots & \lambda_n \end{bmatrix}.$$

The matrix of operator $g(T)$ in (6.25) is

(6.26) $$(U - \lambda_1 I) \cdots (U - \lambda_n I).$$

We will show that this product annihilates every column vector $X = [x_1, \ldots, x_n]'$ by observing the effect of each factor.

The matrix $U - \lambda_n I$ appearing in the last factor has the following form:

$$U - \lambda_n I = \begin{bmatrix} \lambda_1 - \lambda_n & u_{12} & \cdots & u_{1n} \\ 0 & \lambda_2 - \lambda_n & \cdots & u_{2n} \\ 0 & 0 & \ddots & \vdots \\ 0 & 0 & \cdots & 0 \end{bmatrix},$$

with the last row consisting entirely of zeros. The product of this matrix with any column vector $X = [x_1, \ldots, x_n]'$ is another column vector with last entry equal to 0. That is,

$$(U - \lambda_n I)X = (U - \lambda_n I) \begin{bmatrix} x_1 \\ x_2 \\ \vdots \\ x_{n-1} \\ x_n \end{bmatrix} = \begin{bmatrix} * \\ * \\ \vdots \\ * \\ 0 \end{bmatrix},$$

where each $*$ indicates a place holder for the corresponding entry. The next to last factor $U - \lambda_{n-1}I$ in (6.26) has the form

$$U - \lambda_{n-1}I = \begin{bmatrix} \lambda_1 - \lambda_{n-1} & u_{12} & \cdots & u_{1n} \\ 0 & \lambda_2 - \lambda_{n-1} & \cdots & u_{2}n \\ \vdots & \vdots & \ddots & \vdots \\ 0 & \cdots & 0 & * \\ 0 & \cdots & 0 & \lambda_n - \lambda_{n-1} \end{bmatrix}$$

so the product of this matrix with any column vector of the form $[*, *, \ldots, *, 0]'$ is another column vector $[*, *, \ldots, *, 0, 0]'$ with zeros in the last two entries. In other words,

$$(U - \lambda_{n-1}I)(U - \lambda_n I)X = \begin{bmatrix} * \\ * \\ \vdots \\ * \\ 0 \\ 0 \end{bmatrix}.$$

Continuing in this manner we find that each new factor $U - \lambda_k I$ introduces a new zero in the resulting column vector, so the product of all the factors in (6.26) annihilates X. This proves the Cayley-Hamilton theorem for linear operators, and it also shows how each factor $U - \lambda_k I$ contributes to the annihilation of X.

Theorem 6.13 expresses the Cayley-Hamilton theorem in terms of operators. It can also be stated in an equivalent form in terms of matrices.

COROLLARY 6.14. CAYLEY-HAMILTON THEOREM FOR MATRICES. *Let A be an $n \times n$ matrix and let*

(6.27) $$f(\lambda) = \det(\lambda I - A) = \lambda^n + c_{n-1}\lambda^{n-1} + \cdots + c_1\lambda + c_0$$

be its characteristic polynomial. Then $f(A) = O$. In other words, A satisfies the equation

(6.28) $$A^n + c_{n-1}A^{n-1} + \cdots + c_1 A + c_0 I = O.$$

EXAMPLE 1. The 2×2 matrix $A = \begin{bmatrix} 2 & 2 \\ 2 & 5 \end{bmatrix}$ has the characteristic polynomial

$$\det(\lambda I - A) = \lambda^2 - 7\lambda + 6.$$

The Cayley-Hamilton theorem states that $A^2 - 7A + 6I = O$, from which we find

(6.29) $$A^2 = 7A - 6I.$$

Many consequences flow from this equation. First, we can use it to calculate all higher powers of A. For example,

$$A^3 = AA^2 = A(7A - 6I) = 7A^2 - 6A = 7(7A - 6I) - 6A = 43A - 42I,$$
$$A^4 = AA^3 = A(43A - 42I) = 43A^2 - 42A = 43(7A - 6I) - 42A = 259A + 258I.$$

We can also use it to calculate A^{-1}. If we rewrite (6.29) in the form $A(A - 7I) = -6I$ and multiply by the scalar factor $-\frac{1}{6}$ we find that $A\{\frac{1}{6}(7I - A)\} = I$, so $A^{-1} = \frac{1}{6}(7I - A)$.

EXAMPLE 2. The 3×3 matrix $A = \begin{bmatrix} 5 & 4 & 0 \\ 1 & 2 & 0 \\ 1 & 2 & 2 \end{bmatrix}$ has the characteristic polynomial

$$\det(\lambda I - A) = \lambda^3 - 9\lambda^2 + 20\lambda - 12 = (\lambda - 1)(\lambda - 2)(\lambda - 6),$$

so A satisfies the equation

(6.30) $$A^3 - 9A^2 + 20A - 12I = O.$$

This equation can be used to express A^3 and all higher powers of A in terms of I, A, and A^2. For example, we have

$$A^3 = 9A^2 - 20A + 12I.$$

$$A^4 = 9A^3 - 20A^2 + 12A = 9(9A^2 - 20A + 12I) - 20A^2 + 12A$$

$$= 61A^2 - 168A + 108I.$$

It can also be used to express A^{-1} as a polynomial in A. From (6.30) we find

$$A(A^2 - 9A + 20I) = 12I \qquad \text{and hence} \qquad A^{-1} = \frac{1}{12}(A^2 - 9A + 20I).$$

6.14   Exercises

In each of Exercises 1 through 4, express A^{-1}, A^2 and all higher powers of A as a linear combination of I and A.

1. $A = \begin{bmatrix} 1 & 0 \\ 1 & 1 \end{bmatrix}$.   2. $A = \begin{bmatrix} 1 & 0 \\ 1 & 2 \end{bmatrix}$.   3. $A = \begin{bmatrix} 0 & 1 \\ 1 & 0 \end{bmatrix}$.   4. $A = \begin{bmatrix} -1 & 0 \\ 0 & 1 \end{bmatrix}$.

In each of Exercises 5 through 7, calculate A^n and express A^3 in terms of I, A, and A^2.

5. $A = \begin{bmatrix} 0 & 1 & 1 \\ 0 & 0 & 1 \\ 0 & 0 & 0 \end{bmatrix}$.   6. $A = \begin{bmatrix} 0 & 1 & 1 \\ 0 & 1 & 1 \\ 0 & 0 & 0 \end{bmatrix}$.   7. $A = \begin{bmatrix} 2 & 0 & 0 \\ 0 & 1 & 0 \\ 0 & 1 & 1 \end{bmatrix}$.

6.15   The Jordan normal form

Theorem 6.5 shows that every linear transformation of a finite-dimensional linear space into itself can be triangularized if the scalars are complex. This section shows that there is a triangular matrix representation that can be decomposed into special triangular matrices called *Jordan blocks*. Here are some typical examples of Jordan blocks of orders 2, 3, and 4:

$$J = \begin{bmatrix} \lambda & 1 \\ 0 & \lambda \end{bmatrix}, \qquad J = \begin{bmatrix} \mu & 1 & 0 \\ 0 & \mu & 1 \\ 0 & 0 & \mu \end{bmatrix}, \qquad J = \begin{bmatrix} \nu & 1 & 0 & 0 \\ 0 & \nu & 1 & 0 \\ 0 & 0 & \nu & 1 \\ 0 & 0 & 0 & \nu \end{bmatrix}.$$

DEFINITION.   *A Jordan block is any upper triangular square matrix $[a_{ij}]$ with the following properties: the diagonal elements a_{ii} are equal; each superdiagonal entry $a_{i,i+1}$ is 1; all other entries are zero. Also, every 1×1 matrix is called a Jordan block.*

The next theorem describes all the eigenvalues and eigenvectors of a Jordan block.

THEOREM 6.15.   *Let J be a Jordan block of order k. Then J has exactly one eigenvalue, which is equal to the scalar on the main diagonal. The corresponding eigenvectors are the nonzero scalar multiples of the k-dimensional unit coordinate vector $[1, 0, \ldots, 0]$.*

Proof.   Suppose the diagonal entries of J are equal to λ. A column vector $X = [x_1, x_2, \ldots, x_k]^t$ satisfies the equation $JX = \lambda X$ if and only if its components satisfy the following k scalar equations:

$$\lambda x_1 + x_2 = \lambda x_1$$

$$\lambda x_2 + x_3 = \lambda x_2$$

$$\vdots$$

$$\lambda x_{k-1} + x_k = \lambda x_{k-1}$$

$$\lambda x_k = \lambda x_k.$$

Moving down through the first $k - 1$ equations in this list we find

$$x_2 = x_3 = \cdots = x_k = 0,$$

so λ is an eigenvalue for J and all eigenvectors have the form $x_1[1, 0, \ldots, 0]$ with $x_1 \neq 0$.

To show that λ is the *only* eigenvalue for J, assume that $JX = \mu X$ for some scalar $\mu \neq \lambda$. Then the components satisfy the following k scalar equations:

$$\lambda x_1 + x_2 = \mu x_1$$

$$\lambda x_2 + x_3 = \mu x_2$$

$$\vdots$$

$$\lambda x_{k-1} + x_k = \mu x_{k-1}$$

$$\lambda x_k = \mu x_k.$$

Because $\mu \neq \lambda$ the last equation implies $x_k = 0$ and, moving up through the remaining equations, we find $x_{k-1} = \cdots = x_2 = x_1 = 0$. Hence only the zero vector satisfies $JX = \mu X$, so no scalar different from λ can be an eigenvalue for J.

The importance of Jordan blocks is described in the next theorem.

THEOREM 6.16.   JORDAN NORMAL FORM.   *Let V be an n-dimensional linear space with complex scalars, and let $T : V \to V$ be a linear transformation of V into itself. Then there is a basis for V relative to which T has a block-diagonal matrix representation* diag$(J_1, \ldots, J_m)$, *with each J_k being a Jordan block.*

Proof.   The proof is by induction on the dimension n. The result is trivial if $n = 1$, so we assume it holds for all linear transformations acting on linear spaces of dimension $< n$ and prove it also holds for spaces of dimension n.

Let λ be a fixed eigenvalue of T (known to exist by Theorem 6.4), and let $S = T - \lambda I$. If we show that S has a matrix representation in Jordan form then $T = S + \lambda I$ will also.

Because S maps an eigenvector of T onto O, S has nullity $\nu \geq 1$ and rank $r = n - \nu < n$, where $\nu = \dim N(S)$ is the dimension of the null space of S, and r is the dimension of the range $S(V)$.

The plan of the proof is to apply the induction hypothesis to S', the restriction of S to the subspace $S(V)$. This gives a basis E for $S(V)$, relative to which S' has a matrix representation in Jordan form. Then we show that E can be extended to a basis E' for V, relative to which S has a matrix representation in Jordan form.

Because S' maps $S(V)$ into itself, there is a basis E of $S(V)$ consisting of r elements, relative to which S' has a matrix in Jordan form, say

$$m_E(S') = \text{diag}(J_1, \ldots, J_m),$$

where each J_i is a Jordan block of order n_i and the sum of the orders $n_1 + \cdots + n_m = r$.

Denote the elements of E by $e_{i,j}$ and arrange them into m disjoint ordered sets E_i corresponding to the m Jordan blocks, say,

$$E_i = \{e_{i,1}, e_{i,2}, \ldots, e_{i,n_i}\} \qquad \text{for} \qquad i = 1, 2, \ldots, m.$$

If the order n_i of block J_i is 1 with single entry λ_i, then we have

$$(6.31) \qquad S(e_{i,1}) = \lambda_i e_{i,1}.$$

If the order $n_i > 1$ with diagonal entries λ_i then Eq. (6.31) holds for $e_{i,1}$; for the remaining elements we have

$$(6.32) \qquad S(e_{i,j}) = e_{i,j-1} + \lambda_i e_{i,j} \qquad \text{for} \qquad j = 2, \ldots, n_i.$$

Some of the λ_i might be zero. If so, we reorder the basis elements of E if necessary so they correspond to the blocks $J_1, \ldots, J_q$. Then J_i has rank $n_i - 1$ if $i \le q$ and rank n_i if $i > q$. The rank of $m_E(S')$ is the sum of the ranks of the J_i so

$$\text{rank } S' = \text{rank } m_E(S') = \sum_{i=1}^{q}(n_i - 1) + \sum_{i=q+1}^{m} n_i = \sum_{i=1}^{m} n_i - q = r - q.$$

Therefore $q = r - \text{rank } S' = $ nullity of S'. But Eq. (6.31) shows that $S'(e_{i,1}) = 0$ for $i = 1, \ldots, q$, so the q elements $e_{1,1}, \ldots, e_{q,1}$ form a basis for the null space $N(S')$.

Now $N(S') \subseteq N(S)$ so $q = \dim N(S') \le \dim N(S) = \nu$. Let $k = \nu - q$. There exist k elements $w_1, \ldots, w_k$ in $N(S)$ such that the elements

$$\{e_{1,1}, e_{2,1}, \ldots, e_{q,1}, w_1, \ldots, w_k\}$$

form a basis for $N(S)$. Let v_i denote an element in V such that

$$(6.33) \qquad S(v_i) = e_{i,n_i} \qquad \text{for} \qquad i = 1, 2, \ldots, q.$$

We will show that the set

$$E' = E \cup \{w_1, \ldots, w_k\} \cup \{v_1, \ldots, v_q\}$$

is a basis for V.

To show that E' is independent, assume that some linear combination is zero, say,

$$(6.34) \qquad \sum_{i=1}^{m} \sum_{j=1}^{n_i} c_{ij} e_{i,j} + \sum_{i=1}^{k} a_i w_i + \sum_{i=1}^{q} b_i v_i = O.$$

Applying S and using (6.31), (6.32), and (6.33) we find

$$(6.35) \qquad \sum_{i=1}^{m} c_{i1} \lambda_i e_{i,1} + \sum_{i=1}^{m} \sum_{j=2}^{n_i} c_{ij}\{e_{i,j-1} + \lambda_i e_{i,j}\} + \sum_{i=1}^{q} b_i e_{i,n_i} = O,$$

since $S(w_i) = O$. But $\lambda_i = 0$ for $i \leq q$ so Eq. (6.34) can be written as follows:

$$(6.36) \qquad \sum_{i=q+1}^{m} c_{i1} \lambda_i e_{i,1} + \sum_{i=1}^{m} \sum_{j=2}^{n_i} c_{ij} e_{i,j-1} + \sum_{i=q+1}^{m} \sum_{j=2}^{n_i} c_{ij} \lambda_i e_{i,j} + \sum_{i=1}^{q} b_i e_{i,n_i} = O.$$

The elements e_{i,n_i} in the last sum with $i \leq q$ do not occur in the other sums. Since they are independent, each b_i must be zero. Using this in (6.34) we find

$$(6.37) \qquad \sum_{i=1}^{m} \sum_{j=1}^{n_i} c_{ij} e_{i,j} = - \sum_{i=1}^{k} a_i w_i.$$

The left member is in $S(V)$, the right member is in $N(S)$, so each belongs to $S(V) \cap N(S) = N(S')$, hence each is spanned by $\{e_{1,1}, e_{2,1}, \ldots, e_{q,1}\}$. Therefore each $c_{ij} = 0$ for $j \geq 2$, and Eq. (6.35) shows that $c_{i1} = 0$ for each $i \geq q + 1$. Thus Eq. (6.37) becomes

$$\sum_{i=1}^{q} c_{i1} e_{i,1} + \sum_{i=1}^{k} a_i w_i = O.$$

But $\{e_{1,1}, e_{2,1}, \ldots, e_{q,1}, w_1, \ldots, w_k\}$ is a basis for $N(S)$, so $c_{i1} = a_i = 0$ for all i. This proves that E' is independent. The number of elements in E' is $r + k + q = r + v = n$, so E' is a basis for V.

Now S maps the elements of E according to Eqs. (6.31) and (6.32), it maps each w_i onto O, and it maps v_i onto e_{i,n_i} for $i \leq q$. Therefore if we relabel v_i as e_{i,n_i+1} for $i = 1, \ldots, q$ we see that (6.32) also holds for $j = n_i + 1$ with $\lambda_i = 0$ if $i \leq q$. Now let

$$E'_i = \{e_{i,1}, \ldots, e_{i,n_i}, e_{i,n_i+1}\} \qquad \text{for} \qquad i = 1, 2, \ldots, q.$$

If we rearrange the order of the elements of E' according to the scheme

$$E' = \{w_1, \ldots, w_k\} \cup E'_1 \cup \cdots \cup E'_q \cup E_{q+1} \cup \cdots \cup E_m$$

we see that the matrix of S relative to E' has Jordan form, and the proof is complete.

6.16 Miscellaneous exercises on eigenvalues and eigenvectors

1. (a) Calculate the eigenvalues and eigenvectors of the matrix $A = \begin{bmatrix} -1 & -10 & -19 \\ 0 & 4 & 8 \\ 0 & 0 & 0 \end{bmatrix}$.

   (b) Find a nonsingular matrix C with integer entries such that $C^{-1}AC$ is a diagonal matrix.

2. Verify that the matrix $A = \begin{bmatrix} 5 & 4 & 3 \\ -1 & 0 & -3 \\ 1 & -2 & 1 \end{bmatrix}$ has eigenvalues $-2, 4, 4$. Consequently, by

Theorem 6.16 there is a nonsingular matrix C that transforms A to Jordan normal form:

$$C^{-1}AC = \begin{bmatrix} -2 & 0 & 0 \\ 0 & 4 & 1 \\ 0 & 0 & 4 \end{bmatrix}.$$

(Matrix C is not unique because the same equation is satisfied if C is replaced by any nonzero scalar multiple.) Determine such a matrix C with initial entry $c_{11} = 1$. [*Suggestion*: The eigenvectors of A can be used to determine two columns of C. Use undetermined entries for the third column of C, choosing them to satisfy $AC = CB$, where B is the Jordan normal form.]

3. (a) Verify that the matrix $A = \begin{bmatrix} -11 & -7 & -5 \\ 16 & 11 & 6 \\ 12 & 6 & 7 \end{bmatrix}$ has eigenvalues 1, 3, 3.

   (b) Find a nonsingular matrix C with initial entry $c_{11} = 1$ that transforms A to Jordan normal form:

$$C^{-1}AC = \begin{bmatrix} 1 & 0 & 0 \\ 0 & 3 & 1 \\ 0 & 0 & 3 \end{bmatrix}.$$

4. (a) Verify that the matrix $A = \begin{bmatrix} -1 & 3 & 0 \\ 0 & 2 & 0 \\ 2 & 1 & -1 \end{bmatrix}$ has eigenvalues 2, -1, -1.

   (b) Find a nonsingular matrix C with initial entry $c_{11} = 1$ that transforms A to the following Jordan normal form:

$$C^{-1}AC = \begin{bmatrix} 2 & 0 & 0 \\ 0 & -1 & 1 \\ 0 & 0 & -1 \end{bmatrix}.$$

   (c) Find all nonsingular matrices D that transform A to the following Jordan normal form:

$$D^{-1}AD = \begin{bmatrix} -1 & 1 & 0 \\ 0 & -1 & 0 \\ 0 & 0 & 2 \end{bmatrix}.$$

5. Verify that the matrix $\begin{bmatrix} 2 & 1 \\ -1 & 2 \end{bmatrix}$ has two complex eigenvalues. Find two orthogonal eigenvectors.

6. Find a 3×3 matrix whose eigenvalues are 1, 2, 3 with corresponding eigenvectors $(1, 2, 3)$, $(1, 2, 1)$, and $(3, 2, 1)$.

7. Let J denote the $n \times n$ matrix all of whose entries are equal to 1.
   (a) Calculate $J^2, J^3, \ldots, J^n$ in terms of J.
   (b) Prove that one eigenvalue is n and all others are zero.
   (c) When $n = 4$, exhibit an orthogonal set of eigenvectors, with one of them $(1, 1, 1, 1)$.
   (d) Determine the eigenvalues of the matrix tJ, where t is any scalar.

8. Given an $n \times n$ matrix with diagonal entries equal to a and all other entries equal to t. Prove that all eigenvalues are equal to $a - t$ except one, and express that one in terms of a and t.

9. Given an $n \times n$ matrix $A = (a_{ij})$, where $a_{ij} = n$ if $i + j = n + 1$, and $a_{ij} = 0$ otherwise. Thus, if $n = 2$ we have $A = \begin{bmatrix} 0 & 2 \\ 2 & 0 \end{bmatrix}$, and if $n = 3$ we have $A = \begin{bmatrix} 0 & 0 & 3 \\ 0 & 3 & 0 \\ 3 & 0 & 0 \end{bmatrix}$. Calculate A^2, A^3, and A^{-1} in terms of n, A, and the identity matrix I.

10. Let $A = \begin{bmatrix} 0 & 0 & 0 & 1 \\ 0 & 0 & 1 & 0 \\ 0 & 1 & 0 & 0 \\ 1 & 0 & 0 & 0 \end{bmatrix}$.

   (a) For each integer $n \geq 2$ find scalars a and b (depending on n) such that $A^n = aI + bA$.

(b) Determine all the eigenvalues of A (with multiplicities).

(c) Find C such that $C^{-1}AC$ is a diagonal matrix, or else explain why no such matrix C exists.

11. Let $A = \begin{bmatrix} 1 & 1 & 1 & -1 \\ 1 & 1 & -1 & 1 \\ 1 & -1 & 1 & 1 \\ -1 & 1 & 1 & 1 \end{bmatrix}$.

(a) Calculate A^4.

(b) Determine the characteristic polynomial of A and all the eigenvalues (with multiplicities).

12. Let $A = (a_{ij})$ be the 4×4 matrix that has 0 at each diagonal entry and 1 at all other entries; that is, $a_{ij} = 0$ if $i = j$, and $a_{ij} = 1$ if $i \neq j$.

(a) Find constants a and b such that $A^2 = aA + bI$.

(b) Prove that A is nonsingular and calculate A^{-1} explicitly.

(c) Determine all the eigenvalues of A (with multiplicities).

(d) Find 4 independent eigenvectors of A, or else prove that they do not exist.

7

EIGENVALUES OF OPERATORS ACTING ON EUCLIDEAN SPACES

7.1 Eigenvalues and inner products

This chapter describes properties of eigenvalues and eigenvectors of linear transformations that operate on Euclidean spaces, that is, on linear spaces having an inner product. We recall the fundamental properties of inner products.

In a *real* Euclidean space an inner product of two elements x and y is real number (x, y) satisfying the following properties:

(1) $(x, y) = (y, x)$ (symmetry),
(2) $(x, y + z) = (x, y) + (x, z)$ (linearity),
(3) $c(x, y) = (cx, y)$ (homogeneity),
(4) $(x, x) > 0$ if $x \neq O$ (positivity).

In a *complex* linear space, the inner product is a complex number satisfying the same properties, with the exception that symmetry is replaced by *Hermitian symmetry*,

$$(1') \qquad\qquad\qquad (x, y) = \overline{(y, x)},$$

where $\overline{(y, x)}$ denotes the complex conjugate of (y, x). In (3), the scalar c is real or complex. From (3) and $(1')$, we get the companion relation

$$(3') \qquad\qquad\qquad (x, cy) = \bar{c}(x, y),$$

which tells us that scalars are conjugated when taken out of the second factor. Taking $x = y$ in $(1')$ we see that (x, x) is real, so property (4) is meaningful if the inner product is complex.

When we use the term *Euclidean space* without further designation it is to be understood that the space can be real or complex. Although most of our applications will be to finite-dimensional spaces, we do not require this restriction at the outset.

The first theorem shows that eigenvalues (if they exist) can be expressed in terms of the inner product.

THEOREM 7.1. *Let E be a Euclidean space, let V be a subspace of E, and let $T : V \to E$ be a linear transformation having an eigenvalue λ with a corresponding eigenvector x. Then we have*

$$(7.1) \qquad\qquad\qquad \lambda = \frac{(T(x), x)}{(x, x)}.$$

Proof.   Since $T(x) = \lambda x$ we have

$$(T(x), x) = (\lambda x, x) = \lambda(x, x).$$

But $x \neq O$ so we can divide by (x, x) to get (7.1).

Several properties of eigenvalues are easily deduced from Equation (7.1). For example, from the Hermitian symmetry of the inner product we have the companion formula

(7.2)
$$\overline{\lambda} = \frac{(x, T(x))}{(x, x)}$$

for the complex conjugate of λ. From (7.1) and (7.2) we see that λ is real ($\lambda = \overline{\lambda}$) if and only if $(T(x), x)$ is real, that is, if and only if

$$(T(x), x) = (x, T(x)) \qquad \text{for every eigenvector } x \text{ belonging to } \lambda.$$

(This condition is trivially satisfied in a real Euclidean space.) Also, the eigenvalue λ is pure imaginary ($\lambda = -\overline{\lambda}$) if and only if $(T(x), x)$ is pure imaginary, that is, if and only if

$$(T(x), x) = -(x, T(x)) \qquad \text{for every eigenvector } x \text{ belonging to } \lambda.$$

7.2   Hermitian and skew-Hermitian transformations

In this section we introduce two important types of linear operators that act on Euclidean spaces. They have two categories of names, depending on whether the underlying space has a real or complex inner product. In the real case the transformations are called *symmetric* and *skew-symmetric*. In the complex case they are called *Hermitian* and *skew-Hermitian*. These operators occur in many different applications. For example, Hermitian operators on infinite-dimensional spaces play an important role in quantum mechanics. We shall discuss primarily the complex case since it presents no added difficulties.

Throughout this section and the next, E denotes a Euclidean space and V denotes a subspace of E. Neither V nor E is assumed to be finite-dimensional.

DEFINITION.   *A linear transformation $T : V \to E$ is called Hermitian on V if*

$$(T(x), y) = (x, T(y)) \qquad \textit{for all x and y in V.}$$

Operator T is called skew-Hermitian on V if

$$(T(x), y) = -(x, T(y)) \qquad \textit{for all x and y in V.}$$

In other words, a Hermitian operator T can be shifted from one factor of an inner product to the other without changing the value of the product. Shifting a skew-Hermitian operator changes the sign of the product.

Note. As already mentioned, if E is a *real* Euclidean space, Hermitian transformations are also called *symmetric*; skew-Hermitian transformations are called *skew-symmetric*.

Regarding eigenvalues we have the following theorem:

THEOREM 7.2.   *Assume T has an eigenvalue* λ. *Then we have:*
(a) *If T is Hermitian,* λ *is real:* $\lambda = \overline{\lambda}$.
(b) *If T is skew-Hermitian,* λ *is pure imaginary:* $\lambda = -\overline{\lambda}$.

Proof.   Let x be an eigenvector corresponding to λ. Then we have

$$\lambda = \frac{(T(x), x)}{(x, x)} \qquad \text{and} \qquad \overline{\lambda} = \frac{(x, T(x))}{(x, x)}.$$

If T is Hermitian we have $(T(x), x) = (x, T(x))$ so $\lambda = \overline{\lambda}$. If T is skew-Hermitian we have $(T(x), x) = -(x, T(x))$ so $\lambda = -\overline{\lambda}$.

Note. If T is *symmetric*, Theorem 7.2 tells us nothing new about the eigenvalues of T, because (by Theorem 7.1) all the eigenvalues must be real if the inner product is real. If T is *skew-symmetric*, the eigenvalues of T must be both real and pure imaginary. Hence all the eigenvalues of a skew-symmetric operator must be zero (if any exist).

7.3 Orthogonality of eigenvectors corresponding to distinct eigenvalues

We know from Theorem 6.2 that distinct eigenvalues of any linear transformation correspond to independent eigenvectors. For Hermitian and skew-Hermitian transformations more is true.

THEOREM 7.3.   *Let T be a Hermitian or skew-Hermitian transformation, and let* λ *and* μ *be distinct eigenvalues of T with corresponding eigenvectors x and y. Then x and y are orthogonal; that is* $(x, y) = 0$.

Proof.   Write $T(x) = \lambda x$, $T(y) = \mu y$ and compute the two inner products $(T(x), y)$ and $(x, T(y))$. We have

$$(T(x), y) = (\lambda x, y) = \lambda(x, y) \qquad \text{and} \qquad (x, T(y)) = (x, \mu y) = \overline{\mu}(x, y).$$

If T is Hermitian this gives us $\lambda(x, y) = \overline{\mu}(x, y) = \mu(x, y)$ because $\mu = \overline{\mu}$. Therefore $(x, y) = 0$ since $\lambda \neq \mu$. If T is skew-Hermitian we obtain $\lambda(x, y) = -\overline{\mu}(x, y) = \mu(x, y)$ which again implies $(x, y) = 0$.

Examples of symmetric and skew-symmetric operators acting on function spaces are given in Section 7.21 at the end of this chapter. (They require a knowledge of calculus.)

7.4 Exercises

1. Let E be a Euclidean space, let V be a subspace, and let $T : V \rightarrow E$ be a given linear transformation. Prove that a scalar λ is an eigenvalue of T with x as an eigenvector if and only if $x \neq O$ and $(T(x), y) = \lambda(x, y)$ for every y in E.
2. Let $T(x) = cx$ for every x in a linear space V, where c is a fixed scalar. Prove that T is symmetric if V is a real Euclidean space.

3. Assume $T : V \to V$ is a Hermitian transformation.
   (a) Prove that T^n is Hermitian for every integer $n > 0$, as is T^{-1} if T is invertible.
   (b) What can you conclude about T^n and T^{-1} if T is skew-Hermitian?
4. Let $T_1 : V \to E$ and $T_2 : V \to E$ be two Hermitian transformations.
   (a) Prove that $aT_1 + bT_2$ is Hermitian for all real scalars a and b.
   (b) Prove that the composition $T_1 T_2$ is Hermitian if T_1 and T_2 commute, i.e., if $T_1 T_2 = T_2 T_1$.
5. Let $V = \mathbf{R}^3$ with the usual dot product as inner product. Let T be a reflection in the xy plane; that is let $T(i) = i$, $T(j) = j$, and $T(k) = -k$. Prove that T is symmetric.
6. Let V be a subspace of a complex Euclidean space E. Let $T : V \to E$ be a linear transformation and define a scalar-valued function Q on V as follows:

$$Q(x) = (T(x), x) \qquad \text{for all } x \text{ in } V.$$

   (a) If T is Hermitian on V, prove that $Q(x)$ is real for all x.
   (b) If T is skew-Hermitian, prove that $Q(x)$ is pure imaginary for all x.
   (c) Prove that $Q(tx) = |t|^2 Q(x)$ for every scalar t.
   (d) Prove that $Q(x + y) = Q(x) + Q(y) + (T(x), y) + (T(y), x)$, and find a corresponding formula for $Q(x + ty)$.
   (e) Assume $E = V$. If $Q(x) = 0$ for all x prove that $T(x) = O$ for all x.
   (f) If $Q(x)$ is real for all x prove that T is Hermitian. [*Hint:* Use the fact that $Q(x + ty)$ equals its conjugate for every scalar t.]

7.5   Existence of an orthonormal set of eigenvectors for Hermitian and skew-Hermitian operators acting on finite-dimensional spaces

Both Theorems 7.2 and 7.3 are based on the assumption that T has an eigenvalue. As we know, some transformations T need not have eigenvalues. However, if T acts on a *finite-dimensional* complex space, then eigenvalues always exist (by Theorem 6.4) and, moreover, they are roots of the characteristic polynomial. If T is Hermitian, all the eigenvalues are real. If T is skew-Hermitian, all the eigenvalues are pure imaginary.

We just learned that distinct eigenvalues belong to orthogonal eigenvectors if T is Hermitian or skew-Hermitian. Using this property we can prove that T has an orthonormal set of eigenvectors that span the whole space. (We recall that an orthogonal set is called orthonormal if each of its elements has norm 1.)

THEOREM 7.4.   *Assume V is an n-dimensional Euclidean space with complex scalars. Let $T : V \to V$ be Hermitian or skew-Hermitian. Then there exist n eigenvectors $u_1, \ldots, u_n$ of T that form an orthonormal basis for V. Hence the matrix of T relative to this basis is the diagonal matrix $\Lambda = \mathrm{diag}(\lambda_1, \ldots, \lambda_n)$, where λ_k is the eigenvalue belonging to u_k.*

Proof.   We use induction on the dimension n. If $n = 1$, then T has exactly one eigenvalue, and any eigenvector u_1 of norm 1 is an orthonormal basis for V.

Now assume the theorem is true for every Euclidean space of dimension $n - 1$. To prove it is also true for V we choose an eigenvalue λ_1 for T and a corresponding eigenvector u_1 of norm 1. Then $T(u_1) = \lambda_1 u_1$ and $\|u_1\| = 1$. Let S be the subspace spanned by u_1. We shall apply the induction hypothesis to the subspace $S^\perp$ consisting of all element in V that are orthogonal to u_1,

$$S^\perp = \{x : x \in V \text{ and } (x, u_1) = 0\}.$$

To do this we need to know that $\dim S^\perp = n - 1$ and that T maps $S^\perp$ into itself.

From Theorem 3.7(a) we know that u_1 is part of a basis for V, say the basis $(u_1, v_2, \ldots, v_n)$. We can assume, without loss in generality, that this is an orthonormal basis. (If not, we apply the Gram-Schmidt process to convert it into an orthonormal basis, keeping u_1 as the first basis element.) Now take any x in $S^\perp$ and write it as a linear combination of the basis elements, say

$$x = x_1 u_1 + x_2 v_2 + \cdots + x_n v_n.$$

Taking the inner product with u_1 we find $x_1 = (x, u_1) = 0$ because the basis is orthonormal. Therefore, x lies in the space spanned by $v_2, \ldots, v_n$, and hence dim $S^\perp = n - 1$.

Next we show that T maps $S^\perp$ into itself. Assume T is Hermitian. If $x \in S^\perp$ we have

$$(T(x), u_1) = (x, T(u_1)) = (x, \lambda_1 u_1) = \lambda_1(x, u_1) = 0,$$

so $T(x) \in S^\perp$. Since T is Hermitian on $S^\perp$ we can apply the induction hypothesis to conclude that T has $n - 1$ eigenvectors $u_2, \ldots, u_n$ that form an orthonormal basis for $S^\perp$. Therefore the orthogonal set of eigenvectors $u_1, \ldots, u_n$ is an orthonormal basis for V. This proves the theorem if T is Hermitian, and a similar argument works if T is skew-Hermitian.

7.6 Matrix representations for Hermitian and skew-Hermitian operators

In this section we assume that V is a finite-dimensional Euclidean space. A Hermitian or skew-Hermitian transformation can be characterized in terms of its action on the elements of any basis.

THEOREM 7.5. *Let $(u_1, \ldots, u_n)$ be a basis for V and let $T : V \to V$ be a linear transformation. Then we have:*
(a) *T is Hermitian if and only if $(T(u_j), u_i) = (u_j, T(u_i))$ for all i and j.*
(b) *T is skew-Hermitian if and only if $(T(u_j), u_i) = -(u_j, T(u_i))$ for all i and j.*

Proof. Take any two elements x and y in V and express each in terms of the basis elements, say $x = \sum_{j=1}^n x_j u_j$ and $y = \sum_{i=1}^n y_i u_i$. Then we have

$$(T(x), y) = \left(\sum_{j=1}^n x_j T(u_j), y \right) = \sum_{j=1}^n x_j \left(T(u_j), \sum_{i=1}^n y_i u_i \right) = \sum_{j=1}^n \sum_{i=1}^n x_j \bar{y}_i (T(u_j), u_i).$$

Similarly, we find

$$(x, T(y)) = \sum_{j=1}^n \sum_{i=1}^n x_j \bar{y}_i (u_j, T(u_i)).$$

Statements (a) and (b) follow immediately from these equations.

Now we characterize these concepts in terms of a matrix representation of T.

THEOREM 7.6. *Let $(u_1, \ldots, u_n)$ be an orthonormal basis for V, and let $A = (a_{ij})$ be the matrix representation of a linear transformation $T : V \to V$ relative to this basis. Then we have:*

(a) *T is Hermitian if and only if $a_{ij} = \bar{a}_{ji}$ for all i and j.*
(b) *T is skew-Hermitian if and only if $a_{ij} = -\bar{a}_{ji}$ for all i and j.*

Proof. Since A is the matrix of T we have $T(u_j) = \sum_{k=1}^{n} a_{kj} u_k$. Taking the inner product of $T(u_j)$ with u_i and using the linearity of the inner product we obtain

$$(T(u_j), u_i) = \left(\sum_{k=1}^{n} a_{kj} u_k, u_i \right) = \sum_{k=1}^{n} a_{kj}(u_k, u_i).$$

But $(u_k, u_i) = 0$ unless $k = i$, so the last sum simplifies to $a_{ij}(u_i, u_i) = a_{ij}$ because $(u_i, u_i) = 1$. Hence we have

$$a_{ij} = (T(u_j), u_i) \qquad \text{for all } i, j.$$

Interchanging i and j, taking conjugates, and using the Hermitian symmetry of the inner product, we find

$$\bar{a}_{ji} = (u_j, T(u_i)) \qquad \text{for all } i, j.$$

Now we apply Theorem 7.5 to complete the proof.

7.7 Hermitian and skew-Hermitian matrices. The adjoint of a matrix

The following definition is suggested by Theorem 7.6.

DEFINITION. *A square matrix $A = (a_{ij})$ is called Hermitian if $a_{ij} = \bar{a}_{ji}$ for all i and j. Matrix A is called skew-Hermitian if $a_{ij} = -\bar{a}_{ji}$ for all i and j.*

Theorem 7.6 states that a transformation T on a finite-dimensional space V is Hermitian or skew-Hermitian according as its matrix relative to an orthonormal basis is Hermitian or skew-Hermitian.

These matrices can be described in another way. Let $\bar{A}$ denote the matrix obtained by replacing each entry of A by its complex conjugate. Matrix $\bar{A}$ is called the *conjugate* of A. Therefore A is Hermitian if and only if it is equal to the transpose of its conjugate, $A = \bar{A}^t$. It is skew-Hermitian if $A = -\bar{A}^t$. The transpose of the conjugate is given a special name.

DEFINITION. ADJOINT OF A MATRIX. *For any matrix A, the transpose of the conjugate, $\bar{A}^t$, is also called the adjoint of A and is denoted by A^*.*

Thus, a square matrix A is Hermitian if $A = A^*$, and skew-Hermitian if $A = -A^*$. Because a Hermitian matrix is equal to its adjoint, it is also called *self-adjoint*.

Note. Much of the older matrix literature uses the term *adjoint* for the transpose of the cofactor matrix, an entirely different object. The definition given here conforms to the current nomenclature in the theory of linear operators.

7.8 Diagonalization of a Hermitian or skew-Hermitian matrix

THEOREM 7.7. *Every $n \times n$ Hermitian or skew-Hermitian matrix A is similar to the diagonal matrix $\Lambda = \text{diag}(\lambda_1, \ldots, \lambda_n)$ of its eigenvalues. Moreover, we have*

$$\Lambda = C^{-1}AC,$$

where C is a nonsingular matrix whose inverse is its adjoint, $C^{-1} = C^*$.

Proof. Let V be the space $\mathbf{C}^n$ of n-tuples of complex numbers, and let $(e_1, \ldots, e_n)$ be the orthonormal basis of unit coordinate vectors. If $x = \sum_{i=1}^n x_i e_i$ and $y = \sum_{i=1}^n y_i e_i$, let the inner product be given by $(x, y) = \sum_{i=1}^n x_i \bar{y}_i$. For the given matrix A, let T be the transformation represented by A relative to the chosen basis. Theorem 7.4 tells us that V has an orthonormal basis of eigenvectors $(u_1, \ldots, u_n)$, relative to which T has the diagonal matrix representation

$$\Lambda = \text{diag}(\lambda_1, \ldots, \lambda_n),$$

where λ_k is the eigenvalue belonging to u_k. Since both A and Λ represent T they are similar, so we have $\Lambda = C^{-1}AC$, where $C = (c_{ij})$ is the nonsingular transition matrix relating the two bases:

$$[u_1, \ldots, u_n] = [e_1, \ldots, e_n]C.$$

This matrix equation shows that the jth column of C consists of the components of u_j relative to $(e_1, \ldots, e_n)$. Therefore, c_{ij} is the ith component of u_j. The inner product of u_j and u_i is given by

$$(u_j, u_i) = \sum_{k=1}^n c_{kj} \bar{c}_{ki}.$$

Since $(u_1, \ldots, u_n)$ is an orthonormal set, this implies that $CC^* = I$, so $C^{-1} = C^*$.

Note. The proof of Theorem 7.7 also tells us how to determine the diagonalizing matrix C. We find an orthonormal set of eigenvectors $u_1, \ldots, u_n$ and then use the components of u_j (relative to the basis of unit coordinate vectors) as the entries of the jth column of C.

EXAMPLE 1. The real Hermitian matrix

$$A = \begin{bmatrix} 2 & 2 \\ 2 & 5 \end{bmatrix}$$

has eigenvalues $\lambda_1 = 1$ and $\lambda_2 = 6$. The eigenvectors belonging to 1 are $t(2, -1)$, $t \neq 0$. Those belonging to 6 are $t(1, 2)$, $t \neq 0$. The two eigenvectors $u_1 = t(2, -1)$ and $u_2 = t(1, 2)$ with $t = 1/\sqrt{5}$ form an orthonormal set. Therefore the matrix

$$C = \frac{1}{\sqrt{5}} \begin{bmatrix} 2 & 1 \\ -1 & 2 \end{bmatrix}$$

is a diagonalizing matrix for A. In this case $C^* = C^t$ because C is real. It is easily verified that

$$C^t AC = \begin{bmatrix} 1 & 0 \\ 0 & 6 \end{bmatrix}.$$

EXAMPLE 2.   If A is already a diagonal matrix, then the diagonalizing matrix C of Theorem 7.7 either leaves A unchanged or merely rearranges the diagonal entries.

7.9   Unitary matrices. Orthogonal matrices

DEFINITION.   *A square matrix A is called unitary if $AA^* = I$. It is called orthogonal if $AA^t = I$.*

Note. Every real unitary matrix is orthogonal because $A^* = A^t$.

Unitary and orthogonal matrices are pleasant to work with because their inverses can be found with little effort. All that is required is conjugation and transposition. The inverse of a unitary matrix is its adjoint, and the inverse of an orthogonal matrix is its transpose.

If A is unitary its determinant is a complex number of absolute value 1. And if A is orthogonal its determinant is ± 1. The reason for this is that $\det A = \det A^t$ and $\det \overline{A} = \overline{(\det A)}$, so if A is unitary we have $\det(AA^*) = \det I = 1$. But $\det(AA^*) = (\det A)(\det A^*) = (\det A)(\det \overline{A}) = |\det A|^2$, so $\det A$ is a complex number of absolute value 1. If A is orthogonal the same type of argument shows that $(\det A)^2 = 1$ so $\det A = \pm 1$.

Note. Linear transformations represented by unitary and orthogonal matrices are discussed in Section 7.19.

Theorem 7.7 tells us that a Hermitian or skew-Hermitian matrix can always be diagonalized by a unitary matrix. A real Hermitian matrix has real eigenvalues and the corresponding eigenvectors can be taken real. Therefore we have:

COROLLARY OF THEOREM 7.7.   *A real Hermitian matrix can be diagonalized by a real orthogonal matrix.*

This statement is *not* true for real skew-Hermitian matrices. (See Exercise 11 of Section 7.10.)

We also have the following related concepts.

DEFINITION.   *A square matrix A with real or complex entries is called symmetric if $A = A^t$; it is called skew-symmetric if $A = -A^t$.*

EXAMPLE 1.   If $A = \begin{bmatrix} 1 + i & 2 \\ 3 - i & 4i \end{bmatrix}$, then

$$\overline{A} = \begin{bmatrix} 1 - i & 2 \\ 3 + i & -4i \end{bmatrix}, \qquad A^t = \begin{bmatrix} 1 + i & 3 - i \\ 2 & 4i \end{bmatrix}, \qquad \text{and } A^* = \begin{bmatrix} 1 - i & 3 + i \\ 2 & -4i \end{bmatrix}.$$

EXAMPLE 2. If A is real, its adjoint is equal to its transpose, $A^* = A^t$. Thus, every *real* Hermitian matrix is symmetric, but a symmetric matrix need not be Hermitian.

EXAMPLE 3. Both matrices $\begin{bmatrix} 1 & 2 \\ 2 & 3 \end{bmatrix}$ and $\begin{bmatrix} 1 & 2+i \\ 2-i & 3 \end{bmatrix}$ are Hermitian. The first is symmetric, the second is not.

EXAMPLE 4. Both matrices $\begin{bmatrix} 0 & -2 \\ 2 & 0 \end{bmatrix}$ and $\begin{bmatrix} i & -2 \\ 2 & 3i \end{bmatrix}$ are skew-Hermitian. The first is skew-symmetric, the second is not.

EXAMPLE 5. All the diagonal elements of a Hermitian matrix are real. All the diagonal elements of a skew-Hermitian matrix are pure imaginary. All the diagonal elements of a skew-symmetric matrix are zero.

EXAMPLE 6. For any square matrix A, the matrix $B = \frac{1}{2}(A + A^*)$ is Hermitian, and the matrix $C = \frac{1}{2}(A - A^*)$ is skew-Hermitian. Their sum is A. Thus, every square matrix A can be expressed as a sum $A = B + C$, where B is Hermitian and C is skew-Hermitian. It is an easy exercise to verify that this decomposition is unique. Also, every square matrix A can be expressed uniquely as the sum of a symmetric matrix, $\frac{1}{2}(A + A^t)$, and a skew-symmetric matrix, $\frac{1}{2}(A - A^t)$.

7.10 Exercises

1. Determine which of the following matrices are symmetric, skew-symmetric, Hermitian, skew-Hermitian.

(a) $\begin{bmatrix} 0 & 1 & 2 \\ 1 & 0 & 3 \\ 2 & 3 & 4 \end{bmatrix}$.

(c) $\begin{bmatrix} 0 & i & 2 \\ -i & 0 & 3 \\ -2 & -3 & 0 \end{bmatrix}$.

(b) $\begin{bmatrix} 0 & i & 2 \\ i & 0 & 3 \\ -2 & -3 & 4i \end{bmatrix}$.

(d) $\begin{bmatrix} 0 & 1 & 2 \\ -1 & 0 & 3 \\ -2 & -3 & 0 \end{bmatrix}$.

2. (a) Verify that the 2×2 matrix $\begin{bmatrix} \cos\theta & -\sin\theta \\ \sin\theta & \cos\theta \end{bmatrix}$ is an orthogonal matrix.

   (b) Let T be the linear transformation with the above matrix A relative to the usual basis $(\boldsymbol{i}, \boldsymbol{j})$. Prove that T maps each point in the plane with polar coordinates (r, α) onto the point with polar coordinates $(r, \alpha + \theta)$. Thus, T rotates the plane about the origin through an angle θ.

3. Let $V = \mathbf{R}^3$ with the usual basis vectors $\boldsymbol{i}, \boldsymbol{j}, \boldsymbol{k}$. Prove that each of the following matrices is orthogonal and represents the transformation indicated.

(a) $\begin{bmatrix} 1 & 0 & 0 \\ 0 & 1 & 0 \\ 0 & 0 & -1 \end{bmatrix}$ (reflection in the xy plane).

(b) $\begin{bmatrix} 1 & 0 & 0 \\ 0 & -1 & 0 \\ 0 & 0 & -1 \end{bmatrix}$ (reflection through the x axis).

(c) $\begin{bmatrix} -1 & 0 & 0 \\ 0 & -1 & 0 \\ 0 & 0 & -1 \end{bmatrix}$ (reflection through the origin).

(d) $\begin{bmatrix} 1 & 0 & 0 \\ 0 & \cos\theta & -\sin\theta \\ 0 & \sin\theta & \cos\theta \end{bmatrix}$ (rotation about the x axis).

(e) $\begin{bmatrix} -1 & 0 & 0 \\ 0 & \cos\theta & -\sin\theta \\ 0 & \sin\theta & \cos\theta \end{bmatrix}$ (rotation about x axis followed by reflection in the yz plane).

4. A real orthogonal matrix A is called *proper* if $\det A = 1$, and *improper* if $\det A = -1$.

   (a) If A is a proper 2×2 matrix, prove that $A = \begin{bmatrix} \cos\theta & -\sin\theta \\ \sin\theta & \cos\theta \end{bmatrix}$ for some θ. This represents a rotation through an angle θ.

   (b) Prove that $\begin{bmatrix} 1 & 0 \\ 0 & -1 \end{bmatrix}$ and $\begin{bmatrix} -1 & 0 \\ 0 & 1 \end{bmatrix}$ are improper matrices. The first matrix represents reflection of the xy plane through the x axis; the second represents a reflection through the y axis. Find all improper 2×2 matrices.

   In each of Exercises 5 through 8, find (a) an orthogonal set of eigenvectors for A, and (b) a unitary matrix C such that $C^{-1}AC$ is a diagonal matrix.

5. $A = \begin{bmatrix} 9 & 12 \\ 12 & 16 \end{bmatrix}$.

6. $A = \begin{bmatrix} 0 & -2 \\ 2 & 0 \end{bmatrix}$.

7. $A = \begin{bmatrix} 1 & 3 & 0 \\ 3 & -2 & -1 \\ 0 & -1 & 1 \end{bmatrix}$.

8. $A = \begin{bmatrix} 1 & 3 & 4 \\ 3 & 1 & 0 \\ 4 & 0 & 1 \end{bmatrix}$.

9. Determine which of the following matrices are unitary, and which are orthogonal (a, b, θ real).

   (a) $\begin{bmatrix} e^{ia} & 0 \\ 0 & e^{ib} \end{bmatrix}$.

   (b) $\begin{bmatrix} \cos\theta & 0 & \sin\theta \\ 0 & 1 & 0 \\ -\sin\theta & 0 & \cos\theta \end{bmatrix}$.

   (c) $\begin{bmatrix} \frac{1}{2}\sqrt{2} & -\frac{1}{3}\sqrt{3} & \frac{1}{6}\sqrt{6} \\ 0 & \frac{1}{3}\sqrt{3} & \frac{1}{3}\sqrt{6} \\ \frac{1}{2}\sqrt{2} & \frac{1}{3}\sqrt{3} & -\frac{1}{6}\sqrt{6} \end{bmatrix}$.

10. The special theory of relativity makes use of the equations

$$x' = a(x - vt), \qquad y' = y, \qquad z' = z, \qquad t' = a(t - vx/c^2).$$

   Here v is the velocity of a moving object, c is the speed of light, and $a = c/\sqrt{c^2 - v^2}$. The linear operator that maps (x, y, z, t) onto (x', y', z', t') is called a *Lorentz transformation*.

   (a) Let $(x_1, x_2, x_3, x_4) = (x, y, z, ict)$ and $(x_1', x_2', x_3', x_4') = (x', y', z', ict')$. Show that the four equations can be written as one matrix equation,

$$[x_1', x_2', x_3', x_4'] = [x_1, x_2, x_3, x_4] \begin{bmatrix} a & 0 & 0 & -iav/c \\ 0 & 1 & 0 & 0 \\ 0 & 0 & 1 & 0 \\ iav/c & 0 & 0 & a \end{bmatrix}.$$

   (b) Prove that the 4×4 matrix in (a) is orthogonal but not unitary.

11. Let a be a nonzero real number and let A be the skew-symmetric matrix $A = \begin{bmatrix} 0 & a \\ -a & 0 \end{bmatrix}$.

   (a) Find an orthonormal set of eigenvectors for A.

   (b) Find a unitary matrix C such that $C^{-1}AC$ is a diagonal matrix.

   (c) Prove that there is no real orthogonal matrix C such that $C^{-1}AC$ is a diagonal matrix.

12. If the all the eigenvalues of a Hermitian or skew-Hermitian matrix A are equal to c, prove that $A = cI$.

13. If A is a real skew-symmetric matrix, prove that both $I - A$ and $I + A$ are nonsingular and that $(I - A)(I + A)^{-1}$ is orthogonal.

14. Prove each of the following statements about $n \times n$ matrices, or exhibit a counter example.

   (a) If A and B are unitary, then $A + B$ is unitary.

(b) If A and B are unitary, then $A + B$ is not unitary.

(c) If A and B are unitary, then AB is unitary.

(d) If A and AB are unitary, then B is unitary.

7.11 Quadratic forms

We turn now to the study of quadratic forms, one of the most important applications of real symmetric operators.

Let V be a *real* Euclidean space and let $T : V \rightarrow V$ be a symmetric operator. This means that T can be shifted from one factor of an inner product to the other,

$$(T(x), y) = (x, T(y)) \qquad \text{for all } x \text{ and } y \text{ in } V.$$

We keep T fixed and define a real-valued function Q on V by the equation

$$Q(x) = (T(x), x).$$

The function Q is called the *quadratic form* associated with T. The terminology is suggested by the following theorem, which shows that, in the finite-dimensional case, $Q(x)$ is a quadratic polynomial in the components of x.

THEOREM 7.8. *Let $(u_1, \ldots, u_n)$ be an orthonormal basis for a real Euclidean space V. Let $T : V \rightarrow V$ be a symmetric transformation, and let $A = (a_{ij})$ be the matrix representation of T relative to this basis. Then the quadratic form $Q(x) = (T(x), x)$ is related to A as follows:*

$$(7.3) \qquad Q(x) = \sum_{i=1}^{n} \sum_{j=1}^{n} a_{ij} x_i x_j \qquad \text{if } x = \sum_{i=1}^{n} x_i u_i.$$

Proof. By linearity we have $T(x) = \sum_{i=1}^{n} x_i T(u_i)$. Therefore,

$$Q(x) = \left(\sum_{i=1}^{n} x_i T(u_i), \sum_{j=1}^{n} x_j u_j \right) = \sum_{i=1}^{n} \sum_{j=1}^{n} x_i x_j (T(u_i), u_j).$$

This proves (7.3) because by Theorem 7.6 we have $a_{ij} = a_{ji} = (T(u_i), u_j)$.

The sum appearing in Eq. (7.3) is meaningful even if matrix A is not symmetric.

DEFINITION. *Let V be any real Euclidean space with an orthonormal basis $(u_1, \ldots, u_n)$, and let $A = (a_{ij})$ be any $n \times n$ matrix of real scalars. The real-valued function Q defined at each element $x = \sum_{i=1}^{n} x_i u_i$ of V by the formula*

$$(7.4) \qquad Q(x) = \sum_{i=1}^{n} \sum_{j=1}^{n} a_{ij} x_i x_j$$

is called the quadratic form associated with A.

If A is a diagonal matrix, then $a_{ij} = 0$ if $i \neq j$ so the sum in (7.4) contains only squared terms and can be written more simply as

$$Q(x) = \sum_{i=1}^{n} a_{ii} x_i^2 .$$

In this case the quadratic form is called a *diagonal form*.

The double sum appearing in (7.4) can also be expressed as the product of three matrices.

THEOREM 7.9. *Let* $X = [x_1, \ldots, x_n]$ *be a* $1 \times n$ *row matrix, and let* $A = (a_{ij})$ *be an* $n \times n$ *matrix. Then* XAX^t *is a* 1×1 *matrix with entry*

(7.5)
$$\sum_{i=1}^{n} \sum_{j=1}^{n} a_{ij} x_i x_j .$$

Proof. The product XA is a $1 \times n$ matrix, $XA = [y_1, \ldots, y_n]$, where entry y_j is the dot product of X with the jth column of A,

$$y_j = \sum_{i=1}^{n} x_i a_{ij} .$$

Therefore the product XAX^t is a 1×1 matrix whose single entry is the dot product

$$\sum_{j=1}^{n} y_j x_j = \sum_{j=1}^{n} \left(\sum_{i=1}^{n} x_i a_{ij} \right) x_j = \sum_{i=1}^{n} \sum_{j=1}^{n} a_{ij} x_i x_j .$$

Note. It is customary to identify the 1×1 matrix XAX^t with the sum in (7.5) and to call the product XAX^t a quadratic form. Equation (7.4) is written more simply as

$$Q(x) = XAX^t .$$

EXAMPLE 1.   If $X = [x_1, x_2]$ and $A = \begin{bmatrix} 1 & -1 \\ -3 & 5 \end{bmatrix}$ then we have

$$XA = [x_1, x_2] \begin{bmatrix} 1 & -1 \\ -3 & 5 \end{bmatrix} = [x_1 - 3x_2, -x_1 + 5x_2],$$

hence,

$$XAX^t = [x_1 - 3x_2, -x_1 + 5x_2] \begin{bmatrix} x_1 \\ x_2 \end{bmatrix} = x_1^2 - 3x_2 x_1 - x_1 x_2 + 5x_2^2 .$$

EXAMPLE 2.   If $X = [x_1, x_2]$ and $B = \begin{bmatrix} 1 & -2 \\ -2 & 5 \end{bmatrix}$ then we have

$$XBX^t = [x_1, x_2] \begin{bmatrix} 1 & -2 \\ -2 & 5 \end{bmatrix} \begin{bmatrix} x_1 \\ x_2 \end{bmatrix} = x_1^2 - 2x_2 x_1 - 2x_1 x_2 + 5x_2^2 .$$

In both Examples 1 and 2 the two mixed product terms add up to $-4x_1x_2$ so $XAX^t = XBX^t$. These examples show that different matrices can lead to the same quadratic form. Note that one of these matrices, B, is symmetric. This illustrates the next theorem.

THEOREM 7.10. *For any $n \times n$ matrix A and any $1 \times n$ row matrix X we have $XAX^t = XBX^t$, where B is the symmetric matrix $B = \frac{1}{2}(A + A^t)$.*

Proof. Because XAX^t is a 1×1 matrix it is equal to its transpose, $XAX^t = (XAX^t)^t$. But the transpose of a product is the product of transposes in reversed order, (see Exercise 7(d) in Section 4.21) so we have $(XAX^t)^t = XA^tX^t$. Therefore $XAX^t = \frac{1}{2}XAX^t + \frac{1}{2}XA^tX^t = XBX^t$.

7.12 Reduction of a real quadratic form to a diagonal form

A real symmetric matrix A is Hermitian. Therefore, by Theorem 7.7, it is similar to the diagonal matrix $\Lambda = \mathrm{diag}(\lambda_1, \ldots, \lambda_n)$ of its eigenvalues. Moreover, we have $\Lambda = C^tAC$, where C is an orthogonal matrix (by the Corollary to Theorem 7.7 in Section 7.9). Now we show that C can be used to convert the quadratic form XAX^t to a diagonal form.

THEOREM 7.11. *Let XAX^t be the quadratic form associated with a real symmetric matrix A, and let C be an orthogonal matrix that converts A to a diagonal matrix $\Lambda = C^tAC$. Then we have*

$$XAX^t = Y\Lambda Y^t = \sum_{i=1}^{n} \lambda_i y_i^2,$$

where $Y = [y_1, \ldots, y_n]$ is the row matrix $Y = XC$, and $\lambda_1, \ldots, \lambda_n$ are the eigenvalues of A.

Proof. Because C is orthogonal we have $C^{-1} = C^t$, so the equation $Y = XC$ implies $X = YC^t$, and we obtain

$$XAX^t = (YC^t)A(YC^t)^t = Y(C^tAC)Y^t = Y\Lambda Y^t.$$

Note. Theorem 7.11 is described by saying that the linear transformation $Y = XC$ reduces the quadratic form XAX^t to a diagonal form $Y\Lambda Y^t$.

EXAMPLE 1. The quadratic form belonging to the identity matrix is

$$XIX^t = \sum_{i=1}^{n} x_i^2 = \|X\|^2,$$

the square of the length of the vector $X = (x_1, \ldots, x_n)$. A linear transformation $Y = XC$, where C is an orthogonal matrix, gives a new quadratic form $Y\Lambda Y^t$ with $\Lambda = CIC^t = CC^t = I$. Since $XIX^t = YIY^t$ we have $\|X\|^2 = \|Y\|^2$, so Y has the same length as X. A linear transformation that preserves the length of each vector is called an *isometry*. These transformations are discussed in further detail in Section 7.19.

EXAMPLE 2.   Determine an orthogonal matrix C that reduces the quadratic form

$$Q(x) = 2x_1^2 + 4x_1x_2 + 5x_2^2$$

to a diagonal form.

Solution. We write the mixed product term as $2x_1x_2 + 2x_2x_1$ so that $Q(x) = XAX^t$, where $A = \begin{bmatrix} 2 & 2 \\ 2 & 5 \end{bmatrix}$. This symmetric matrix was diagonalized in Example 1 following Theorem 7.7. It has the eigenvalues $\lambda_1 = 1, \lambda_2 = 6$, and an orthonormal set of eigenvectors u_1, u_2, where $u_1 = \frac{1}{\sqrt{5}}(2, -1)$, and $u_2 = \frac{1}{\sqrt{5}}(1, 2)$. An orthogonal diagonalizing matrix is $C = \frac{1}{\sqrt{5}} \begin{bmatrix} 2 & 1 \\ -1 & 2 \end{bmatrix}$. The corresponding diagonal form is

$$Y\Lambda Y^t = \lambda_1 y_1^2 + \lambda_2 y_2^2 = y_1^2 + 6y_2^2.$$

The result of Example 2 has a simple geometric interpretation, illustrated in Figure 7.1. The linear transformation $Y = XC$ can be regarded as a rotation that maps the basis i, j onto the new basis u_1, u_2. A point with coordinates (x_1, x_2) relative to the first basis has a new pair of coordinates (y_1, y_2) relative to the second basis. Since $XAX^t = Y\Lambda Y^t$, the set of points (x_1, x_2) satisfying an equation of the form $XAX^t = c$ for some c is identical with the set of points (y_1, y_2) satisfying $Y\Lambda Y^t = c$. The second equation, written as $y_1^2 + 6y_2^2 = c$, is the Cartesian equation of an ellipse if $c > 0$. Therefore the equation $XAX^t = c$, written as $2x_1^2 + 4x_1x_2 + 5x_2^2 = c$, represents the same ellipse in the original coordinate system. Figure 7.1 shows the ellipse when $c = 9$.

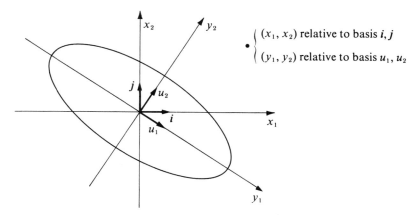

FIGURE 7.1   Rotation of axes by an orthogonal matrix. The ellipse has Cartesian equation $XAX^t = 9$ in the x_1x_2 system, and equation $Y\Lambda Y^t = 9$ in the y_1y_2 system.

7.13 Applications to conic sections

The reduction of a quadratic form to a diagonal form can be used to identify the set of all points (x, y) in the plane that satisfy a Cartesian equation of the form

(7.6) $$ax^2 + bxy + cy^2 + dx + ey + f = 0.$$

We shall find that this set is always a conic section, that is, an ellipse, hyperbola, parabola, or one of the degenerate cases (the empty set, a single point, or one or two straight lines). The type of conic is governed by the second-degree terms, that is, by the quadratic form $ax^2 + bxy + cy^2$. To conform with the notation used in the foregoing sections, we write x_1 for x, and x_2 for y, and express this quadratic form as a matrix product,

$$XAX^t = ax_1^2 + bx_1x_2 + cx_2^2,$$

where $X = [x_1, x_2]$ and where

$$A = \begin{bmatrix} a & b/2 \\ b/2 & c \end{bmatrix},$$

is a symmetric matrix. A rotation $Y = XC$ (or perhaps a rotation followed by a reflection in one of the coordinate axes) reduces this form to a diagonal form $\lambda_1 y_1^2 + \lambda_2 y_2^2$, where λ_1 and λ_2 are the eigenvalues of A. An orthonormal set of eigenvectors u_1, u_2 determines a new set of coordinate axes, relative to which the Cartesian equation (7.6) becomes

(7.7) $$\lambda_1 y_1^2 + \lambda_2 y_2^2 + d'y_1 + e'y_2 + f = 0,$$

with new coefficients d' and e' in the linear terms. In this equation there is no mixed product term y_1y_2, so the type of conic is easily identified by examining the eigenvalues λ_1 and λ_2.

If the conic is not degenerate, Equation (7.7) represents an ellipse if λ_1 and λ_2 have the same sign, a *hyperbola* if λ_1 and λ_2 have opposite sign, and a *parabola* if either λ_1 or λ_2 is zero. The three cases correspond to $\lambda_1\lambda_2 > 0$, $\lambda_1\lambda_2 < 0$, and $\lambda_1\lambda_2 = 0$. We illustrate with some specific examples.

EXAMPLE 1. $2x^2 + 4xy + 5y^2 + 4x + 13y - \frac{1}{4} = 0$. We rewrite this as

(7.8) $$2x_1^2 + 4x_1x_2 + 5x_2^2 + 4x_1 + 13x_2 - \frac{1}{4} = 0.$$

The quadratic form $2x_1^2 + 4x_1x_2 + 5x_2^2$ is the one treated in Example 2 of the foregoing section. Its matrix has eigenvalues $\lambda_1 = 1$, $\lambda_2 = 6$, and an orthonormal set of eigenvectors u_1, u_2, where $u_1 = \frac{1}{\sqrt{5}}(2, -1)$ and $u_2 = \frac{1}{\sqrt{5}}(1, 2)$. An orthogonal diagonalizing matrix is $C = \frac{1}{\sqrt{5}} \begin{bmatrix} 2 & 1 \\ -1 & 2 \end{bmatrix}$. This reduces the quadratic part of (7.8) to the form $y_1^2 + 6y_2^2$. To determine the effect on the linear part we write the equation of rotation $Y = XC$ in the form $X = YC^t$ and obtain

$$[x_1, x_2] = \frac{1}{\sqrt{5}}[y_1, y_2] \begin{bmatrix} 2 & -1 \\ 1 & 2 \end{bmatrix}, \qquad x_1 = \frac{1}{\sqrt{5}}(2y_1 + y_2), \qquad x_2 = \frac{1}{\sqrt{5}}(-y_1 + 2y_2).$$

Therefore the linear part $4x_1 + 13x_2$ is transformed to

$$\frac{4}{\sqrt{5}}(2y_1 + y_2) + \frac{13}{\sqrt{5}}(-y_1 + 2y_2) = -\sqrt{5}y_1 + 6\sqrt{5}y_2.$$

The transformed Cartesian equation becomes

$$y_1^2 + 6y_2^2 - \sqrt{5}y_1 + 6\sqrt{5}y_2 - \frac{1}{4} = 0.$$

By completing the squares in y_1 and y_2, we rewrite this as follows:

$$\left(y_1 - \frac{1}{2}\sqrt{5}\right)^2 + 6\left(y_2 + \frac{1}{2}\sqrt{5}\right)^2 = 9.$$

This is the equation of an ellipse with its center at the point $(\frac{1}{2}\sqrt{5}, -\frac{1}{2}\sqrt{5})$ in the $y_1 y_2$ system. The positive directions of the y_1 and y_2 axes are determined by the eigenvectors u_1 and u_2, as indicated in Figure 7.2.

We can simplify the equation further by writing

$$z_1 = y_1 - \frac{1}{2}\sqrt{5}, \qquad z_2 = y_2 + \frac{1}{2}\sqrt{5}.$$

Geometrically, this is the same as introducing a new system of coordinate axes parallel to the $y_1 y_2$ axes but with the new origin at the center of the ellipse. In the $z_1 z_2$ system the equation of the ellipse is simply

$$z_1^2 + 6z_2^2 = 9, \qquad \text{or} \qquad \frac{z_1^2}{9} + \frac{z_2^2}{3/2} = 1.$$

The ellipse and all three coordinate systems are shown in Figure 7.2.

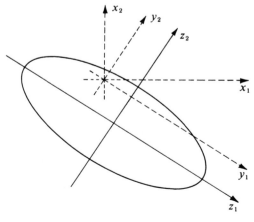

FIGURE 7.2   Rotation and translation of coordinate axes. The rotation $Y = XC$ is followed by the translation $z_1 = y_1 - \frac{1}{2}\sqrt{5}$, $z_2 = y_2 + \frac{1}{2}\sqrt{5}$.

EXAMPLE 2. $2x^2 - 4xy - y^2 - 4x + 10y - 13 = 0$. We rewrite this as

$$2x_1^2 - 4x_1x_2 - x_2^2 - 4x_1 + 10x_2 - 13 = 0.$$

The quadratic part is XAX^t, where $A = \begin{bmatrix} 2 & -2 \\ -2 & -1 \end{bmatrix}$. This matrix has the eigenvalues $\lambda_1 = 3$, and $\lambda_2 = -2$. An orthonormal set of eigenvectors is u_1, u_2, where $u_1 = \frac{1}{\sqrt{5}}(2, -1)$ and $u_2 = \frac{1}{\sqrt{5}}(1, 2)$. An orthogonal diagonalizing matrix is $C = \frac{1}{\sqrt{5}} \begin{bmatrix} 2 & 1 \\ -1 & 2 \end{bmatrix}$. The equation of rotation $X = YC^t$ gives us

$$x_1 = \frac{1}{\sqrt{5}}(2y_1 + y_2), \qquad x_2 = \frac{1}{\sqrt{5}}(-y_1 + 2y_2).$$

Therefore the transformed equation becomes

$$3y_1^2 - 2y_2^2 - \frac{4}{\sqrt{5}}(2y_1 + y_2) + \frac{10}{\sqrt{5}}(-y_1 + 2y_2) - 13 = 0,$$

or

$$3y_1^2 - 2y_2^2 - \frac{18}{\sqrt{5}}y_1 + \frac{16}{\sqrt{5}}y_2 - 13 = 0.$$

By completing the squares in y_1 and y_2 we obtain the equation

$$3\left(y_1 - \frac{3}{5}\sqrt{5}\right)^2 - 2\left(y_2 - \frac{4}{5}\sqrt{5}\right)^2 = 12,$$

which represents a hyperbola with its center at $(\frac{3}{5}\sqrt{5}, \frac{4}{5}\sqrt{5})$ in the y_1y_2 system. The translation $z_1 = y_1 - \frac{3}{5}\sqrt{5}$, $z_2 = y_2 - \frac{4}{5}\sqrt{5}$ simplifies this equation further to

$$3z_1^2 - 2z_2^2 = 12, \qquad \text{or} \qquad \frac{z_1^2}{4} - \frac{z_2^2}{6} = 1.$$

The hyperbola is shown in Figure 7.3(a). The eigenvectors u_1 and u_2 determine the directions of the positive y_1 and y_2 axes.

EXAMPLE 3. $9x^2 + 24xy + 16y^2 - 20x + 15y = 0$. We rewrite this as

$$9x_1^2 + 24x_1x_2 + 16x_2^2 - 20x_1 + 15x_2 = 0.$$

The symmetric matrix for the quadratic part is $A = \begin{bmatrix} 9 & 12 \\ 12 & 16 \end{bmatrix}$. Its eigenvalues are $\lambda_1 = 25$, and $\lambda_2 = 0$. An orthonormal set of eigenvectors is $u_1 = \frac{1}{5}(3, 4)$, $u_2 = \frac{1}{5}(-4, 3)$. An orthogonal diagonalizing matrix is $C = \frac{1}{5} \begin{bmatrix} 3 & -4 \\ 4 & 3 \end{bmatrix}$. The equation of rotation $X = YC^t$

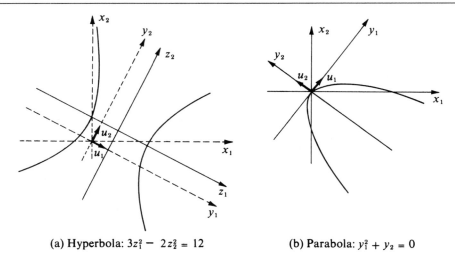

(a) Hyperbola: $3z_1^2 - 2z_2^2 = 12$ (b) Parabola: $y_1^2 + y_2 = 0$

FIGURE 7.3   The curves in Examples 2 and 3.

gives us

$$x_1 = \frac{1}{5}(3y_1 - 4y_2), \qquad x_2 = \frac{1}{5}(4y_1 + 3y_2).$$

The transformed Cartesian equation becomes

$$25y_1^2 - \frac{20}{5}(3y_1 - 4y_2) + \frac{15}{5}(4y_1 + 3y_2) = 0.$$

This simplifies to $y_1^2 + y_2 = 0$, the equation of a parabola with its vertex at the origin. The parabola is shown in Figure 7.3(b).

EXAMPLE 4.   *Degenerate cases.*   A knowledge of the eigenvalues alone does not reveal whether the Cartesian equation represents a degenerate conic section. For example, the three equations

$$x^2 + 2y^2 = 1, \qquad x^2 + 2y^2 = 0, \qquad x^2 + 2y^2 = -1,$$

all have the same eigenvalues; the first represents a nondegenerate ellipse, the second is satisfied only by $(x, y) = (0, 0)$, and the third represents the empty set. The last two can be regarded as degenerate cases of the ellipse.

The graph of the equation $y^2 = 0$ is the x axis. The equation $y^2 - 1 = 0$ represents the two parallel lines $y = 1$ and $y = -1$. These can be regarded as degenerate cases of the parabola. The equation $x^2 - 4y^2 = 0$ represents two intersecting lines because it is satisfied if either $x - 2y = 0$ or $x + 2y = 0$. This can be regarded as a degenerate case of the hyperbola.

However, if the Cartesian equation $ax^2 + bxy + cy^2 + dx + ey + f = 0$ represents a nondegenerate conic section, then the *type* of conic can be determined quite easily. The characteristic polynomial of the matrix of the quadratic form $ax^2 + bxy + cy^2$ is

$$\det \begin{bmatrix} \lambda - a & -b/2 \\ -b/2 & \lambda - c \end{bmatrix} = \lambda^2 - (a + c)\lambda + \left(ac - \frac{1}{4}b^2 \right) = (\lambda - \lambda_1)(\lambda - \lambda_2).$$

Therefore the product of the eigenvalues is the constant term,

$$\lambda_1 \lambda_2 = ac - \frac{1}{4}b^2 = \frac{1}{4}(4ac - b^2).$$

Since the type of conic is determined by the algebraic sign of the product $\lambda_1 \lambda_2$, we see that the conic is an *ellipse, hyperbola,* or *parabola,* according as $4ac - b^2$ is *positive, negative,* or *zero*. The number $4ac - b^2$ is called the *discriminant* of the quadratic form $ax^2 + bxy + cy^2$. This is the same discriminant we encountered in our treatment of conics in Theorem 2.19. There, the discriminant was related to the eccentricity of the conic. Now we have related it to the product of the eigenvalues of the matrix representing the quadratic form. In Examples 1, 2, and 3 the discriminant has the values 34, −24, and 0, respectively.

7.14 Exercises

In each of Exercises 1 through 7, find (a) a symmetric matrix A for the quadratic form; (b) the eigenvalues of A; (c) an orthonormal set of eigenvectors; (d) an orthogonal diagonalizing matrix C.

1. $4x_1^2 + 4x_1x_2 + x_2^2$.
2. x_1x_2.
3. $x_1^2 + 2x_1x_2 - x_2^2$.
4. $34x_1^2 - 24x_1x_2 + 41x_2^2$.

5. $x_1^2 + x_1x_2 + x_1x_3 + x_2x_3$.
6. $2x_1^2 + 4x_1x_3 + x_2^2 - x_3^2$.
7. $3x_1^2 + 4x_1x_2 + 8x_1x_3 + 4x_2x_3 + 3x_3^2$.

In each of Exercises 8 through 18, identify and make a sketch of the conic section represented by the Cartesian equation.

8. $y^2 - 2xy + 2x^2 - 5 = 0$.
9. $y^2 - 2xy + 5x = 0$.
10. $y^2 - 2xy + x^2 - 5x = 0$.
11. $5x^2 - 4xy + 2y^2 - 6 = 0$.
12. $19x^2 + 4xy + 16y^2 - 212x + 104y = 356$.
13. $9x^2 + 24xy + 16y^2 - 52x + 14y = 6$.
14. $5x^2 + 6xy + 5y^2 - 2 = 0$.
15. $x^2 + 2xy + y^2 - 2x + 2y + 3 = 0$.
16. $2x^2 + 4xy + 5y^2 - 2x - y - 4 = 0$.
17. $x^2 + 4xy - 2y^2 - 12 = 0$.
18. $xy + y - 2x - 2 = 0$.
19. For what value (or values) of c will the graph of the Cartesian equation $2xy - 4x + 7y + c = 0$ be a pair of lines?
20. The region enclosed by an ellipse with semi-major axis of length A and semi-minor axis of length B has area πAB. If the equation $ax^2 + bxy + cy^2 = 1$ represents an ellipse, prove that the area of the region it encloses is $2\pi/\sqrt{4ac - b^2}$. This gives a geometric meaning to the discriminant $4ac - b^2$.

7.15 Positive definite quadratic forms

As another application of Theorem 7.11 we establish a connection between the algebraic sign of a quadratic form and its eigenvalues.

THEOREM 7.12. *Let V be any real Euclidean space with an orthonormal basis* $(u_1, \ldots, u_n)$, *let* $A = (a_{ij})$ *be an* $n \times n$ *real symmetric matrix, and let*

$$Q(x) = \sum_{i=1}^{n} \sum_{j=1}^{n} a_{ij} x_i x_j$$

where $x = \sum_{i=1}^{n} x_i u_i$. *Then we have:*
   (a) $Q(x) > 0$ *for all* $x \neq O$ *if and only if all the eigenvalues of A are positive.*
   (b) $Q(x) < 0$ *for all* $x \neq O$ *if and only if all the eigenvalues of A are negative.*

   Note. In case (a), the quadratic form is called *positive definite*; in case (b) it is called *negative definite*.

   Proof. Use the components of x to form a row matrix $X = [x_1, \ldots, x_n]$. By Theorem 7.11, there is an orthogonal matrix C that reduces the quadratic form $Q(x) = XAX^t$ to a diagonal form:

$$Q(x) = \sum_{i=1}^{n} \lambda_i y_i^2,$$

where $Y = [y_1, \ldots, y_n]$ is the row matrix $Y = XC$, and $\lambda_1, \ldots, \lambda_n$ are the eigenvalues of A. The eigenvalues are real because A is symmetric.

   If all the eigenvalues are positive, the sum in the diagonal form is positive whenever $Y \neq O$. But since $Y = XC$ we have $X = YC^{-1}$, so $Y \neq O$ if and only if $x \neq O$. Therefore $Q(x) > 0$ for all $x \neq O$.

   Conversely, if $Q(x) > 0$ for all $x \neq O$ we can choose x so that $Y = XC$ is the kth basis vector u_k. For this x the diagonal form for $Q(x)$ reduces to a single term, $Q(x) = \lambda_k$, so each $\lambda_k > 0$. This proves part (a). The proof of (b) is entirely analogous.

★ 7.16★ Eigenvalues of a symmetric transformation obtained as values of its quadratic form

   Now we drop the requirement that V be finite-dimensional and we find a relation between the eigenvalues of a symmetric operator and numerical values of its quadratic form.
   Suppose x is an eigenvector with norm 1 belonging to an eigenvalue λ. Then $T(x) = \lambda x$ so we have

(7.9) $$Q(x) = (T(x), x) = (\lambda x, x) = \lambda(x, x) = \lambda$$

because $(x, x) = 1$. The set of all x in V satisfying $(x, x) = 1$ is called the *unit sphere* in V. Equation (7.9) proves the following theorem.

   THEOREM 7.13. *Let* $T : V \to V$ *be a symmetric transformation on a real Euclidean space V, and let* $Q(x) = (T(x), x)$. *Then the eigenvalues of T (if any exist) are to be found among the values that Q takes on the unit sphere of V.*

---

*Starred sections can be omitted or postponed without loss in continuity.

EXAMPLE 1. Let $V = \mathbf{R}^2$ with the usual basis (i, j) and the usual dot product as inner product. Let T be the symmetric transformation with matrix $A = \begin{bmatrix} 4 & 0 \\ 0 & 8 \end{bmatrix}$. Then the quadratic form of T is a diagonal form given by

$$Q(x) = \sum_{i=1}^{2} \sum_{j=1}^{2} a_{ij} x_i x_j = 4x_1^2 + 8x_2^2.$$

The eigenvalues of T are $\lambda_1 = 4$, $\lambda_2 = 8$. It is easy to see that these eigenvalues are, respectively, the minimum and maximum values that Q takes on the unit circle $x_1^2 + x_2^2 = 1$. In fact, on this circle we have

$$Q(x) = 4(x_1^2 + x_2^2) + 4x_2^2 = 4 + 4x_2^2, \qquad \text{where } -1 \le x_2 \le 1.$$

This has its smallest value, 4, when $x_2 = 0$, and its largest value, 8, when $x_2 = \pm 1$.

Figure 7.4 shows the unit circle and two ellipses. The inner ellipse has the Cartesian equation $4x_1^2 + 8x_2^2 = 4$. It consists of all points $x = (x_1, x_2)$ in the plane satisfying $Q(x) = 4$. The outer ellipse has Cartesian equation $4x_1^2 + 8x_2^2 = 8$ and consists of all points satisfying $Q(x) = 8$. The points $(\pm 1, 0)$ where the inner ellipse touches the unit circle are eigenvectors belonging to the eigenvalue $\lambda_1 = 4$. The points $(0, \pm 1)$ on the outer ellipse are eigenvectors belonging to the eigenvalue $\lambda_2 = 8$.

The foregoing examples illustrate extremal properties of eigenvalues that hold more generally. In the next section we will prove that the smallest and largest eigenvalues (if they exist) are always the minimum and maximum values that Q takes on the unit sphere. Our discussion of these extremal properties will make use of the following theorem on quadratic forms. It should be noted that this theorem does not require that V be finite-dimensional.

THEOREM 7.14. *Let $T : V \to V$ be a symmetric transformation on a real Euclidean space V with quadratic form $Q(x) = (T(x), x)$. Assume that Q does not change sign on V. Then if $Q(x) = 0$ for some x in V we also have $T(x) = O$. In other words, if Q does not change sign, then Q vanishes only on the null space of T.*

Proof. Assume $Q(x) = 0$ for some x in V and let y be an arbitrary element in V. Choose any real t and consider $Q(x + ty)$ as a function of t. Using linearity of T, linearity

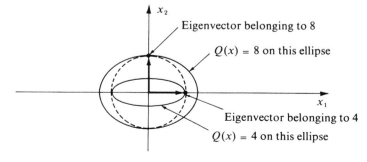

FIGURE 7.4   Geometric relation between the eigenvalues of T and the values of Q on the unit sphere, illustrated with a two-dimensional example.

of the inner product, and symmetry of T, we have

$$Q(x + ty) = (T(x + ty), x + ty) = (T(x) + tT(y), x + ty)$$

$$= (T(x), x) + t(T(x), y) + t(T(y), x) + t^2(T(y), y)$$

$$= Q(x) + 2t(T(x), y) + t^2 Q(y) = 0 + 2at + bt^2,$$

where $a = (T(x), y)$ and $b = Q(y)$. If Q is nonnegative on V we have the inequality

$$2at + bt^2 \geq 0 \qquad \text{for all real } t.$$

Now we show that this implies $a = 0$. If $b = 0$ this inequality cannot hold for all t unless $a = 0$. If $b > 0$ the graph of the quadratic polynomial $p(t) = 2at + bt^2$ is a parabola that opens upward and does not intersect the t axis at more than one point. Since $p(0) = 0$, we must have $a = 0$, otherwise $p(t)$ would have another root at $t = -b/(2a)$. But $a = (T(x), y)$, so we have shown that $(T(x), y) = 0$ for all y. In particular, when $y = T(x)$ we find $(T(x), T(x)) = 0$ so $T(x) = O$.

If Q is nonpositive on V, we get $p(t) = 2at + bt^2 \leq 0$ for all t. Applying the foregoing argument to $-p(t)$ we again find that $a = 0$.

★ 7.17 Extremal properties of eigenvalues of a symmetric transformation

Now we shall prove that the extreme values of a quadratic form on the unit sphere are eigenvalues.

THEOREM 7.15. *Let $T : V \to V$ be a symmetric transformation on a real Euclidean space V with quadratic form $Q(x) = (T(x), x)$. Among all values that Q takes on the unit sphere, assume there is an extremum[†] (maximum or minimum) at a point u with $(u, u) = 1$. Then u is an eigenvector for T and the corresponding eigenvalue is $Q(u)$, the extreme value of Q on the unit sphere.*

Proof. Assume Q has a minimum at u, so that

(7.10) $$Q(x) \geq Q(u) \qquad \text{for all } x \text{ with } (x, x) = 1.$$

Let $\lambda = Q(u)$. If $(x, x) = 1$ we have $Q(u) = \lambda(x, x) = (\lambda x, x)$. Therefore inequality (7.10) can be written as

(7.11) $$(T(x), x) \geq (\lambda x, x) \qquad \text{for all } x \text{ with } (x, x) = 1.$$

Now we prove that (7.11) is valid for *all* x in V. Suppose $\|x\| = a$. Then $x = ay$, where $\|y\| = 1$. Hence

$$(T(x), x) = (T(ay), ay) = a^2(T(y), y) \qquad \text{and} \qquad (\lambda x, x) = (\lambda ay, ay) = a^2(\lambda y, y).$$

[†] If V is infinite-dimensional, the quadratic form Q need not have an extremum on the unit sphere. This will be the case when T has no eigenvalues. In the finite-dimensional case, Q always has a maximum and a minimum somewhere on the unit sphere. This follows as a consequence of a more general theorem on extreme values of continuous functions.

But by (7.11) we have $(T(y), y) \geq (\lambda y, y)$ since $(y, y) = 1$. Multiplying each member of this inequality by a^2 we get (7.11) for $x = ay$.

Since $(T(x), x) - (\lambda x, x) = (T(x) - \lambda x, x)$, we can rewrite inequality (7.11) in the form

$$(7.12) \qquad\qquad (S(x), x) \geq 0, \qquad \text{where} \qquad S = T - \lambda I.$$

When $x = u$ we have equality in (7.10) and hence also in (7.12). The linear transformation S is symmetric, and inequality (7.12) states that the quadratic form Q_1 given by $Q_1(x) = (S(x), x)$ is nonnegative on V. When $x = u$ we have $Q_1(u) = 0$. Therefore, by Theorem 7.14 we must have $S(u) = 0$. In other words, $T(u) = \lambda u$, so u is an eigenvector for T, and $\lambda = Q(u)$ is the corresponding eigenvector. This completes the proof if Q has a minimum at u.

If there is a maximum at u all the inequalities in the foregoing proof are reversed and we apply Theorem 7.14 to the nonpositive quadratic form Q_1.

★ 7.18 The finite-dimensional case

Suppose now that $\dim V = n$. Then T has n real eigenvalues that can be arranged in increasing order, say

$$\lambda_1 \leq \lambda_2 \leq \cdots \leq \lambda_n.$$

According to Theorem 7.15, the smallest eigenvalue λ_1 is the minimum of Q on the unit sphere, and the largest eigenvalue is the maximum of Q on the unit sphere. Now we will show that the intermediate eigenvalues also occur as extreme values of Q on certain subsets of the unit sphere.

Let u_1 be an eigenvector on the unit sphere that minimizes Q. Then $\lambda_1 = Q(u_1)$. Any eigenvector belonging to an eigenvalue $\lambda \neq \lambda_1$ must be orthogonal to u_1, and it is natural to search for such an eigenvector in the orthogonal complement of the subspace S spanned by u_1.

The orthogonal complement $S^\perp$ consists of all elements in V orthogonal to u_1. In particular, $S^\perp$ contains all eigenvectors belonging to eigenvalues $\lambda \neq \lambda_1$. It is easily verified that $\dim S^\perp = n - 1$ and that T maps $S^\perp$ into itself.[‡] Let S_{n-1} denote the unit sphere in the $(n - 1)$-dimensional subspace $S^\perp$. (The unit sphere S_{n-1} is a subset of the unit sphere in V.) Applying Theorem 7.15 to the subspace $S^\perp$ we find that $\lambda_2 = Q(u_2)$, where u_2 is a point that minimizes Q on S_{n-1}.

The next eigenvalue λ_3 can be obtained in a similar way as the minimum value of Q on the unit sphere S_{n-2} in the $(n - 2)$-dimensional space consisting of those elements orthogonal to both u_1 and u_2. Continuing in this manner we find that each eigenvalue λ_k is the minimum value that Q takes on a unit sphere S_{n-k+1} in a subspace of dimension $n - k + 1$. The largest of these minima, λ_n, is also the *maximum* value that Q takes on each of the spheres S_{n-k+1}. The corresponding set of eigenvectors $u_1, \ldots, u_n$ is an orthonormal basis for V.

[‡]This was done in the proof of Theorem 7.4, Section 7.5.

7.19   Unitary transformations

We turn now to another important class of transformations called unitary transformations. In the finite-dimensional case they are represented by unitary matrices. But in this section the underlying space need not be finite-dimensional.

DEFINITION.   *Let E be a Euclidean space and let V be a subspace of E. A linear transformation T : V → E is called unitary on V if we have*

$$(7.13) \qquad\qquad (T(x), T(y)) = (x, y) \qquad \text{for all x and y in V.}$$

When E is a real Euclidean space, a unitary transformation is also said to be orthogonal.

Equation (7.13) is described by saying that T preserves inner products. Therefore it is natural to expect that T also preserves orthogonality and norms, because these are derived from the inner product.

THEOREM 7.16.   *If T : V → E is a unitary operator on V, then for all x and y in V we have:*
  (a) $(x, y) = 0$ *implies* $(T(x), T(y)) = 0$ *(T preserves orthogonality).*
  (b) $\|T(x)\| = \|x\|$ *(T preserves norms).*
  (c) $\|T(x) - T(y)\| = \|x - y\|$ *(T preserves distances).*
  (d) *T is invertible, and* T^{-1} *is unitary on* $T(V)$.

Proof.   Part (a) follows at once from Equation (7.13). Part (b) follows by taking $y = x$ in (7.13). Part (c) follows from (b) because $T(x) - T(y) = T(x - y)$.
   To prove (d) we use (b), which shows that $T(x) = O$ implies $x = O$, so T is invertible. If $x \in T(V)$ and $y \in T(V)$ we can write $x = T(u)$, $y = T(v)$ for some u and v in V, so we have

$$(T^{-1}(x), T^{-1}(y)) = (u, v) = (T(u), T(v)) = (x, y).$$

Therefore T^{-1} satisfies (7.13), so T^{-1} is unitary on $T(V)$.

Regarding eigenvalues and eigenvectors we have the following theorem:

THEOREM 7.17.   *Let T : V → E be a unitary transformation on V.*
  (a) *If T has an eigenvalue* λ, *then* $|\lambda| = 1$.
  (b) *Eigenvectors belonging to distinct eigenvalues are orthogonal.*
  (c) *If V = E and dim V = n, and if V is a complex Euclidean space, then there exist eigenvectors* $u_1, \ldots, u_n$ *of T that form an orthonormal basis for V. The matrix of T relative to this basis is the diagonal matrix* $\Lambda = \text{diag}(\lambda_1, \ldots, \lambda_n)$, *where* λ_k *is the eigenvalue belonging to* u_k.

Proof.   To prove (a), let x be an eigenvector belonging to λ. Then $x \neq O$ and $T(x) = \lambda x$. Taking $y = x$ in Equation (7.13) we get

$$(\lambda x, \lambda x) = (x, x) \qquad \text{or} \qquad \lambda \overline{\lambda}(x, x) = (x, x).$$

Since $(x, x) > 0$ and $\lambda \overline{\lambda} = |\lambda|^2$, this implies $|\lambda| = 1$.

To prove (b), let x and y be eigenvectors belonging to distinct eigenvalues λ and μ, so that $T(x) = \lambda x$ and $T(y) = \mu y$, where $\lambda \neq \mu$. We compute the inner product $(T(x), T(y))$ in two ways. Because T is unitary we have

$$(T(x), T(y)) = (x, y),$$

and because x and y are eigenvectors we also have

$$(T(x), T(y)) = (\lambda x, \mu y) = \lambda \overline{\mu}(x, y).$$

Therefore $\lambda \overline{\mu}(x, y) = (x, y)$, so $(x, y) = 0$ unless $\lambda \overline{\mu} = 1$. But $\lambda \overline{\lambda} = 1$ by (a), so if we had $\lambda \overline{\mu} = 1$ we would also have $\lambda \overline{\lambda} = \lambda \overline{\mu}$, hence $\overline{\lambda} = \overline{\mu}$, which contradicts the assumption that $\lambda \neq \mu$. Therefore $\lambda \overline{\mu} \neq 1$ and $(x, y) = 0$.

Part (c) is proved by induction on n in much the same way that we proved Theorem 7.4, the corresponding result for Hermitian operators. The only change required is in that part of the proof which shows that T maps $S^{\perp}$ into itself, where

$$S^{\perp} = \{x : x \in V \text{ and } (x, u_1) = 0\}.$$

Here u_1 is an eigenvector of T with eigenvalue λ_1. From the equation $T(u_1) = \lambda_1 u_1$ we find

$$u_1 = \lambda_1^{-1} T(u_1) = \overline{\lambda}_1 T(u_1)$$

because $\lambda_1 \overline{\lambda}_1 = |\lambda_1|^2 = 1$. Now choose any x in $S^{\perp}$ and note that

$$(T(x), u_1) = (T(x), \overline{\lambda}_1 T(u_1)) = \lambda_1 (T(x), T(u_1)) = \lambda_1 (x, u_1) = 0.$$

Hence $T(x) \in S^{\perp}$ if $x \in S^{\perp}$, so T maps $S^{\perp}$ into itself. The rest of the proof is identical with that of Theorem 7.4, so we shall not repeat the details.

The next two theorems describe properties of unitary transformations on a finite-dimensional space. We give only a brief outline of each proof.

THEOREM 7.18. *Assume* $\dim V = n$ *and let* $U = (u_1, \ldots, u_n)$ *be a fixed basis for* V. *Then a linear transformation* $T : V \to V$ *is unitary if and only if*

(7.14) $$(T(u_i), T(u_j)) = (u_i, u_j) \qquad \textit{for all } i \textit{ and } j.$$

In particular, if U *is orthonormal, then* T *is unitary if and only* T *maps* U *onto an orthonormal basis.*

Sketch of Proof. Write $x = \sum_{i=1}^{n} x_i u_i$ and $y = \sum_{j=1}^{n} y_j u_j$. Then we have

$$(x, y) = \left(\sum_{i=1}^{n} x_i u_i, \sum_{j=1}^{n} y_j u_j \right) = \sum_{i=1}^{n} \sum_{j=1}^{n} x_i \overline{y}_j (u_i, u_j)$$

and

$$(T(x), T(y)) = \left(\sum_{i=1}^{n} x_i T(u_i), \sum_{j=1}^{n} y_j T(u_j) \right) = \sum_{i=1}^{n} \sum_{j=1}^{n} x_i \bar{y}_j (T(u_i), T(u_j)).$$

Now compare (x, y) with $(T(x), T(y))$.

THEOREM 7.19.   *Assume* $\dim V = n$ *and let* $(u_1, \ldots, u_n)$ *be an orthonormal basis for V. Let* $A = (a_{ij})$ *be the matrix representation of a linear transformation* $T : V \to V$ *relative to this basis. Then T is unitary if and only if A is unitary, that is, if and only if*

(7.15) $$A^* A = I.$$

Sketch of Proof.   Since (u_i, u_j) is the ij-entry of the identity matrix, Equation (7.15) implies

(7.16) $$(u_i, u_j) = \sum_{k=1}^{n} \bar{a}_{ki} a_{kj} = \sum_{k=1}^{n} a_{ki} \bar{a}_{kj}.$$

Since A is the matrix of T we have $T(u_i) = \sum_{k=1}^{n} a_{ki} u_k$, $T(u_j) = \sum_{r=1}^{n} a_{rj} u_r$, so

$$(T(u_i), T(u_j)) = \left(\sum_{k=1}^{n} a_{ki} u_k, \sum_{r=1}^{n} a_{rj} u_r \right) = \sum_{k=1}^{n} \sum_{r=1}^{n} a_{ki} \bar{a}_{rj} (u_k, u_r) = \sum_{k=1}^{n} a_{ki} \bar{a}_{kj}.$$

Now compare this with (7.16) and use Theorem 7.18.

THEOREM 7.20.   *Every unitary matrix A has the following properties:*
(a) *A is nonsingular and* $A^{-1} = A^*$.
(b) *Each of* A^t, *$\bar{A}$, and A^* is a unitary matrix.*
(c) *The eigenvalues of A are complex numbers of absolute value 1.*
(d) $|\det A| = 1$; *if A is real, then* $\det A = \pm 1$.

The proof of Theorem 7.20 is left as an exercise for the reader.

7.20   Exercises

1. (a) Let $T : V \to V$ be the transformation given by $T(x) = cx$, where c is a fixed scalar. Prove that T is unitary if and only if $|c| = 1$.
   (b) If V is one-dimensional, prove that the only unitary transformations on V are those described in (a). In particular, if V is a real one-dimensional space, there are only two orthogonal transformations, $T(x) = x$ and $T(x) = -x$.
2. Prove each of the following statements about a real orthogonal $n \times n$ matrix A.
   (a) If λ is a real eigenvalue of A, then $\lambda = 1$ or $\lambda = -1$.
   (b) If λ is a complex eigenvalue of A, then the complex conjugate $\bar{\lambda}$ is also an eigenvalue of A. In other words, the nonreal eigenvalues of A occur in conjugate pairs.
   (c) If n is odd, then A has at least one real eigenvalue.
3. Let V be a real Euclidean space of dimension n. An orthogonal transformation $T : V \to V$ with determinant 1 is called a *rotation*. If n is odd, prove that 1 is an eigenvalue for T. This shows that every rotation of an odd-dimensional space has a fixed axis. [*Hint:* Use Exercise 2.]

4. Given a real orthogonal matrix with -1 as an eigenvalue of multiplicity k. Prove that $\det A = (-1)^k$.
5. If T is linear and norm-preserving, prove that T is unitary.
6. If $T : V \to V$ is both unitary and Hermitian, prove that $T^2 = I$.
7. Given two orthonormal bases for a Euclidean space of dimension n. Prove that there is a unitary transformation T that maps one of these bases onto the other.
8. Find a real number a such that the following matrix is unitary:

$$\begin{bmatrix} a & \frac{1}{2}i & \frac{1}{2}a(2i-1) \\ ia & \frac{1}{2}(1+i) & \frac{1}{2}a(1-i) \\ a & -\frac{1}{2} & \frac{1}{2}a(2-i) \end{bmatrix}.$$

9. If A is a skew-Hermitian matrix, prove that both $I - A$ and $I + A$ are nonsingular and that the product $(I - A)(I + A)^{-1}$ is unitary.
10. If A is a unitary matrix and if $I + A$ is nonsingular, prove that $(I - A)(I + A)^{-1}$ is skew-Hermitian.
11. If A is Hermitian, prove that $A - iI$ is nonsingular and that $(A - iI)^{-1}(A + iI)$ is unitary.
12. Prove that any unitary matrix can be diagonalized by a unitary matrix.
13. A square matrix is called *normal* if $AA^* = A^*A$. Determine which of the following types of matrices are normal.
    (a) Hermitian matrices.   (d) Skew-symmetric matrices.
    (b) Skew-Hermitian matrices.   (e) Unitary matrices.
    (c) Symmetric matrices.   (f) Orthogonal matrices.
14. If A is normal ($AA^* = A^*A$) and if U is unitary, prove that U^*AU is normal.

★ 7.21   Examples of symmetric and skew-symmetric operators acting on function spaces

We conclude this chapter with some examples of symmetric and skew-symmetric operators acting on real function spaces. These examples require a knowledge of calculus.

Recall the definitions given in Section 7.2. If E is a Euclidean space and if V is a subspace, a linear transformation $T : V \to E$ is called *Hermitian* on V if

$$(T(x), y) = (x, T(y)) \qquad \text{for all } x \text{ and } y \text{ in } V.$$

If the inner product is real, a Hermitian operator is called *symmetric*. Operator T is called *skew-Hermitian* on V if

$$(T(x), y) = -(x, T(y)) \qquad \text{for all } x \text{ and } y \text{ in } V.$$

When the inner product is real, skew-Hermitian operators are called *skew-symmetric*.

EXAMPLE 1.   *Symmetry and skew-symmetry in the space $C(a, b)$.*   Let $C(a, b)$ denote the linear space of all real functions continuous on a closed interval $[a, b]$, with real inner product

$$(f, g) = \int_a^b f(t)g(t)\, dt.$$

Let V be a subspace of $C(a, b)$. If $T : V \to C(a, b)$ is a linear transformation, then

$$(f, T(g)) = \int_a^b f(t)Tg(t)\, dt,$$

where we have written $Tg(t)$ for $T(g)(t)$. In this case the conditions for symmetry and skew-symmetry become

(7.17) $$\int_a^b \{f(t)Tg(t) - g(t)Tf(t)\}\, dt = 0 \qquad \text{if } T \text{ is symmetric,}$$

and

(7.18) $$\int_a^b \{f(t)Tg(t) + g(t)Tf(t)\}\, dt = 0 \qquad \text{if } T \text{ is skew-symmetric.}$$

EXAMPLE 2.   *Multiplication by a fixed function.*   In the space $C(a, b)$ of Example 1, choose a fixed function p and define $T(f) = pf$, the product of p and f. For this T, Equation (7.17) is satisfied for all f and g in $C(a, b)$ because the integrand is zero. Therefore, multiplication by a fixed function is a symmetric operator.

EXAMPLE 3.   *The differentiation operator.*   In the space $C(a, b)$ of Example 1, let V be the subspace consisting of all functions f that have a continuous derivative in the open interval (a, b) and that also satisfy the *boundary condition* $f(a) = f(b)$. Let $D : V \to C(a, b)$ be the differentiation operator given by $D(f) = f'$. It is easy to prove that D is skew-symmetric. In this case the integrand in (7.18) is the derivative of the product fg, so the integral is equal to

$$\int_a^b (fg)'(t)\, dt = f(b)g(b) - f(a)g(a).$$

Since both f and g satisfy the boundary condition, we have $f(b)g(b) - f(a)g(a) = 0$. Thus, the boundary condition implies skew-symmetry of D. The only eigenfunctions in the subspace V are the nonzero constant functions. They belong to the eigenvalue 0.

EXAMPLE 4.   *Sturm-Liouville operators.*   This example is important in the theory of linear second-order differential equations. We use the space $C(a, b)$ of Example 1 once more and let V be the subspace consisting of all f that have a continuous second derivative in $[a, b]$ and that also satisfy the two boundary conditions

(7.19) $$p(a)f(a) = 0, \qquad p(b)f(b) = 0,$$

where p is a fixed function in $C(a, b)$ with a continuous derivative on $[a, b]$. Let q be another fixed function in $C(a, b)$ and let $T : V \to C(a, b)$ be the operator defined by the equation

$$T(f) = (pf')' + qf.$$

This is called a *Sturm-Liouville operator.* To test for symmetry we note that

$$fT(g) - gT(f) = f \cdot (pg')' - g \cdot (pf')'.$$

Using this in (7.17) and integrating both $\int_a^b f \cdot (pg')' \, dt$ and $\int_a^b g \cdot (pf')' \, dt$ by parts, we find

$$\int_a^b \{f(t)Tg(t) - g(t)Tf(t)\} \, dt = fpg' \Big|_a^b - \int_a^b pg'f' \, dt - gpf' \Big|_a^b + \int_a^b pg'f' \, dt = 0,$$

because both f and g satisfy the boundary conditions (7.19). Hence T is symmetric on V. The eigenfunctions of T are those nonzero f that satisfy, for some real λ, a differential equation of the form

$$(pf')' + qf = \lambda f$$

on $[a, b]$, and also satisfy the boundary conditions (7.19).

EXAMPLE 5.   Theorem 7.3 tells us that distinct eigenvalues of a symmetric transformation correspond to orthogonal eigenvectors. This example applies this result to those nonzero functions that satisfy a differential equation of the form

(7.20) $$(pf')' + qf = \lambda f$$

on an interval $[a, b]$, and that also satisfy the boundary conditions $p(a)f(a) = p(b)f(b) = 0$. The conclusion is that any two solutions f and g corresponding to two distinct values of λ are orthogonal. For example, consider the differential equation of simple harmonic motion,

$$f'' + k^2 f = 0$$

on the interval $[0, \pi]$, where $k \neq 0$. This has the form (7.20) with $p = 1$, $q = 0$, and $\lambda = -k^2$. All solutions are given by $f(t) = c_1 \cos kt + c_2 \sin kt$. The boundary condition $f(0) = 0$ implies $c_1 = 0$. The second boundary condition, $f(\pi) = 0$, implies $c_2 \sin k\pi = 0$. Since $c_2 \neq 0$ for a nonzero solution, we must have $\sin k\pi = 0$, which means that k is an integer. In other words, nonzero solutions that satisfy the boundary conditions exist if and only if k is an integer. These solutions are $f(t) = \sin nt, n = \pm 1, \pm 2, \dots$. The orthogonality condition implied by Theorem 7.3 now becomes the familiar relation

$$\int_0^\pi \sin nt \sin mt \, dt = 0,$$

valid whenever m^2 and n^2 are distinct integers.

7.22 Exercises

The exercises in this section require a knowledge of calculus.

1. Let $C(0, 1)$ be the real linear space of all real functions continuous on $[0, 1]$ with inner product $(f, g) = \int_0^1 f(t)g(t) \, dt$. Let V be the subspace of all f such that $\int_0^1 f(t) \, dt = 0$. Let $T : V \to C(0, 1)$ be the integration operator defined by $Tf(x) = \int_0^x f(t) \, dt$. Prove that T is skew-symmetric.
2. Let V be the Euclidean space of all real polynomials with inner product $(f, g) = \int_{-1}^1 f(t)g(t) \, dt$. Determine which of the following transformations $T : V \to V$ is symmetric or skew-symmetric:
   (a) $Tf(x) = f(-x)$.   (c) $Tf(x) = f(x) + f(-x)$.
   (b) $Tf(x) = f(x)f(-x)$.   (d) $Tf(x) = f(x) - f(-x)$.

3. Refer to Example 4 in Section 7.21. Modify the inner product as follows:

$$(f, g) = \int_a^b f(t)g(t)w(t)\,dt,$$

where w is a fixed positive function in $C(a, b)$. Modify the Sturm-Liouville operator T by writing

$$T(f) = \frac{(pf')' + qf}{w}.$$

Prove that the modified operator is symmetric on the subspace V.

4. This exercise shows that the Legendre polynomials (introduced in Section 3.14) are eigenfunctions of a Sturm-Liouville operator. The Legendre polynomials are defined by the equation

$$P_n(t) = \frac{1}{2^n n!} f_n^{(n)}(t), \text{ where } f_n(t) = (t^2 - 1)^n.$$

(a) Verify that $(t^2 - 1)f_n'(t) = 2nt f_n(t)$.

(b) Differentiate the equation in (a) $n + 1$ times to obtain

$$(t^2 - 1)f_n^{(n+2)}(t) + 2t(n + 1)f_n^{(n+1)}(t) + n(n + 1)f_n^{(n)}(t) = 2nt f_n^{(n+1)}(t) + 2n(n + 1)f_n^{(n)}(t).$$

You may use Leibniz's formula for the nth derivative $h^{(n)}(t)$ of a product $h(t) = f(t)g(t)$:

$$h^{(n)}(t) = \sum_{k=0}^n \binom{n}{k} f^{(k)}(t)g^{(n-k)}(t).$$

(c) Show that the equation in part (b) can be rewritten in the form

$$[(t^2 - 1)P_n'(t)]' = n(n + 1)P_n(t).$$

This shows that $P_n(t)$ is an eigenfunction of the Sturm-Liouville operator T given on the interval $[-1, 1]$ by $T(f) = (pf')'$, where $p(t) = t^2 - 1$. The eigenfunction $P_n(t)$ belongs to the eigenvalue $\lambda = n(n + 1)$. In this example the boundary conditions for symmetry are automatically satisfied because $p(1) = p(-1) = 0$.

8

APPLICATIONS TO LINEAR DIFFERENTIAL EQUATIONS

8.1 Introduction

One of the most important applications of linear algebra is to the study of differential equations. Differential equations arise naturally in the physical world when one tries to determine something from a knowledge of its rate of change. For example, we can try to find the position of a moving particle from a knowledge of its velocity (first derivative of position) or its acceleration (second derivative of position). Or, a radioactive substance disintegrates at a known rate and we try to determine the amount of material present at a given time. In examples like these, the problem is to determine an unknown function $y = f(x)$ from prescribed information about its derivatives y' and/or y'' expressed as an equation, called a *differential equation*. Any function satisfying the differential equation is called a *solution*.

The *order* of a differential equation is the order of the highest derivative that appears in the equation. Thus, the differential equation $y' = y$ is of first order, while $y'' + 16y = 0$ is of second order. All solutions of $y' = y$ have the form $y = ce^x$, where c is any constant. All solutions of $y'' + 16y = 0$ have the form $y = c_1 \cos 4x + c_2 \sin 4x$, where c_1 and c_2 are arbitrary constants.

The history of differential equations began in the 17th century when Newton, Leibniz, and the Bernoullis solved some simple differential equations of first and second order arising from problems in geometry and mechanics. These early discoveries seemed to suggest that solutions of differential equations arising from geometric or physical problems could be expressed in terms of the familiar elementary functions of calculus. Much of the early work was aimed at developing ingenious techniques for solving differential equations by elementary means, that is to say, by addition, subtraction, multiplication, division, composition, and integration, applied only a finite number of times to the familiar functions of calculus.

Special methods such as separation of variables and the use of integrating factors were devised more or less haphazardly before the end of the 17th century. During the 18th century, more systematic procedures were developed, primarily by Euler, Lagrange, and Laplace. It soon became apparent that relatively few differential equations could be solved by elementary means. Gradually, mathematicians began to realize that it was hopeless to seek methods for solving all differential equations. Instead, they found it more fruitful to ask whether or not a given differential equation has any solution at all and, when it has, to try to deduce properties of the solution from the differential equation itself. In this framework, mathematicians began to think of differential equations as new sources of functions.

Experience has shown that it is difficult to obtain results of great generality about solutions of differential equations, except for a few types. Among these are the so-called *linear* differential equations that occur in a great variety of physical problems. The next few sections summarize some of the principal results concerning these equations.

8.2 Review of results concerning linear differential equations of first and second orders

A linear differential equation of first order is one of the form

$$(8.1) \qquad y' + P(x)y = Q(x),$$

where P and Q are given (known) functions and y is the unknown function. This is one of the few general types for which all solutions can be obtained by elementary means. An explicit formula for determining all solutions is given in the following theorem.

THEOREM 8.1. *Assume P and Q are continuous on an open interval J. Choose any point a in J and let b be any real number. Then there is one and only one function $y = f(x)$ that satisfies the differential equation (8.1) and the initial condition $f(a) = b$. This function is given by the explicit formula*

$$(8.2) \qquad f(x) = be^{-A(x)} + e^{-A(x)} \int_a^x Q(t)e^{A(t)}\, dt,$$

where $A(x) = \int_a^x P(t)\, dt$.

Proof. We multiply each member of Equation (8.1) by $e^{A(x)}$ and note that $A'(x) = P(x)$. This converts the left member of (8.1) to $y'e^{A(x)} + ye^{A(x)}A'(x)$, the derivative of the product $ye^{A(x)}$, so the differential equation can now be written as

$$(ye^{A(x)})' = Q(x)e^{A(x)}.$$

The factor $e^{A(x)}$ that we used to convert the left member of (8.1) to the derivative of a product is called an integrating factor. Integrating this last equation over the interval $[a, x]$ we find

$$f(x)e^{A(x)} - f(a)e^{A(a)} = \int_a^x Q(t)e^{A(t)}\, dt.$$

Now $f(a) = b$ and $A(a) = 0$, and when we solve this equation for $f(x)$ we obtain (8.2).

Linear differential equations of the second order are those of the form

$$y'' + P_1(x)y' + P_2(x)y = R(x).$$

If the coefficients P_1, P_2 and the right-hand member R are continuous on some interval J, an existence theorem guarantees that solutions always exist over the interval J. Nevertheless, there is no general formula analogous to (8.2) for expressing these solutions in terms of P_1, P_2 and R. Thus, even in this relatively simple generalization of (8.1), the theory is far from complete, except in special cases. But if the coefficients are *constants* and if R is zero, all solutions can be determined explicitly in terms of polynomials, exponential and trigonometric functions as explained by the following theorem. It is easy to verify that the functions described in the theorem are, indeed, solutions of the differential equation. The fact that all solutions are contained in Equation (8.4) is a consequence of Theorem 8.5, given below.

THEOREM 8.2.   *Consider the differential equation*

$$(8.3) \qquad\qquad y'' + ay' + by = 0,$$

where a and b are given real constants. Let $d = a^2 - 4b$. Then every solution of (8.3) on the interval $(-\infty, +\infty)$ has the form

$$(8.4) \qquad\qquad y = e^{-ax/2}\big[c_1 u_1(x) + c_2 u_2(x)\big],$$

where c_1 and c_2 are constants, and the functions u_1 and u_2 are determined according to the algebraic sign of d as follows:
   (a) *If $d = 0$, then $u_1(x) = 1$ and $u_2(x) = x$.*
   (b) *If $d > 0$, then $u_1(x) = e^{kx}$ and $u_2(x) = e^{-kx}$, where $k = \frac{1}{2}\sqrt{d}$.*
   (c) *If $d < 0$, then $u_1(x) = \cos kx$ and $u_2(x) = \sin kx$, where $k = \frac{1}{2}\sqrt{-d}$.*

The number $d = a^2 - 4b$ is called the *discriminant* of the quadratic equation

$$(8.5) \qquad\qquad r^2 + ar + b = 0,$$

which is known as the *characteristic equation* of the differential equation in (8.3). Its roots are given by

$$r_1 = \frac{-a + \sqrt{d}}{2}, \qquad r_2 = \frac{-a - \sqrt{d}}{2}.$$

The algebraic sign of d determines the nature of these roots. If $d > 0$, both roots are real and the solution in (8.4) is a linear combination of exponential functions given by

$$y = c_1 e^{r_1 x} + c_2 e^{r_2 x}.$$

If $d < 0$, the roots r_1 and r_2 are conjugate complex numbers given by

$$r_1 = -\frac{1}{2}a + ik, \text{ and } r_2 = -\frac{1}{2}a - ik, \text{ where } k = \frac{1}{2}\sqrt{-d}.$$

In this case, the exponential functions $e^{r_1 x}$ and $e^{r_2 x}$ are complex conjugates, given by

$$e^{r_1 x} = e^{-ax/2} e^{ikx} = e^{-ax/2} \cos kx + i e^{-ax/2} \sin kx$$

and

$$e^{r_2 x} = e^{-ax/2} e^{-ikx} = e^{-ax/2} \cos kx - i e^{-ax/2} \sin kx.$$

The solution appearing in Equation (8.4) is a linear combination of the real and imaginary parts of these two complex exponentials.

8.3 Exercises

The exercises in this section are intended as review of introductory material on linear differential equations of first and second orders.

Linear equations of first order. In each of Exercises 1, 2, 3 use Theorem 8.1 to solve the initial-value problem on the specified interval.

1. $y' - 3y = e^{2x}$ on $(-\infty, +\infty)$, with $y = 0$ when $x = 0$.
2. $xy' - 2y = x^5$ on $(0, +\infty)$, with $y = 1$ when $x = 1$.
3. $y' + y \tan x = \sin 2x$ on $(-\frac{1}{2}\pi, \frac{1}{2}\pi)$, with $y = 2$ when $x = 0$.
4. If a strain of bacteria grows at a rate proportional to the amount present and if the population doubles in one hour, by how much will it increase at the end of two hours?
5. A curve with Cartesian equation $y = f(x)$ passes through the origin. Lines drawn parallel to the coordinate axes through an arbitrary point of the curve form a rectangle with two sides on the axes. The curve divides every such rectangle into two regions A and B, one of which has an area equal to n times the other. Find the function f.
6. (a) Let u be a nonzero solution of the second-order equation $y'' + P(x)y' + Q(x)y = 0$. Show that the substitution $y = uv$ converts the equation

$$y'' + P(x)y' + Q(x)y = R(x)$$

   into a first-order linear equation for v'.
   (b) Obtain a nonzero solution of the equation $y'' - 4y' + x^2(y' - 4y) = 0$ by inspection, and use the method of part (a) to find a solution of

$$y'' - 4y' + x^2(y' - 4y) = 2xe^{-x^3/3}$$

   such that $y = 0$ and $y' = 4$ when $x = 0$.
7. The change of variable $v = (y + 1)^2$ converts the nonlinear differential equation

$$(y + 1)y' + x(y^2 + 2y) = x$$

   into a first-order linear equation for v. Use this information to find all solutions $y = f(x)$ of the nonlinear equation.

Linear equations of second order with constant coefficients. In each of Exercises 8 through 11, find all solutions on $(-\infty, +\infty)$.

8. $y'' - 4y = 0$.
9. $y'' + 4y = 0$.
10. $y'' - 2y' + 5y = 0$.
11. $y'' + 2y' + y = 0$.
12. (a) Find that solution $y = f(x)$ of the differential equation $y'' - 2y' - 3y = 0$ such that $f(0) = 5$ and $f'(0) = 35$.
    (b) Find constants A, B, C such that the function $y = Ae^x + B \sin x + C \cos x$ satisfies the equation $y'' - 2y' - 3y = 2e^x - 10 \sin x$.
13. Find all values of the constant k such that the differential equation $y'' + ky = 0$ has a non-trivial solution $y = f_k(x)$ for which $f_k(0) = f_k(1) = 0$. For each permissible k, determine the corresponding solution $y = f_k(x)$. Consider both positive and negative values of k.
14. If (a, b) is a given point in the plane and if m is a given real number, prove that the differential equation $y'' + k^2y = 0$ has exactly one solution whose graph passes through (a, b) and has slope m there. Discuss separately the case $k = 0$.
15. In each case, find a linear differential equation of second order satisfied by u_1 and u_2.
    (a) $u_1(x) = e^x$, $\quad u_2(x) = e^{-x}$.
    (b) $u_1(x) = e^{2x}$, $\quad u_2(x) = xe^{2x}$.
    (c) $u_1(x) = e^{-x/2} \cos x$, $\quad u_2(x) = e^{-x/2} \sin x$.

(d) $u_1(x) = \sin(2x + 1)$, $u_2(x) = \sin(2x + 2)$.

(e) $u_1(x) = \cosh x$, $u_2(x) = \sinh x$. *Note:* $\cosh x = \frac{1}{2}(e^x + e^{-x})$, $\sinh x = \frac{1}{2}(e^x - e^{-x})$.

16. A particle undergoes simple harmonic motion. Initially it has displacement 1, velocity 1, and acceleration -12. Compute its displacement and acceleration when the velocity is $\sqrt{8}$.

17. Given the differential equation $y'' + P(x)y' + Q(x)y = 0$, where P and Q are real functions, continuous everywhere.

   (a) If the equation has two solutions $y_1 = e^{ax}$ and $y_2 = e^{bx}$, where a, b are distinct real constants prove that both P and Q are constants and that, in fact, $P(x) = -(a + b)$, and $Q(x) = ab$.

   (b) Now assume the differential equation has the two solutions $y_1 = e^{ax}$ and $y_2 = xe^{ax}$, where a is a real constant. Prove or disprove: Both P and Q are constants.

8.4   Linear differential equations of order n

The discussion of linear differential equations of higher order can be simplified by the use of operator notation. Let $C(J)$ denote the linear space of all real-valued functions continuous on an interval J, where J can be either a bounded or an unbounded interval. Let $C^{(n)}(J)$ denote the subspace consisting of all functions f whose first n derivatives f', $f'', \ldots, f^{(n)}$ exist and are continuous on J. Let $P_1, \ldots, P_n$ be n given functions in $C(J)$. Consider the operator $L : C^{(n)}(J) \to C(J)$ defined by

$$L(f) = f^{(n)} + P_1 f^{(n-1)} + \cdots + P_n f.$$

The operator L itself is sometimes written as

$$L = D^n + P_1 D^{n-1} + \cdots + P_n,$$

where D^k denotes the kth derivative operator. In operator notation, a linear differential equation of order n is one of the form

(8.6) $$L(y) = R,$$

where R is a given function defined on J. A solution of this equation is any function y in $C^{(n)}(J)$ that satisfies (8.6) on the interval J.

It is easy to verify that $L(y_1 + y_2) = L(y_1) + L(y_2)$, and that $L(cy) = cL(y)$ for every constant c. Therefore L is a *linear operator*. That is why the differential equation $L(y) = R$ is referred to as a linear equation. The operator L is called a *linear differential operator of order n*.

With each linear equation $L(y) = R$ we can associate the equation

$$L(y) = 0$$

in which the right-hand side has been replaced by zero. This is called the *homogeneous equation* corresponding to $L(y) = R$. When R is not identically zero, the equation $L(y) = R$ is called a *nonhomogeneous equation*. It turns out that we can always solve the nonhomogeneous equation whenever we can solve the corresponding homogeneous equation. Therefore we treat first the homogeneous case.

The set of solutions of the homogeneous equation is the null space $N(L)$ of the operator L. This is also called the *solution space* of the equation. The solution space is a subspace of $C^{(n)}(J)$. Although $C^{(n)}(J)$ is infinite-dimensional, the solution space $N(L)$ is always finite-dimensional, and in fact, its dimension is the order n of the operator L. This result is a consequence of the following uniqueness-existence theorem.

8.5 The existence-uniqueness theorem

THEOREM 8.3.   EXISTENCE-UNIQUENESS THEOREM FOR LINEAR EQUATIONS OF ORDER n. *Let $P_1, \ldots, P_n$ be continuous functions on an open interval J, and let L be the linear differential operator*

$$L = D^n + P_1 D^{n-1} + \cdots + P_n.$$

If $x_0 \in J$ and if $k_0, k_1, \ldots, k_{n-1}$ are n given real numbers, then there exists one and only one function $y = f(x)$ that satisfies the homogeneous differential equation $L(y) = 0$ on J and that also satisfies the initial conditions

$$f(x_0) = k_0, f'(x_0) = k_1, \ldots, f^{(n-1)}(x_0) = k_{n-1}.$$

Note. The vector in $\mathbf{R}^n$ given by $(f(x_0), f'(x_0), \ldots, f^{(n-1)}(x_0))$ is called the *initial-value vector* of f at x_0. Theorem 8.3 states that if we choose a point x_0 in J and a vector in $\mathbf{R}^n$, then the homogeneous equation $L(y) = 0$ has exactly one solution $y = f(x)$ on J with this vector as initial-value vector at x_0. In particular, only the zero solution has initial-value vector O at any point. When $n = 2$ there is exactly one solution with prescribed value $f(x_0)$ and prescribed derivative $f'(x_0)$ at a prescribed point x_0. For example, if $f(t)$ represents the position of a particle at time t undergoing simple harmonic motion, then $f(t)$ is completely determined by specifying its initial position at time t_0 and its initial velocity $f'(t_0)$.

A proof of the existence-uniqueness theorem will be obtained as a corollary of more general existence-uniqueness theorems for systems of differential equations discussed in Chapter 10. An alternate proof for equations with constant coefficients is given in Section 9.9.

8.6 The dimension of the solution space of a homogeneous linear differential equation

THEOREM 8.4.   DIMENSIONALITY THEOREM.   *Let $L : C^{(n)}(J) \to C(J)$ be a linear differential operator of order n given by*

(8.7) $$L = D^n + P_1 D^{n-1} + \cdots + P_n.$$

Then the solution space of the equation $L(y) = 0$ has dimension n.

Proof.   To illustrate how the theorems of linear algebra come into play, we shall deduce the dimensionality theorem from the existence-uniqueness theorem.

Let $\mathbf{R}^n$ denote the linear space of n-tuples of real scalars. Let T be the linear transformation that maps each function f in the solution space $N(L)$ onto the initial-value vector of f at x_0; that is,

$$T(f) = (f(x_0), f'(x_0), \ldots, f^{(n-1)}(x_0)),$$

where x_0 is a fixed point in J. The uniqueness theorem tells us that $T(f) = O$ implies $f = 0$. Therefore, by Theorem 4.10, T is one-to-one on $N(L)$. Hence T^{-1} is also one-to-one and maps $\mathbf{R}^n$ onto $N(L)$. Therefore, by Theorem 4.11, $\dim N(L) = \dim \mathbf{R}^n = n$.

Once we know that the solution space has dimension n, any set of n independent solutions will serve as a basis. Therefore, as a corollary of the dimensionality theorem we have:

THEOREM 8.5.   *Let $L : C^{(n)}(J) \to C(J)$ be a linear differential operator of order n. If $u_1, \ldots, u_n$ are n independent solutions of the homogeneous differential equation $L(y) = 0$ on J, then every solution $y = f(x)$ on J can be expressed in the form*

$$(8.8) \qquad f(x) = \sum_{k=1}^{n} c_k u_k(x),$$

where $c_1, \ldots, c_n$ are constants.

> *Note.* Because all solutions of the differential equation $L(y) = 0$ are contained in formula (8.8), the linear combination on the right with arbitrary constants $c_1, \ldots, c_n$ is called the *general solution* of the differential equation.

The dimensionality theorem tells us that the solution space of a homogeneous linear differential equation of order n always has a basis of n solutions, but it does not tell us how to determine such a basis. In fact, no simple method is known for determining a basis of solutions for every linear equation. However, special methods have been devised for special equations. Among these are differential equations with constant coefficients. In this case, the problem of determining a basis can be reduced to an algebraic problem of finding the roots of a polynomial of degree n.

8.7   The algebra of constant-coefficient operators

A constant-coefficient operator A is a linear operator of the form

$$(8.9) \qquad A = a_0 D^n + a_1 D^{n-1} + \cdots + a_{n-1} D + a_n,$$

where D is the derivative operator and $a_0, a_1, \ldots, a_n$ are real constants. If $a_0 \neq 0$ the operator is said to have order n. The operator A can be applied to any function y with derivatives $y', y'', \ldots, y^{(n)}$ in some interval, the result being a function $A(y)$ given by

$$A(y) = a_0 y^{(n)} + a_1 y^{(n-1)} + \cdots + a_{n-1} y' + a_n y.$$

In this section we restrict our attention to functions having derivatives of every order on $(-\infty, +\infty)$. The set of all such functions is denoted by C^∞ and is referred to as the class of *infinitely differentiable functions*. If $y \in C^\infty$ then $A(y)$ is also in C^∞.

The usual algebraic operations on linear transformations (*addition, multiplication by scalars,* and *multiplication by composition*) are applicable to constant-coefficient operators. The sum of two constant-coefficient operators A and B (not necessarily of the same order) is a constant-coefficient operator, as is the scalar multiple λA, so the set of all constant-ceofficient operators is a linear space. The product (composition) of A and B is also a constant-coefficient operator. Therefore sums, products, and scalar multiples of constant-coefficient operators satisfy the usual commutative, associative, and distributive laws satisfied by all linear transformations. Moreover, because we have $D^r D^s = D^s D^r$ for all positive integers r and s, any two constant-coefficient operators commute: $AB = BA$.

With each constant-coefficient operator A, as given by Equation (8.9), we associate a polynomial p_A, called the *associated polynomial* of A, that has the same coefficients as A. Thus, we have

$$p_A(r) = a_0 r^n + a_1 r^{n-1} + \cdots + a_{n-1} r + a_n.$$

Conversely, given any polynomial p with real coefficients, there is a corresponding operator A whose coefficients are the same as those of p. The next theorem shows that this association between operators and polynomials is a one-to-one correspondence. Moreover, the correspondence associates with sums, compositions, and scalar multiples of operators the respective sums, products, and scalar multiples of their associated polynomials.

THEOREM 8.6.   *Let A and B denote constant-coefficient operators with associated polynomials p_A and p_B, respectively, and let λ be any real number. Then we have:*

(a) $A = B$ *if and only if* $p_A = p_B$.
(b) $p_{A+B} = p_A + p_B$.
(c) $p_{AB} = p_A \cdot p_B$, *the product of p_A and p_B.*
(d) $p_{\lambda A} = \lambda p_A$.

Proof.   We consider part (a) first. Assume $p_A = p_B$. Since both polynomials p_A and p_B have the same degree and the same coefficients, the operators A and B have the same order and the same coefficients, so $A = B$.

Next we prove that $A = B$ implies $p_A = p_B$. The equation $A = B$ means that $A(y) = B(y)$ for every y in C^∞. Take $y = e^{rx}$, where r is a constant. Then $y^{(k)} = r^k e^{rx}$ for every $k \geq 0$, so

$$A(y) = p_A(r)e^{rx} \quad \text{and} \quad B(y) = p_B(r)e^{rx}.$$

Put these in the equation $A(y) = B(y)$ and cancel the nonzero factor e^{rx} to find $p_A(r) = p_B(r)$. Since r is arbitrary, we must have $p_A = p_B$. This completes the proof of part (a).

Parts (b), (c) and (d) follow at once from the definition of the associated polynomial.

From Theorem 8.6 we see that every algebraic relation involving sums, products and scalar multiples of polynomials p_A and p_B also holds for the operators A and B. In particular, if the associated polynomial p_A can be factored as a product of two or more polynomials, each factor must be the associated polynomial of some constant-coefficient operator, so, by Theorem 8.6, there is a corresponding factorization of the operator A. For example, if $p_A(r) = p_B(r)p_C(r)$, then $A = BC$. If $p_A(r)$ can be factored as a product of n linear factors, say,

(8.10) $$p_A(r) = a_0(r - r_1)(r - r_2) \cdots (r - r_n),$$

the corresponding factorization of A takes the form

$$A = a_0(D - r_1)(D - r_2) \cdots (D - r_n).$$

The fundamental theorem of algebra tells us that every polynomial $p_A(r)$ of degree $n \geq 1$ has a factorization of the form (8.10), where $r_1, \ldots, r_n$ are its roots. Each root is written as often as its multiplicity indicates (double roots are written twice, triple roots three times, etc.). The roots may be real or complex. Because $p_A(r)$ has real coefficients, the complex roots occur in conjugate pairs, $\alpha \pm i\beta$ if $\beta \neq 0$. The two linear factors corresponding to each such pair can be combined to give one quadratic factor $r^2 - 2\alpha r + \alpha^2 + \beta^2$ whose coefficients are real. Therefore, every polynomial $p_A(r)$ can be factored as a product of linear and quadratic polynomials with *real coefficients*. This gives a corresponding factorization of the operator A as a product of first-order and second-order constant-coefficient operators with real coefficients.

EXAMPLE 1. Let $A = D^2 - 5D + 6$. The associated polynomial $p_A(r)$ has the factorization $r^2 - 5r + 6 = (r - 2)(r - 3)$, so operator A has the factorization

$$D^2 - 5D + 6 = (D - 2)(D - 3).$$

EXAMPLE 2. Let $A = D^4 - 2D^3 + 2D^2 - 2D + 1$. The associated polynomial $p_A(r)$ has the factorization

$$r^4 - 2r^3 + 2r^2 - 2r + 1 = (r - 1)(r - 1)(r^2 + 1),$$

so A has the factorization $A = (D - 1)(D - 1)(D^2 + 1)$.

8.8 Determination of a basis of solutions for linear equations with constant coefficients by factorization of operators

The next theorem shows how factorization of constant-coefficient operators helps us solve linear differential equations with constant coefficients.

THEOREM 8.7. *Let L be a constant-coefficient operator that can be factored as a product of constant-coefficient operators, say*

$$L = A_1 A_2 \cdots A_k.$$

Then the solution space of the linear differential equation $L(y) = 0$ contains the solution space of each differential equation $A_i(y) = 0$. In other words,

(8.11) $\qquad\qquad N(A_i) \subseteq N(L) \qquad \textit{for each } i = 1, 2, \ldots, k.$

Proof. If u is in the null space of the last factor A_k we have $A_k(u) = O$. Therefore the null space of L contains the null space of the last factor A_k. Constant-coefficient operators commute, so we can rearrange the factors to make any one of them the last factor. This proves (8.11).

If $L(u) = O$ the operator L is said to *annihilate* u. Theorem 8.7 tells us that if a factor A_i of L annihilates u, the L also annihilates u.

The following examples shows how the theorem can be used to solve homogeneous differential equations with constant coefficients. The examples have been chosen to illustrate different features, depending on the nature of the roots of the associated polynomial.

CASE 1.   Real distinct roots.

EXAMPLE 1.   Find a basis of solutions for the differential equation

(8.12) $(D^3 - 7D + 6)y = 0.$

Solution. This has the form $L(y) = 0$ with

$$L = D^3 - 7D + 6 = (D - 1)(D - 2)(D + 3).$$

The null space of $(D - 1)$ contains $u_1(x) = e^x$, that of $(D - 2)$ contains $u_2(x) = e^{2x}$, and that of $(D + 3)$ contains $u_3(x) = e^{-3x}$. In Section 3.7, Example 7, we proved that the exponential functions u_1, u_2, u_3 are independent. Any three independent solutions of a third-order equation form a basis for the solution space, so the general solution of (8.12) is given by

$$y = c_1 e^x + c_2 e^{2x} + c_3 e^{-3x}.$$

The method used to solve Example 1 can be used to prove the following theorem, which tells how to find a basis for the solution space of any constant-coefficient operator that can be factored into distinct linear factors.

THEOREM 8.8.   *Let L be a constant-coefficient operator whose associated polynomial $p_L(r)$ has n distinct real roots $r_1, \ldots, r_n$. Then the general solution of the differential equation $L(y) = 0$ on the interval $(-\infty, +\infty)$ is given by a linear combination of exponential functions,*

(8.13) $$y = \sum_{k=1}^{n} c_k e^{r_k x}.$$

Proof.   We have the factorization

$$L = a_0(D - r_1) \cdots (D - r_n).$$

Because the null space of $(D - r_k)$ contains the exponential function $u_k(x) = e^{r_k x}$, the null space of L contains the n exponential functions

(8.14) $u_1(x) = e^{r_1 x}, \ldots, u_n(x) = e^{r_n x}.$

In Section 3.7, Example 7, we proved that these functions are independent. Therefore they form a basis for the solution space of the differential equation $L(y) = 0$, so the general solution is given by the linear combination in (8.13).

CASE 2.   Real roots, some of which are repeated. If all the roots are real but not distinct, the functions in (8.14) are not independent and therefore do not form a basis for the solution space. If a root r occurs with multiplicity m then $(D - r)^m$ is a factor of L. The next theorem shows how to obtain m independent solutions in the null space of this factor.

THEOREM 8.9.   *The m functions*

$$u_1(x) = e^{rx}, u_2(x) = xe^{rx}, \ldots, u_m(x) = x^{m-1}e^{rx}$$

are m independent elements annihilated by the operator $(D - r)^m$.

Proof.   Independence of these functions follows from that of the polynomials 1, x, $x^2, \ldots, x^{m-1}$. To prove that $u_1, \ldots, u_m$ are annihilated by $(D - r)^m$ we use induction on m.

If $m = 1$ there is only one function, $u_1(x) = e^{rx}$, which is clearly annihilated by $(D - r)$. Suppose, then, that the theorem is true for $m - 1$. This means that the functions $u_1, \ldots, u_{m-1}$ are annihilated by $(D - r)^{m-1}$. Since

$$(D - r)^m = (D - r)(D - r)^{m-1}$$

the functions $u_1, \ldots, u_{m-1}$ are also annihilated by $(D - r)^m$. To complete the proof we must show that $(D - r)^m$ annihilates u_m as well. Therefore we consider

$$(D - r)^m u_m = (D - r)^{m-1}(D - r)(x^{m-1}e^{rx}).$$

We have

$$(D - r)(x^{m-1}e^{rx}) = D(x^{m-1}e^{rx}) - rx^{m-1}e^{rx}$$

$$= (m - 1)x^{m-2}e^{rx} + x^{m-1}re^{rx} - rx^{m-1}e^{rx}$$

$$= (m - 1)x^{m-2}e^{rx} = (m - 1)u_{m-1}(x).$$

When we apply $(D - r)^{m-1}$ to both members of this last equation we get 0 on the right because $(D - r)^{m-1}$ annihilates u_{m-1}. Hence $(D - r)^m u_m = 0$, so u_m is annihilated by $(D - r)^m$. This completes the proof.

EXAMPLE 2.   Find the general solution of the differential equation $L(y) = 0$, where

$$L = D^3 - D^2 - 8D + 12.$$

Solution. The operator L has the factorization

$$L = (D - 2)^2(D + 3).$$

By Theorem 8.9, the two functions

$$u_1(x) = e^{2x}, \qquad u_2(x) = xe^{2x}$$

are in the null space of $(D - 2)^2$. The function $u_3(x) = e^{-3x}$ is in the null space of $(D + 3)$. Since u_1, u_2, u_3 are independent (see Exercise 17 of Section 8.9) they form a basis for the null space of L, so the general solution of the differential equation is

$$y = c_1e^{2x} + c_2xe^{2x} + c_3e^{-3x}.$$

Theorem 8.9 tells us how to find a basis of solutions for any nth order linear equation with constant coefficients whose associated polynomial has only real roots, some of which are repeated. If the *distinct* roots are $r_1, \ldots, r_k$, and if they occur with respective multiplicities $m_1, \ldots, m_k$, that part of the basis corresponding to r_p is given by the m_p functions

$$u_{q,p}(x) = x^{q-1} e^{r_p x}, \qquad \text{where} \qquad q = 1, 2, \ldots, m_p.$$

As p takes the values $1, 2, \ldots, k$ we get $m_1 + \cdots + m_k$ functions altogether. In Exercise 17 of Section 8.9 we outline a proof showing that all these functions are independent. Since the sum of the multiplicities $m_1 + \cdots + m_k$ is equal to n (the order of the equation), the n functions $u_{p,q}$ form a basis for the solution space of the equation.

EXAMPLE 3.   Solve the equation $(D^6 + 2D^5 - 2D^3 - D^2)y = 0$.

Solution. We have $D^6 + 2D^5 - 2D^3 - D^2 = D^2(D - 1)(D + 1)^3$. The part of the basis corresponding to the factor D^2 is $u_1(x) = 1$, $u_2(x) = x$; the part corresponding to the factor $(D - 1)$ is $u_3(x) = e^x$; and the part corresponding to the factor $(D + 1)^3$ consists of three functions: $u_4(x) = e^{-x}$, $u_5(x) = xe^{-x}$, and $u_6(x) = x^2 e^{-x}$. The six functions $u_1, \ldots, u_6$ are independent, so the general solution of the equation is

$$y = c_1 + c_2 x + c_3 e^x + (c_4 + c_5 x + c_6 x^2) e^{-x}.$$

CASE 3.   *Complex Roots.* If complex exponentials are used, there is no need to distinguish between real and complex roots of the associated polynomial of the operator L. If real-valued solutions are desired, we factor L into linear and quadratic factors with real coefficients. Each pair of conjugate complex roots $\alpha \pm i\beta$ corresponds to a quadratic factor

(8.15) $$D^2 - 2\alpha D + \alpha^2 + \beta^2.$$

The null space of this second-order operator contains the two independent functions $u(x) = e^{\alpha x} \cos \beta x$ and $v(x) = e^{\alpha x} \sin \beta x$. If the pair of roots $\alpha \pm i\beta$ occurs with multiplicity m, the quadratic factor occurs to the mth power. The null space of the operator

$$(D^2 - 2\alpha D + \alpha^2 + \beta^2)^m$$

contains the $2m$ independent functions

$$u_q(x) = x^{q-1} e^{\alpha x} \cos \beta x, \qquad v_q(x) = x^{q-1} e^{\alpha x} \sin \beta x, \qquad q = 1, 2, \ldots, m.$$

These facts can be proved by induction on m. The following examples illustrate some of the possibilities.

EXAMPLE 4. $y''' - 4y'' + 13y' = 0$. The associated polynomial, $r^3 - 4r^2 + 13r$, has the roots $0, 2 \pm 3i$; the general solution is

$$y = c_1 + e^{2x}(c_2 \cos 3x + c_3 \sin 3x).$$

EXAMPLE 5. $y''' - 2y'' + 4y' - 8y = 0$. The associated polynomial is

$$r^3 - 2r^2 + 4r - 8 = (r - 2)(r^2 + 4);$$

its roots are 2, $\pm 2i$; so the general solution of the differential equation is

$$y = c_1 e^{2x} + c_2 \cos 2x + c_3 \sin 2x.$$

EXAMPLE 6. $y^{(5)} - 9y^{(4)} + 34y''' - 66y'' + 65y' - 25y = 0$. The associated polynomial has the factorization

$$(r - 1)(r^2 - 4r + 5)^2;$$

its roots are 1, $2 \pm i$, $2 \pm i$, so the general solution of the differential equation is

$$y = c_1 e^x + e^{2x}\left[(c_2 + c_3 x)\cos x + (c_4 + c_5 x)\sin x\right].$$

8.9 Exercises

Find the general solution of each of the differential equations in Exercises 1 through 12.

1. $y''' - 2y'' - 3y' = 0$.
2. $y''' - y' = 0$.
3. $y''' + 4y'' + 4y' = 0$.
4. $y''' - 3y'' + 3y' - y = 0$.
5. $y^{(4)} + 4y''' + 6y'' + 4y' + y = 0$.
6. $y^{(4)} - 16y = 0$.

7. $y^{(4)} + 16y = 0$.
8. $y''' - y = 0$.
9. $y^{(4)} + 4y''' + 8y'' + 8y' + 4y = 0$.
10. $y^{(4)} + 2y'' + y = 0$.
11. $y^{(6)} + 4y^{(4)} + 4y'' = 0$.
12. $y^{(6)} + 8y^{(4)} + 16y'' = 0$.

13. If m is a positive constant, find the particular solution $y = f(x)$ of the differential equation

$$y''' - my'' + m^2 y' - m^3 y = 0$$

that satisfies the conditions $f(0) = f'(0) = 0$, $f''(0) = 1$.

14. A linear differential equation with constant coefficients has associated polynomial $f(r)$, all of whose roots are real and negative. Prove that every solution of the differential equation approaches zero as $x \to +\infty$. What can you conclude about the behavior of all solutions on the interval $[0, +\infty)$ if all the roots of the associated polynomial are nonpositive?

15. In each case, find a linear differential equation with constant coefficients satisfied by all the given functions.
    (a) $u_1(x) = e^x$, $\quad u_2(x) = e^{-x}$, $\quad u_3(x) = e^{2x}$, $\quad u_4(x) = e^{-2x}$.
    (b) $u_1(x) = e^{-2x}$, $\quad u_2(x) = xe^{-2x}$, $\quad u_3(x) = x^2 e^{-2x}$.
    (c) $u_1(x) = 1$, $\quad u_2(x) = x$, $\quad u_3(x) = e^x$, $\quad u_4(x) = xe^x$.
    (d) $u_1(x) = x$, $\quad u_2(x) = e^x$, $\quad u_3(x) = xe^x$.
    (e) $u_1(x) = x^2$, $\quad u_2(x) = e^x$, $\quad u_3(x) = xe^x$.
    (f) $u_1(x) = e^{-2x}\cos 3x$, $\quad u_2(x) = e^{-2x}\sin 3x$, $\quad u_3(x) = e^{-2x}$, $\quad u_4(x) = xe^{-2x}$.
    (g) $u_1(x) = \cosh x$, $\quad u_2(x) = \sinh x$, $\quad u_3(x) = x\cosh x$, $\quad u_4(x) = x\sinh x$.
    (h) $u_1(x) = \cosh x \sin x$, $\quad u_2(x) = \sinh x \cos x$, $\quad u_3(x) = x$.

16. Let $r_1, \ldots, r_n$ be n distinct real numbers and let $Q_1, \ldots, Q_n$ be n polynomials, none of which is the zero polynomial. Prove that the n functions

$$u_1(x) = Q_1(x)e^{r_1 x}, \ldots, u_n(x) = Q_n(x)e^{r_n x},$$

are independent.

Outline of proof. Use induction on n. For $n = 1$ and $n = 2$ the result is easily verified. Assume the statement is true for $n = p$ and let $c_1, \ldots, c_{p+1}$ be $p + 1$ scalars such that

$$\sum_{k=1}^{p+1} c_k Q_k(x) e^{r_k x} = 0.$$

Multiply both sides by $e^{-r_{p+1} x}$ and differentiate the resulting equation. Then use the induction hypothesis to show that all the scalars c_k are zero. An alternate proof can be given based on order of magnitude as $x \rightarrow +\infty$, as was done in Example 7 of Section 3.7.

17. Let $m_1, m_2, \ldots, m_k$ be k positive integers, let $r_1, r_2, \ldots, r_k$ be k distinct real numbers, and let $n = m_1 + \cdots + m_k$. For each pair of integers p, q satisfying $1 \le p \le k$, $1 \le q \le m_p$, let

$$u_{q,p}(x) = x^{q-1} e^{r_p x}.$$

For example, when $p = 1$ the corresponding functions are

$$u_{1,1}(x) = e^{r_1 x}, \qquad u_{2,1}(x) = x e^{r_1 x}, \ldots, u_{m_1,1}(x) = x^{m_1 - 1} e^{r_1 x}.$$

Prove that the n functions $u_{q,p}(x)$ so defined are independent. [*Hint:* Use Exercise 16.]

8.10 The relation between the homogeneous and nonhomogeneous equations

We return now to the general linear differential equation of order n with coefficients that are not necessarily constant. The next theorem describes the relation between solutions of a homogeneous equation $L(y) = 0$ and those of a nonhomogeneous equation $L(y) = R(x)$.

THEOREM 8.10. *Let $L : C^{(n)}(J) \rightarrow C(J)$ be a linear differential operator of order n. Let $u_1, \ldots, u_n$ be n independent solutions of the homogeneous equation $L(y) = 0$, and let y_1 be a particular solution of the nonhomogeneous equation $L(y) = R$, where $R \in C(J)$. Then every solution $y = f(x)$ of the nonhomogeneous equation has the form*

(8.16)
$$f(x) = y_1(x) + \sum_{k=1}^{n} c_k u_k(x),$$

where $c_1, \ldots, c_n$ are constants.

Proof. Let f be any solution of the nonhomogeneous equation. By linearity we have $L(f - y_1) = L(f) - L(y_1) = R - R = 0$. Therefore $f - y_1$ is in the solution space of the homogeneous equation, so $f - y_1$ is a linear combination of $u_1, \ldots, u_n$, from which we get (8.16).

Because all the solutions of $L(y) = R$ are found in (8.16), the sum on the right (with arbitrary constants $c_1, \ldots, c_n$) is called the general solution of the nonhomogeneous equation. Theorem 8.10 tells us that the general solution of the nonhomogeneous equation is obtained by adding a particular solution y_1 to the general solution of the homogeneous equation.

Note. Theorem 8.10 has a simple geometric analogy that helps gives an insight into its meaning. To determine all points on a plane we find a particular point on the plane and add to it all points on the parallel plane through the origin. To find all solutions of $L(y) = R$ we find a particular solution and add to it all solutions of the homogeneous equation $L(y) = 0$. The set of solutions of the nonhomogeneous equation is analogous to a plane through a particular point. The solution space of the homogeneous equation is analogous to a parallel plane through the origin.

When using Theorem 8.10 in practice, we must solve two problems: (1) Find the general solution of the homogeneous equation $L(y) = 0$, and (2) find a particular solution of the nonhomogeneous equation $L(y) = R$. The next section shows that we can always solve problem (2) if we can solve problem (1).

8.11 Determination of a particular solution of the nonhomogeneous equation. The method of variation of parameters

We turn now to the problem of determining one particular solution y_1 of the nonhomogeneous equation $L(y) = R$. We shall describe a powerful method known as *variation of parameters*, which tells us how to determine y_1 if we know n independent solutions $u_1, \ldots, u_n$ of the homogeneous equation $L(y) = 0$. The method provides a particular solution of the form

$$(8.17) \qquad y_1 = v_1 u_1 + \cdots + v_n u_n,$$

where $v_1, \ldots, v_n$ are functions that can be calculated in terms of $u_1, \ldots, u_n$ and the right-hand member R. The method leads to a system of n linear algebraic equations satisfied by the derivatives $v_1', \ldots, v_n'$. This system can always be solved because it has a nonsingular coefficient-matrix. Integration of the derivatives then gives the required functions $v_1, \ldots, v_n$. The method was first used by Johann Bernoulli to solve linear equations of first order, and then by Lagrange in 1774 to solve linear equations of second order.

The details for the nth order case can be simplified by using vector and matrix notation. The right-hand member of (8.17) can be written as an inner product,

$$(8.18) \qquad y_1 = (v, u),$$

where v and u are n-dimensional vector functions given by

$$v = (v_1, \ldots, v_n) \qquad \text{and} \qquad u = (u_1, \ldots, u_n).$$

The derivative of a vector-valued function is defined to be the vector obtained by differentiating each component. For example, $v' = (v_1', \ldots, v_n')$ and $u' = (u_1', \ldots, u_n')$.

The usual rules for derivatives of sums and scalar multiples of real-valued functions also apply to sums and scalar multiples of vector-valued functions. The product rule is applicable to inner products. Thus, the derivative $(v, u)'$ is the sum of two inner products, $(v, u') + (v', u)$.

In the method of variation of parameters we try to choose v so that the inner product defining y_1 will satisfy the nonhomogeneous equation $L(y) = R$, given that $L(u) = O$, where $L(u)$ is the vector with components $L(u_1), \ldots, L(u_n)$. We begin by calculating the first derivative of y_1. By the product rule we find

$$(8.19) \qquad y_1' = (v, u') + (v', u).$$

We have n functions $v_1, \ldots, v_n$ to determine, so we should be able to put n conditions on them. If we impose the condition that the second term on the right of Equation (8.19) should vanish, the formula for y_1' simplifies to

$$y_1' = (v, u'), \qquad \text{provided that} \qquad (v', u) = 0.$$

Differentiating the new relation for y_1' we find

$$y_1'' = (v, u'') + (v', u').$$

If we can choose v so that $(v', u') = 0$ then the formula for y_1'' also simplifies and we get

$$y_1'' = (v, u''), \qquad \text{provided that also } (v', u') = 0.$$

Continuing in this manner for the first $n - 1$ derivatives of y_1 we find

$$y_1^{(n-1)} = (v, u^{(n-1)}), \qquad \text{provided that also } (v', u^{(n-2)}) = 0.$$

So far, we have put $n - 1$ conditions on v. Differentiating once more by the product rule we get

$$y_1^{(n)} = (v, u^{(n)}) + (v', u^{(n-1)}).$$

This time we impose the condition $(v', u^{(n-1)}) = R(x)$, and the last equation becomes

$$y_1^{(n)} = (v, u^{(n)}) + R(x), \qquad \text{provided that also } (v', u^{(n-1)}) = R(x).$$

Suppose, for the moment, that we can satisfy the n conditions imposed on v. Let $L = D^n + P_1(x)D^{n-1} + \cdots + P_n(x)$. When we apply L to y_1 we find

$$
\begin{aligned}
L(y_1) &= y_1^{(n)} + P_1(x)y_1^{(n-1)} + \cdots + P_n(x)y_1 \\
&= \{(v, u^{(n)}) + R(x)\} + P_1(x)(v, u^{(n-1)}) + \cdots + P_n(x)(v, u) \\
&= (v, L(u)) + R(x) = (v, O) + R(x) = R(x).
\end{aligned}
$$

Thus $L(y_1) = R(x)$, so y_1 is a solution of the nonhomogeneous equation.

The method will succeed if we can satisfy the n conditions we have imposed on v. These conditions stated that $(v', u^{(k)}) = 0$ for $k = 0, 1, \ldots, n - 2$, and that $(v', u^{(n-1)}) = R(x)$. We can write these n equations as a single matrix equation,

$$
(8.20) \qquad\qquad W(x)v'(x) = R(x) \begin{bmatrix} 0 \\ \vdots \\ 0 \\ 1 \end{bmatrix},
$$

where $v'(x)$ is regarded as an $n \times 1$ column matrix, and where W is the $n \times n$ matrix function whose rows consist of the components of u and its successive derivatives:

$$
W = \begin{bmatrix}
u_1 & u_2 & \cdots & u_n \\
u_1' & u_2' & \cdots & u_n' \\
\vdots & \vdots & \vdots & \vdots \\
u_1^{(n-1)} & u_2^{(n-1)} & \cdots & u_n^{(n-1)}
\end{bmatrix}.
$$

Matrix W is called the *Wronskian matrix* of $u_1, \ldots, u_n$, after J. M. H. Wronski (1778–1853).

In the next section we shall prove that the Wronskian matrix in this particular situation is nonsingular. Therefore, multiplying both sides of Equation (8.20) by the inverse of $W(x)$ we get

$$v'(x) = R(x)W(x)^{-1} \begin{bmatrix} 0 \\ \vdots \\ 0 \\ 1 \end{bmatrix}.$$

Choose two points c and x in the interval J under consideration and integrate this vector equation (component by component) over the interval from c to x to obtain

$$v(x) - v(c) = \int_c^x R(t)W(t)^{-1} \begin{bmatrix} 0 \\ \vdots \\ 0 \\ 1 \end{bmatrix} dt = z(x),$$

where

$$z(x) = \int_c^x R(t)W(t)^{-1} \begin{bmatrix} 0 \\ \vdots \\ 0 \\ 1 \end{bmatrix} dt.$$

Then $v(x) = v(c) + z(x)$ and the formula $y_1 = (u, v)$ for the particular solution now becomes

$$y_1 = (u, v) = (u, v(c) + z) = (u, v(c)) + (u, z).$$

The first term $(u, v(c))$ is a linear combination of $u_1, \dots, u_n$ so it satisfies the homogeneous equation. Therefore we can omit this term and use the second term (u, z) as a particular solution of the nonhomogeneous equation. In other words, a particular solution of $L(y) = R(x)$ is given by the inner product

$$(u(x), z(x)) = \left(u(x), \int_c^x R(t)W(t)^{-1} \begin{bmatrix} 0 \\ \vdots \\ 0 \\ 1 \end{bmatrix} dt \right).$$

We can summarize the results of the foregoing discussion by the following theorem.

THEOREM 8.11.   *Let $u_1, \dots, u_n$ be n independent solutions of a homogeneous nth order linear differential equation $L(y) = 0$ on an interval J. Then a particular solution y_1 of the nonhomogeneous equation $L(y) = R$ is given by the formula*

$$y_1 = v_1 u_1 + \cdots + v_n u_n,$$

where $v_1, \dots, v_n$ are the entries of the $n \times 1$ column matrix v determined by the equation

$$(8.21) \qquad v(x) = \int_c^x R(t)W(t)^{-1} \begin{bmatrix} 0 \\ \vdots \\ 0 \\ 1 \end{bmatrix} dt.$$

In this formula, W is the Wronskian matrix of $u_1, \ldots, u_n$, and c is any point in J.

EXAMPLE.   Find the general solution of the differential equation

$$y'' - y = \frac{2}{1 + e^x}$$

on the interval $(-\infty, +\infty)$.

Solution. The homogeneous equation $(D^2 - 1)y = 0$ has the two independent solutions $u_1(x) = e^x$ and $u_2(x) = e^{-x}$. The Wronskian matrix of u_1 and u_2 is

$$W(x) = \begin{bmatrix} e^x & e^{-x} \\ e^x & -e^{-x} \end{bmatrix},$$

whose determinant is -2. Therefore $W(x)$ is nonsingular and its inverse is given by

$$W(x)^{-1} = -\frac{1}{2} \begin{bmatrix} -e^{-x} & -e^{-x} \\ -e^x & e^x \end{bmatrix}.$$

Consequently

$$W(x)^{-1} \begin{bmatrix} 0 \\ 1 \end{bmatrix} = -\frac{1}{2} \begin{bmatrix} -e^{-x} \\ e^x \end{bmatrix}$$

and we have

$$R(t)W(t)^{-1} \begin{bmatrix} 0 \\ 1 \end{bmatrix} = -\frac{1}{2} \frac{2}{1 + e^t} \begin{bmatrix} -e^{-t} \\ e^t \end{bmatrix} = \begin{bmatrix} \dfrac{e^{-t}}{1 + e^t} \\ \dfrac{-e^t}{1 + e^t} \end{bmatrix}.$$

Integrating each component of the vector on the right from 0 to x we find

$$v_1(x) = \int_0^x \frac{e^{-t}}{1 + e^t} \, dt = \int_0^x \left(e^{-t} - 1 + \frac{e^t}{1 + e^t} \right) dt = -e^{-x} - x + \log(1 + e^x) + 1 - \log 2$$

and

$$v_2(x) = \int_0^x \frac{-e^t}{1 + e^t} \, dt = -\log(1 + e^x) + \log 2.$$

Neglecting the additive constants (because they contribute to the solution of the homogeneous equation) we find that a particular solution of the nonhomogeneous equation is

$$y_1 = v_1 u_1 + v_2 u_2 = -1 - xe^x + (e^x - e^{-x})\log(1 + e^x).$$

Therefore the general solution of the nonhomogeneous equation is

$$y = y_1 + c_1 e^x + c_2 e^{-x}.$$

8.12 Nonsingularity of the Wronskian matrix of n independent solutions of a homogeneous linear equation

In this section we prove that the Wronskian matrix W of n independent solutions $u_1, \ldots, u_n$ of a homogeneous equation $L(y) = 0$ is nonsingular. We do this by showing that its determinant is an exponential function that never vanishes on the interval J under consideration.

Let $w(x) = \det W(x)$ for each x in J, and assume the differential equation satisfied by $u_1, \ldots, u_n$ has the form

$$(8.22) \qquad y^{(n)} + P_1(x)y^{(n-1)} + \cdots + P_n(x)y = 0.$$

Then we have:

THEOREM 8.12. *The Wronskian determinant $w(x)$ satisfies the first-order equation*

$$(8.23) \qquad w' + P_1(x)w = 0$$

on J. Consequently, if $c \in J$ we have

$$(8.24) \qquad w(x) = w(c)\exp\left(-\int_c^x P_1(t)\,dt\right) \qquad (Abel's\ formula).$$

Moreover, $w(x) \neq 0$ for all x in J.

Proof. Let u be the row vector $u = (u_1, \ldots, u_n)$. Because each component of u satisfies the differential equation (8.22) the same is true of u. The rows of the Wronskian matrix W are the vectors $u, u', \ldots, u^{(n-1)}$. Hence we can write

$$w = \det W = \det(u, u', \ldots, u^{(n-2)}, u^{(n-1)}).$$

The derivative of w is the determinant of the matrix obtained by differentiating the last row of W (see Exercise 7 of Section 5.22). That is,

$$w' = \det(u, u', \ldots, u^{(n-2)}, u^{(n)}).$$

Multiplying the last row of w by $P_1(x)$ we also have

$$P_1(x)w = \det(u, u', \ldots, u^{(n-2)}, P_1(x)u^{(n-1)}).$$

Adding the last two equations we find

$$w' + P_1(x)w = \det(u, u', \ldots, u^{(n-2)}, u^{(n)} + P_1(x)u^{(n-1)}).$$

But the rows of this last determinant are dependent because u satisfies the differential equation (8.22). Therefore the determinant is zero, which means that w satisfies (8.23). Solving the first-order equation (8.23) we obtain Abel's formula (8.24).

To show that $w(x)$ never vanishes on J it suffices to show that $w(c) \neq 0$ for some c in J. We do this by a contradiction argument. Suppose that $w(t) = 0$ for all t in J. Choose a fixed t in J, say $t = t_0$, and consider the linear system of algebraic equations

$$W(t_0)X = O,$$

where X is a column vector. Since $w(t_0) = \det W(t_0) = 0$, the matrix $W(t_0)$ is singular so this system has a nonzero solution vector, say $X = (c_1, \ldots, c_n) \neq (0, \ldots, 0)$. Take the components of this nonzero vector and form the linear combination

$$f(t) = c_1 u_1(t) + \cdots + c_n u_n(t).$$

The function f so defined satisfies $L(f) = 0$ on J because it is a linear combination of $u_1, \ldots, u_n$. The matrix equation $W(t_0)X = O$ implies that

$$f(t_0) = f'(t_0) = \cdots = f^{(n-1)}(t_0) = 0.$$

Therefore f has the initial-value vector O at $t = t_0$ so, by the uniqueness theorem, f is the zero solution. This means that $(c_1, \ldots, c_n) = (0, \ldots, 0)$, which is a contradiction. Therefore $w(t) \neq 0$ for some t in J. Taking c in Abel's formula to be this t we see that $w(x)$ never vanishes on J. This completes the proof of Theorem 8.12.

8.13 Special methods for determining a particular solution of the nonhomogeneous equation. Reduction to a system of first-order linear equations

Because variation of parameters is a general method that always provides a particular solution of the nonhomogeneous equation, it may not be the simplest method to use in specific cases. Special methods exist that are often easier to apply when the equation has a special form. For example, if the equation has constant coefficients we can reduce the problem to that of solving a succession of first-order linear equations. This method is best explained through a simple example.

EXAMPLE. Find a particular solution of the second-order equation

$$(8.25) \qquad\qquad (D - 1)(D - 2)y = xe^{x+x^2}.$$

Solution. Let $u = (D - 2)y$. Then the equation becomes

$$(D - 1)u = xe^{x+x^2}.$$

This is a first-order linear equation in u that can be solved using Theorem 8.1. A particular solution is

$$u = \frac{1}{2}e^{x+x^2}.$$

Substituting this in the equation $u = (D - 2)y$ we obtain

$$(D - 2)y = \frac{1}{2}e^{x+x^2},$$

a first-order linear equation for y. Solving this by Theorem 8.1 we find that a particular solution (with $y_1(0) = 0$) is given by

$$y_1(x) = \frac{1}{2}e^{2x}\int_0^x e^{t^2-t}\,dt.$$

Although the integral cannot be evaluated in terms of elementary functions, we consider the equation as having been solved, because the solution is expressed in terms of integrals of familiar functions. The general solution of (8.25) is

$$y = c_1 e^x + c_2 e^{2x} + \frac{1}{2}e^{2x}\int_0^x e^{t^2-t}\,dt.$$

8.14 The annihilator method for determining a particular solution of the nonhomogeneous equation

We describe next a method that can be used if the equation $L(y) = R$ has constant coefficients and if the right-hand member R is itself annihilated by a constant-coefficient operator, say $A(R) = 0$. In principle, the method is very simple. We apply the operator A to both members of the differential equation $L(y) = R$ and obtain a new equation $AL(y) = 0$ that must be satisfied by all solutions of the original equation. Since AL is another constant-coefficient operator we can determine its null space by calculating the roots of the associated polynomial of AL. Then the problem becomes that of choosing from this null space a particular function y_1 that satisfies $L(y_1) = R$. The following examples illustrate the process.

EXAMPLE 1.   Find a particular solution of the fourth-order equation

$$(D^4 - 16)y = x^4 + x + 1.$$

Solution. The right-hand member, a polynomial of degree 4, is annihilated by the operator D^5. Therefore, any solution of the given equation is also a solution of the ninth-order equation

(8.26) $$D^5(D^4 - 16)y = 0.$$

The roots of the associated polynomial are 0, 0, 0, 0, 0, ± 2, $\pm 2i$, so all the solutions of (8.26) are to be found in the linear combination

$$y = c_1 + c_2 x + c_3 x^2 + c_4 x^3 + c_5 x^4 + c_6 e^{2x} + c_7 e^{-2x} + c_8 \cos 2x + c_9 \sin 2x.$$

We want to choose the c_i so that $L(y) = x^4 + x + 1$, where $L = D^4 - 16$. Since the last four terms are annihilated by L, we can take $c_6 = c_7 = c_8 = c_9 = 0$ and try to find $c_1, \ldots, c_5$ so that

$$L(c_1 + c_2 x + c_3 x^2 + c_4 x^3 + c_5 x^4) = x^4 + x + 1.$$

In other words, we seek a particular solution y_1 that is a polynomial of degree 4 satisfying $L(y_1) = x^4 + x + 1$. To simplify the algebra by avoiding subscripts, we write

$$16 y_1 = ax^4 + bx^3 + cx^2 + dx + e.$$

Differentiating four times we find $16 y_1^{(4)} = 24a$, so $y_1^{(4)} = 3a/2$. Substituting in the differential equation $L(y_1) = x^4 + x + 1$, we must determine a, b, c, d, e to satisfy

$$\frac{3}{2}a - ax^4 - bx^3 - cx^2 - dx - e = x^4 + x + 1.$$

Equating like powers of x we obtain $a = -1$, $b = c = 0$, $d = -1$, $e = -\frac{5}{2}$, so the particular solution y_1 is given by

$$y_1 = -\frac{1}{16}x^4 - \frac{1}{16}x - \frac{5}{32}.$$

This solution can be verified by substituting directly into the differential equation.

EXAMPLE 2. Solve the differential equation $y'' - 5y' + 6y = xe^x$.

Solution. The differential equation has the form

(8.27) $$L(y) = R,$$

where $R(x) = xe^x$ and $L = D^2 - 5D + 6$. The corresponding homogeneous equation is

$$(D - 2)(D - 3)y = 0;$$

it has the independent solutions $u_1(x) = e^{2x}$, $u_2(x) = e^{3x}$. Now we seek a particular solution y_1 of the nonhomogeneous equation. We recognize the right-hand member $R(x) = xe^x$ as a solution of the homogeneous equation

$$(D - 1)^2 y = 0.$$

Therefore, if we operate on both sides of (8.27) with the operator $(D - 1)^2$ we find that any function that satisfies (8.27) must also satisfy the equation

$$(D - 1)^2 (D - 2)(D - 3)y = 0.$$

The roots of the associated polynomial for this differential equation are 1, 1, 2, 3, so all its solutions are to be found in the linear combination

$$y = ae^x + bxe^x + ce^{2x} + de^{3x},$$

where a, b, c, d are constants. We want to choose a, b, c, d so the resulting solution y_1 satisfies $L(y_1) = xe^x$. Operator L annihilates the two exponentials for every choice of c and d, so we need only choose a and b so that $L(ae^x + bxe^x) = xe^x$ and take $c = d = 0$. If we put

$$y_1 = ae^x + bxe^x$$

we have

$$D(y_1) = (a + b)e^x + bxe^x, \qquad D^2(y_1) = (a + 2b)e^x + bxe^x,$$

so the equation $(D^2 - 5D + 6)y_1 = xe^x$ becomes

$$(2a - 3b)e^x + 2bxe^x = xe^x.$$

Canceling e^x and equating coefficients of like powers of x we find $a = \frac{3}{4}$, $b = \frac{1}{2}$. Therefore

$$y_1 = \frac{3}{4}e^x + \frac{1}{2}xe^x,$$

so the general solution of $L(y) = R$ is given by the formula

$$y = c_1 e^{2x} + c_2 e^{3x} + \frac{3}{4}e^x + \frac{1}{2}xe^x.$$

The method used in the foregoing examples is called the *annihilator method*. It will always work if we can find a constant-coefficient operator A that annihilates R. From our knowledge of homogeneous linear differential equations with constant coefficients, we know that the only real-valued functions annihilated by constant-coefficient operators are linear combinations of terms of the form

$$x^{m-1}e^{\alpha x}, \qquad x^{m-1}e^{\alpha x}\cos \beta x, \qquad x^{m-1}e^{\alpha x}\sin \beta x,$$

where m is a positive integer and α and β are real constants. The function $y = x^{m-1}e^{\alpha x}$ is a solution of a differential equation with associated polynomial having root α with multiplicity m. Therefore, this function has the annihilator $(D - \alpha)^m$. Each function $y = x^{m-1}e^{\alpha x}\cos \beta x$ and $y = x^{m-1}e^{\alpha x}\sin \beta x$ is a solution of a differential equation whose associated polynomial has complex roots $\alpha \pm i\beta$, each occuring with multiplicity m, so each is annihilated by the mth power of the operator $D^2 - 2\alpha D + (\alpha^2 + \beta^2)$. For ease of reference, we list these annihilators in Table 8.1, along with some of their special cases.

TABLE 8.1

Function		Annihilator
$y = x^{m-1}$		D^m
$y = e^{\alpha x}$		$D - \alpha$
$y = x^{m-1}e^{\alpha x}$		$(D - \alpha)^m$
$y = \cos \beta x$	or $\quad y = \sin \beta x$	$D^2 + \beta^2$
$y = x^{m-1}\cos \beta x$	or $\quad y = x^{m-1}\sin \beta x$	$(D^2 + \beta^2)^m$
$y = e^{\alpha x}\cos \beta x$	or $\quad y = e^{\alpha x}\sin \beta x$	$D^2 - 2\alpha D + (\alpha^2 + \beta^2)$
$y = x^{m-1}e^{\alpha x}\cos \beta x$	or $\quad y = x^{m-1}e^{\alpha x}\sin \beta x$	$[D^2 - 2\alpha D + (\alpha^2 + \beta^2)]^m$

Although the annihilator method is very efficient when applicable, it is limited to equations whose right-hand members R have a constant-coefficient annihilator. If $R(x)$ has the form e^{x^2}, or $\log x$, or $\tan x$, the method will not work; we must then use variation of parameters or some other method to find a particular solution.

8.15   Exercises

In each of Exercises 1 through 10, find the general solution on the interval $(-\infty, +\infty)$.

1. $y'' - y' = x^2$.
2. $y'' - 4y = e^{2x}$.
3. $y'' + 2y' = 3xe^x$.
4. $y'' + 4y = \sin x$.
5. $y'' - 2y' + y = e^x + e^{2x}$.
6. $y''' - y' = e^x$.
7. $y''' - y' = e^x + e^{-x}$.
8. $y''' + 3y'' + 3y' + y = xe^{-x}$.
9. $y'' + y = xe^x \sin 2x$.
10. $y^{(4)} - y = x^2 e^{-x}$.

11. If a constant-coefficient operator A annihilates f, and if a constant-coefficient operator B annihilates g, show that the product AB annihilates $f + g$.

12. Let A be a constant-coefficient operator with associated polynomial p_A.
    (a) Use the annihilator method to prove that the differential equation $A(y) = e^{\alpha x}$ has a particular solution of the form
    $$y_1 = \frac{e^{\alpha x}}{p_A(\alpha)}$$
    if α is not a zero of the polynomial p_A.
    (b) If α is a simple zero of p_A (multiplicity 1), prove that the equation $A(y) = e^{\alpha x}$ has the particular solution
    $$y_1 = \frac{xe^{\alpha x}}{p_A'(\alpha)}.$$
    (c) Generalize the results of (a) and (b) when α is a zero of p_A of multiplicity m.

13. Given two constant-coefficient operators A and B whose associated polynomials have no zeros in common. Let $C = AB$.
    (a) Prove that every solution of the differential equation $C(y) = 0$ has the form $y = y_1 + y_2$, where $A(y_1) = 0$ and $B(y_2) = 0$.
    (b) Prove that the functions y_1 and y_2 in part (a) are uniquely determined. That is, for a given y satisfying $C(y) = 0$ there is only one pair y_1, y_2 with the properties in part (a).

14. If $L(y) = y'' + ay' + by$, where a and b are constants, let f be that particular solution of $L(y) = 0$ satisfying the conditions $f(0) = 0$ and $f'(0) = 1$. Show that a particular solution of $L(y) = R$ is given by the formula
    $$y_1(x) = \int_c^x f(x - t)R(t)\, dt$$
    for any choice of c. In particular, if the roots of the associated polynomial are equal, say $r_1 = r_2 = m$, show that the formula becomes
    $$y_1(x) = e^{mx} \int_c^x (x - t)e^{-mt} R(t)\, dt.$$

15. Let Q be the operator "multiplication by x." That is, $Q(y)(x) = xy(x)$ for each y in class C^∞ and each real x. Let I denote the identity operator defined by $I(y) = y$ for each y in C^∞.
    (a) Prove that $DQ - QD = I$.
    (b) Show that $D^2Q - QD^2$ is a constant-coefficient operator of first order, and determine this operator explicitly as a linear polynomial in D.
    (c) Show that $D^3Q - QD^3$ is a constant-coefficient operator of second order, and determine this operator explicitly as a quadratic polynomial in D.
    (d) Guess the generalization suggested for $D^nQ - QD^n$ and prove it by induction.

16. The graph of a solution $y = u(x)$ of the differential equation $y'' - 3y' - 4y = 0$ intersects the graph of a solution $y = v(x)$ of the differential equation $y'' + 4y' - 5y = 0$ at the origin. Determine the functions u and v if the two curves have equal slopes at the origin and if

$$\lim_{x \to \infty} \frac{[v(x)]^4}{u(x)} = \frac{5}{6}.$$

17. The graph of a solution $y = u(x)$ of the differential equation $y'' - 4y' + 29y = 0$ intersects the graph of a solution $y = v(x)$ of the differential equation $y'' + 4y' + 13y = 0$ at the origin. The two curves have equal slopes at the origin. Determine u and v if $u'(\pi/2) = 1$.

18. Given that the differential equation $y'' + 4xy' + Q(x)y = 0$ has two solutions of the form $y_1 = u(x)$ and $y_2 = xu(x)$, where $u(0) = 1$. Determine both $u(x)$ and $Q(x)$ explicitly in terms of x.

19. Let $L(y) = y'' + P_1 y' + P_2 y$. To solve the nonhomogeneous equation $L(y) = R$ by variation of parameters, we need to know two linearly independent solutions of the homogeneous equation. This exercise shows that if one solution u_1 of $L(y) = 0$ is known, and if u_1 is never zero on an interval J, a second solution u_2 of the homogeneous equation is given by the formula

$$u_2(x) = u_1(x) \int_c^x \frac{Q(t)}{[u_1(t)]^2} \, dt,$$

where $Q(x) = e^{-A(x)}$ with $A(x) = \int_a^x P_1(t) \, dt$, and where a, c are any two points in J. These two solutions are indepedent on J.
(a) Prove that the function u_2 does, indeed, satisfy $L(y) = 0$.
(b) Differentiate the quotient u_2/u_1 to show that u_1 and u_2 are independent on J.

Exercise 19 can be helpful in solving Exercises 20 through 23.

20. Find the general solution of the equation $xy'' - 2(x + 1)y' + (x + 2)y = x^3 e^{2x}$ for $x > 0$, given that the homogeneous equation has a solution of the form $y = e^{mx}$.

21. Find the general solution of the differential equaton $4x^2 y'' + 4xy' - y = 0$, given that there is a particular solution of the form $y = x^m$ for $x > 0$.

22. Find the general solution of the equation $(2x - 3x^2)y'' + 4y' + 6xy = 0$, given that it has a solution that is a polynomial in x.

23. Find the general solution of the equation $x^2(1 - x)y'' + 2x(2 - x)y' + 2(1 + x)y = x^2$, given that the homogeneous equation has a solution of the form $y = x^c$.

24. The second order differential equation $y'' + P(x)y' + Q(x)y = 0$ has the solution $u(x) = x^3$ on the positive x axis. The Wronskian determinant of two independent solutions has the form $w(x) = cx^5$, where c is a nonzero constant.
(a) Determine the functions $P(x)$ and $Q(x)$ explicitly.
(b) Find the general solution of the nonhomogeneous equation $y'' + P(x)y' + Q(x)y = x^2$ on the positive x axis.

Eigenvalue problems leading to linear differential equations

25. Let V be the linear space of all functions continuous on $(-\infty, +\infty)$ such that $f(t) \to 0$ as $t \to -\infty$ and such that both integrals $\int_{-\infty}^x f(t) \, dt$ and $\int_{-\infty}^x tf(t) \, dt$ exist for all real x. If $f \in V$, let $g = T(f)$ be defined by $g(x) = x \int_{-\infty}^x f(t) \, dt - \int_{-\infty}^x tf(t) \, dt$. Find all real eigenvalues of T and, for each eigenvalue λ, determine the corresponding eigenfunctions.

26. Let V be the linear space of all real functions continuous on the interval $[0, \pi]$. Let S be the subspace of all f that have a continuous second derivative in $[0, \pi]$ and that also satisfy the boundary conditions $f(0) = f(\pi) = 0$. Let $T : S \to V$ be the linear transformation that maps each f onto its second derivative, $T(f) = f''$. Prove that the eigenvalues of T are the numbers of the form $-n^2$, where $n = 1, 2, \ldots$, and that the corresponding eigenfunctions are $f(t) = c_n \sin nt$, where $c_n \neq 0$.

27. Let V be the linear space of all real functions f having derivatives of every order on the real axis, and let S be the subspace consisting of all f that satisfy the boundary conditions $f(0) = f(2\pi)$ and $f'(0) = f'(2\pi) = 0$. Let $T : S \longrightarrow V$ be the second derivative operator, $T(f) = f''$. Determine all the eigenvalues of T (there are infinitely many) and, for each eigenvalue λ, find all the eigenfunctions belonging to λ.

28. Let V be the linear space of all real functions f having derivatives of every order everywhere and such that $f(0) = f''(0)$. Let $T : V \longrightarrow V$ be the second derivative operator, $T(f) = f''$. Find all the eigenvalues of T and, for each eigenvalue λ, find all the eigenfunctions belonging to λ.

29. Let V be the linear space of all real functions having derivatives of every order everywhere. Let S be the subspace of those f in V such that $f(0) = 2f'(0)$. Let $T(f) = 4f''$. Determine all the eigenvalues of T and, for each eigenvalue λ, determine the corresponding eigenfunctions in S.

30. Let V be as in Exercise 29, and let S be the subspace of those f in V such that $f(0) = f(1)$. Let $T(f) = f''$. Determine all the eigenvalues of T and, for each eigenvalue λ, determine the corresponding eigenfunctions in S.

9

APPLICATIONS TO SYSTEMS OF
DIFFERENTIAL EQUATIONS

9.1 Introduction

Although the study of differential equations began in the 17th century, it was not until the 19th century that mathematicians realized that relatively few differential equations could be solved by elementary means. The work of Cauchy, Liouville, and others showed the importance of establishing general theorems to guarantee the existence of solutions to certain specific classes of differential equations. Chapter 8 illustrated the use of an existence-uniqueness theorem in the study of linear differential equations. This chapter develops further topics in linear algebra, such as exponential matrices, and uses them to study systems of linear differential equations.

Existence theory for differential equations of higher order can be reduced to the first-order case by the introduction of *systems*. For example, the second-order equation

$$(9.1) \qquad y'' + 2ty' - y = e^t$$

can be transformed to a system of two first-order equations by introducing two unknown functions y_1 and y_2, where

$$y_1 = y, \qquad y_2 = y_1'.$$

Then we have $y_2' = y_1'' = y''$, so Eq. (9.1) can be written as a system of two first-order equations:

$$(9.2) \qquad \begin{aligned} y_1' &= y_2 \\ y_2' &= y_1 - 2ty_2 + e^t. \end{aligned}$$

The methods of Chapter 8 cannot be used here because each equation involves two unknown functions.

In this chapter we consider systems consisting of n linear differential equations of first order involving n unknown functions $y_1, \ldots, y_n$. These systems have the form

$$(9.3) \qquad \begin{aligned} y_1' &= p_{11}(t)y_1 + p_{12}(t)y_2 + \cdots + p_{1n}(t)y_n + q_1(t) \\ &\vdots \\ y_n' &= p_{n1}(t)y_1 + p_{n2}(t)y_2 + \cdots + p_{nn}(t)y_n + q_n(t). \end{aligned}$$

The functions p_{ik} and q_i that appear in (9.3) are considered as known functions defined on a given interval J. The functions $y_1, \ldots, y_n$ are unknowns to be determined. Systems of this type are called *first-order linear systems*. In general, each equation in the system involves more than one unknown function so the equations cannot be solved separately.

One reason for discussing systems is that every linear differential equation of order n can be transformed to a linear system. For example, consider the nth order linear equation

$$(9.4) \qquad y^{(n)} + a_1 y^{(n-1)} + \cdots + a_n y = R(t),$$

where the coefficients a_i are given functions. To transform this to a system we write y_1 for y, and we introduce a new unknown function for each of the successive derivatives of y. That is, we let

$$y_1 = y, \qquad y_2 = y_1', \qquad y_3 = y_2', \qquad \ldots, \qquad y_n = y_{n-1}',$$

and rewrite (9.4) as a system of n equations:

$$(9.5) \qquad \begin{aligned} y_1' &= y_2 \\ y_2' &= y_3 \\ &\;\;\vdots \\ y_{n-1}' &= y_n \\ y_n' &= -a_n y_1 - a_{n-1} y_2 - \cdots - a_1 y_n + R(t). \end{aligned}$$

The discussion of systems is simplified considerably by using vector and matrix notation. For the general system (9.3) we introduce vector-valued functions $Y = (y_1, \ldots, y_n)$, $Q = (q_1, \ldots, q_n)$, and an $n \times n$ matrix-valued function $P = [p_{ij}]$, defined by the equations

$$Y(t) = (y_1(t), \ldots, y_n(t)), \qquad Q(t) = (q_1(t), \ldots, q_n(t)), \qquad P(t) = [p_{ij}(t)]$$

for each t in J. We regard the vectors Y and Q as $n \times 1$ column matrices and write the system (9.3) in the simpler form

$$(9.6) \qquad Y' = P(t)Y + Q(t).$$

For example, in system (9.2) we have

$$Y = \begin{bmatrix} y_1 \\ y_2 \end{bmatrix}, \qquad P(t) = \begin{bmatrix} 0 & 1 \\ 1 & -2t \end{bmatrix}, \qquad Q(t) = \begin{bmatrix} 0 \\ e^t \end{bmatrix}.$$

In system (9.5) we have

$$Y = \begin{bmatrix} y_1 \\ y_2 \\ \vdots \\ y_n \end{bmatrix}, \qquad P(t) = \begin{bmatrix} 0 & 1 & 0 & \cdots & 0 \\ 0 & 0 & 1 & \cdots & 0 \\ \vdots & \vdots & \vdots & \vdots & \vdots \\ 0 & 0 & 0 & \cdots & 1 \\ -a_n & -a_{n-1} & -a_{n-2} & \cdots & -a_1 \end{bmatrix}, \qquad Q(t) = \begin{bmatrix} 0 \\ 0 \\ \vdots \\ 0 \\ R(t) \end{bmatrix}.$$

The problem of finding a vector-valued function Y that satisfies (9.6) and an initial condition of the form $Y(a) = B$, where $a \in J$ and B is a given n-dimensional constant vector, is called an *initial-value problem* for system (9.6). It takes no extra effort to discuss the more general problem in which $Q(t)$, B and $Y(t)$ are $n \times m$ matrices.

In the scalar case ($m = n = 1$) we know from Theorem 8.1 that if P and Q are continuous on J, all solutions of (9.6) are given by the explicit formula

$$(9.7) \qquad Y(x) = e^{A(x)} Y(a) + e^{A(x)} \int_a^x e^{-A(t)} Q(t) \, dt,$$

where $A(x) = \int_a^x P(t) \, dt$, and a is any point in J. We will show that this formula can be suitably generalized for systems, that is, when $P(t)$ is an $n \times n$ matrix function and when $Y(t)$ and $Q(t)$ are $n \times m$ matrix functions. But first we must define integrals of matrices and exponentials of matrices. Therefore, we digress briefly to discuss the calculus of matrix functions.

9.2   Calculus of matrix functions

Generalizing the concepts of integral and derivative for matrix functions is straightforward. If $P(t) = [p_{ij}(t)]$ is any $n \times m$ matrix of functions, we define the integral $\int_a^b P(t) \, dt$ by the equation

$$\int_a^b P(t) \, dt = \left[\int_a^b p_{ij}(t) \, dt \right].$$

That is, the integral of matrix $P(t)$ is the matrix obtained by integrating each entry of $P(t)$, assuming of course, that each entry is integrable on $[a, b]$. The reader can verify that the linearity property for integrals generalizes to matrix functions.

Continuity and differentiability of matrix functions are also defined in terms of the entries. We say that a matrix function $P = [p_{ij}]$ is continuous at t if each entry p_{ij} is continuous at t. The derivative P' is defined by differentiating each entry,

$$P'(t) = [p'_{ij}(t)],$$

whenever all derivatives $p'_{ij}(t)$ exist. It is easy to verify the basic differentiation rules for sums and products. For example, if P and Q are differentiable matrix functions, we have

$$(P + Q)' = P' + Q'$$

if P and Q are of the same size; and we also have

$$(PQ)' = PQ' + P'Q$$

if the product PQ is meaningful. The chain rule also holds. That is, if $F(t) = P[g(t)]$, where P is a differentiable matrix function and g is a differentiable scalar function, then $F'(t) = g'(t)P'[g(t)]$. The first and second fundamental theorems of calculus and the zero derivative theorem are also valid for matrix functions. Proofs of these properties are requested in the next set of exercises.

The definition of the exponential of a matrix is not so simple and requires further preparation. This is discussed in the next few sections.

9.3 Infinite series of matrices. Norms of matrices

Let $A = [a_{ij}]$ be an $n \times n$ matrix of real or complex entries. We wish to define the exponential e^A in such a way that it possesses some of the fundamental properties of the ordinary real or complex-valued exponential. In particular, we shall require that the definition imply the law of exponents in the form

$$(9.8) \qquad e^{tA}e^{sA} = e^{(s+t)A} \qquad \text{for all real } s \text{ and } t,$$

and the relation

$$(9.9) \qquad e^O = I,$$

where O and I are the $n \times n$ zero and identity matrices, respectively. It might seem natural to define e^A to be the matrix $[e^{a_{ij}}]$ obtained by exponentiating each entry a_{ij}. But this is unacceptable because it satisfies neither of properties (9.8) or (9.9). Instead, we shall define e^A by means of a power series expansion,

$$e^A = \sum_{k=0}^{\infty} \frac{A^k}{k!}.$$

We know that this formula holds if A is a real number (it also holds if A is complex), and we will prove that it implies properties (9.8) and (9.9) if A is a matrix. Before we can do this we need to explain what is meant by a convergent series of matrices.

DEFINITION. CONVERGENT SERIES OF MATRICES. *Given an infinite sequence of $m \times n$ matrices $\{C_k\}$ whose entries are real or complex numbers, denote the ij-entry of C_k by $c_{ij}^{(k)}$. If all mn series*

$$(9.10) \qquad \sum_{k=1}^{\infty} c_{ij}^{(k)} \qquad (i = 1, \ldots, m; j = 1, \ldots, n)$$

are convergent, then we say the series of matrices $\sum_{k=1}^{\infty} C_k$ is convergent, and its sum is defined to be the $m \times n$ matrix whose ij-entry is the series in (9.10).

A simple and useful test for convergence of series of matrices can be given in terms of the *norm* of a matrix, a generalization of the absolute value of a number.

DEFINITION. NORM OF A MATRIX. *If $A = [a_{ij}]$ is an $m \times n$ matrix of real or complex entries, the norm of A, denoted by $\|A\|$, is defined to be the sum of the absolute values of all its entries:*

$$(9.11) \qquad \|A\| = \sum_{i=1}^{m} \sum_{j=1}^{n} |a_{ij}|.$$

Other definitions of norms are sometimes used, but we have chosen this one because of the ease with which we can deduce the following properties.

THEOREM 9.1.   FUNDAMENTAL PROPERTIES OF NORMS.   *For rectangular matrices A and B and all real or complex c we have*

$$\|A + B\| \leq \|A\| + \|B\|, \qquad \|AB\| \leq \|A\|\|B\|, \qquad \|cA\| = |c|\|A\|.$$

Proof.   We prove only the result for $\|AB\|$, assuming that A is $m \times n$ and B is $n \times p$. The proofs of the others are simpler and are left as exercises.

If $A = [a_{ik}]$ and $B = [b_{kj}]$ we have $AB = [\sum_{k=1}^{n} a_{ik}b_{kj}]$, so from (9.11) we obtain

$$\|AB\| = \sum_{i=1}^{m}\sum_{j=1}^{p}\left|\sum_{k=1}^{n} a_{ik}b_{kj}\right| \leq \sum_{i=1}^{m}\sum_{k=1}^{n} |a_{ik}| \sum_{j=1}^{p} |b_{kj}| \leq \sum_{i=1}^{m}\sum_{k=1}^{n} |a_{ik}|\,\|B\| = \|A\|\|B\|.$$

Note that in the special case when $B = A$ (both $n \times n$ matrices) the inequality for $\|AB\|$ becomes $\|A^2\| \leq \|A\|^2$. By induction on k we also have

$$\|A^k\| \leq \|A\|^k \qquad \text{for } k = 2, 3, \ldots.$$

for any $n \times n$ matrix A. These inequalities will be used in the discussion of the exponential matrix.

The next theorem gives a useful sufficient condition for convergence of a series of matrices.

THEOREM 9.2.   TEST FOR CONVERGENCE OF A MATRIX SERIES.   *If $\{C_k\}$ is a sequence of $n \times n$ matrices such that the series of norms $\sum_{k=1}^{\infty} \|C_k\|$ converges, then the matrix series $\sum_{k=1}^{\infty} C_k$ also converges.*

Proof.   Let the ij-entry of C_k be denoted by $c_{ij}^{(k)}$. Since $|c_{ij}^{(k)}| \leq \|C_k\|$, convergence of $\sum_{k=1}^{\infty} \|C_k\|$ implies absolute convergence of each series $\sum_{k=1}^{\infty} c_{ij}^{(k)}$. Hence each series $\sum_{k=1}^{\infty} c_{ij}^{(k)}$ is convergent, so the matrix series $\sum_{k=1}^{\infty} C_k$ also converges.

9.4   Exercises

1. Verify that the linearity property of integrals also holds for integrals of matrix functions.
2. Verify each of the following differentiation rules for matrix functions, assuming P and Q are differentiable. In (a), P and Q must be of the same size so that $P + Q$ is meaningful. In (b) and (d) they need not be of the same size provided the products are meaningful. In (c) and (d), Q is assumed to be nonsingular.
   (a) $(P + Q)' = P' + Q'$.
   (b) $(PQ)' = PQ' + P'Q$.
   (c) $(Q^{-1})' = -Q^{-1}Q'Q^{-1}$.
   (d) $(PQ^{-1})' = -PQ^{-1}Q'Q^{-1} + P'Q^{-1}$.
3. (a) Let P be a differentiable matrix function. Prove that the derivatives of P^2 and P^3 are given by the formulas

   $$(P^2)' = PP' + P'P, \qquad (P^3)' = P^2P' + PP'P + P'P^2.$$

   (b) Guess a general formula for the derivative of P^k and prove it by induction on k.
4. Let P be a differentiable matrix function and let g be a differentiable scalar function whose range is a subset of the domain of P. Let F be the composite function defined by $F(t) = P[g(t)]$, and prove the chain rule: $F'(t) = g'(t)P'[g(t)]$.

5. Prove the zero-derivative theorem for matrix functions: *If $P'(t) = O$ for every t in an open interval (a, b), then the matrix function $P(t)$ is constant on (a, b).*
6. State and prove generalizations of the first and second fundamental theorems of calculus for matrix functions.
7. State and prove a formula for integration by parts in which the integrands are matrix functions.
8. Prove the following properties of matrix norms: $\|A + B\| \leq \|A\| + \|B\|$, $\|cA\| = |c|\|A\|$.
9. If a matrix function P is integrable on an interval $[a, b]$ prove that

$$\left\| \int_a^b P(t)\, dt \right\| \leq \int_a^b \|P(t)\|\, dt.$$

10. Let D be an $n \times n$ diagonal matrix, say $D = \text{diag}(\lambda_1, \ldots, \lambda_n)$. Prove that the matrix series $\sum_{k=0}^{\infty} D^k/k!$ converges and is also a diagonal matrix,

$$\sum_{k=0}^{\infty} \frac{D^k}{k!} = \text{diag}(e^{\lambda_1}, \ldots, e^{\lambda_n}).$$

(The term corresponding to $k = 0$ is understood to be the identity matrix I.)
11. Let D be an $n \times n$ diagonal matrix, $D = \text{diag}(\lambda_1, \ldots, \lambda_n)$. If the matrix series $\sum_{k=0}^{\infty} c_k D^k$ converges, prove that

$$\sum_{k=0}^{\infty} c_k D^k = \text{diag}\left(\sum_{k=0}^{\infty} c_k \lambda_1^k, \ldots, \sum_{k=0}^{\infty} c_k \lambda_n^k \right).$$

12. Assume the matrix series $\sum_{k=1}^{\infty} C_k$ converges, where each C_k is an $n \times n$ matrix. Prove that the matrix series $\sum_{k=1}^{\infty} (AC_k B)$ also converges and that its sum is the matrix $A(\sum_{k=1}^{\infty} C_k)B$, if A and B are matrices for which the products $AC_k B$ are meaningful.

9.5 The exponential matrix

Using the test in Theorem 9.2 it is easy to prove that the matrix series

$$(9.12) \qquad \qquad \sum_{k=0}^{\infty} \frac{A^k}{k!}$$

converges for every square matrix A with real or complex entries. (The term corresponding to $k = 0$ is understood to be the identity matrix I.) The norm of each term satisfies the inequality

$$\left\| \frac{A^k}{k!} \right\| \leq \frac{\|A\|^k}{k!} = \frac{a^k}{k!}$$

where $a = \|A\|$. Because the series $\sum_{k=0}^{\infty} a^k/k!$ converges for every real a, Theorem 9.2 implies that the series in (9.12) converges for every square matrix A.

DEFINITION. EXPONENTIAL MATRIX. *For any $n \times n$ matrix A with real or complex entries we define the exponential e^A to be the $n \times n$ matrix given by the convergent series in (9.12):*

$$e^A = \sum_{k=0}^{\infty} \frac{A^k}{k!}.$$

Note that this definition implies $e^O = I$, where O is the zero matrix. Further properties of the exponential will be developed with the help of differential equations.

9.6   The differential equation satisfied by e^{tA}

Let t be a real number, let A be an $n \times n$ matrix, and let $E(t)$ be the $n \times n$ matrix given by

$$E(t) = e^{tA}.$$

We shall keep A fixed and study this matrix as a function of t. First we obtain a differential equation satisfied by E.

THEOREM 9.3.   *For every real t the matrix function E defined by* $E(t) = e^{tA}$ *satisfies the matrix differential equation*

$$E'(t) = E(t)A = AE(t).$$

Proof.   From the definition of the exponential matrix we have

$$E(t) = \sum_{k=0}^{\infty} \frac{(tA)^k}{k!} = \sum_{k=0}^{\infty} \frac{t^k A^k}{k!}.$$

Let $c_{ij}^{(k)}$ denote the ij-entry of A^k. Then the ij-entry of $t^k A^k/k!$ is $t^k c_{ij}^{(k)}/k!$. Hence, from the definition of a matrix series, we have

$$(9.13) \qquad \sum_{k=0}^{\infty} \frac{t^k A^k}{k!} = \left[\sum_{k=0}^{\infty} \frac{t^k}{k!} c_{ij}^{(k)} \right].$$

Each entry on the right of (9.13) is a power series in t, convergent for all t. Therefore its derivative exists for all t and is given by the differentiated series

$$\sum_{k=1}^{\infty} \frac{k t^{k-1}}{k!} c_{ij}^{(k)} = \sum_{k=0}^{\infty} \frac{t^k}{k!} c_{ij}^{(k+1)}.$$

This shows that the derivative $E'(t)$ exists and is given by the matrix series

$$E'(t) = \sum_{k=0}^{\infty} \frac{t^k A^{k+1}}{k!} = \left(\sum_{k=0}^{\infty} \frac{t^k A^k}{k!} \right) A = E(t)A.$$

In the last equation we used the property $A^{k+1} = A^k A$. Since A commutes with A^k we could also have written $A^{k+1} = AA^k$ to obtain the relation $E'(t) = AE(t)$. This completes the proof.

Note.   The foregoing proof also shows that e^{tA} commutes with A.

9.7   Uniqueness theorem for the matrix differential equation $F'(t) = AF(t)$

In this section we prove a uniqueness theorem that characterizes all solutions of the matrix differential equation $F'(t) = AF(t)$. The proof makes use of the following theorem.

THEOREM 9.4.   NONSINGULARITY OF e^{tA}.   *For any* $n \times n$ *matrix A and any scalar t we have*

$$(9.14) \qquad\qquad e^{tA}e^{-tA} = I.$$

Hence e^{tA} *is nonsingular, and its inverse is* e^{-tA}.

Proof.   Let F be the matrix function defined for all real t by the equation

$$F(t) = e^{tA}e^{-tA}.$$

We shall prove that $F(t)$ is the identity matrix by showing that the derivative $F'(t)$ is the zero matrix. Differentiating F by the product rule, using the result of Theorem 9.3, we find

$$F'(t) = e^{tA}(e^{-tA})' + (e^{tA})'e^{-tA} = e^{tA}(-Ae^{-tA}) + Ae^{tA}e^{-tA}$$
$$= -Ae^{tA}e^{-tA} + Ae^{tA}e^{-tA} = O,$$

because A commutes with e^{tA}. Therefore, by the zero-derivative theorem, F is a constant matrix. But $F(0) = e^{O}e^{-O} = I$, so $F(t) = I$ for all t. This proves (9.14).

THEOREM 9.5.   UNIQUENESS THEOREM.   *Let A and B be given constant matrices, where A is* $n \times n$ *and B is* $n \times m$. *Then the only* $n \times m$ *matrix function F satisfying the initial-value problem*

$$F'(t) = AF(t), \qquad F(0) = B$$

for all real t is

$$(9.15) \qquad\qquad F(t) = e^{tA}B.$$

Proof.   First we note that $e^{tA}B$ is a solution. Now let F be any solution and consider the matrix function

$$G(t) = e^{-tA}F(t).$$

Differentiating this product we obtain

$$G'(t) = e^{-tA}F'(t) - Ae^{-tA}F(t) = e^{-tA}AF(t) - e^{-tA}AF(t) = O.$$

Therefore $G(t)$ is a constant matrix,

$$G(t) = G(0) = F(0) = B.$$

In other words, $e^{-tA}F(t) = B$. Multiplying by e^{tA} and using (9.14) we obtain (9.15).

Note.   The same type of proof shows that $F(t) = Be^{tA}$ is the only solution of the initial-value problem

$$F'(t) = F(t)A, \qquad F(0) = B,$$

where now $F(t)$ and B are $m \times n$ matrices and A is $n \times n$.

9.8   The law of exponents for exponential matrices

The law of exponents $e^A e^B = e^{A+B}$ is not always true for matrix exponentials. A counter example is given in Exercise 13 of Section 9.11. However, it is not difficult to prove that the law of exponents holds for matrices A and B that commute.

THEOREM 9.6.   *Let A and B be two n $\times$ n matrices that commute: AB = BA. Then we have*

(9.16) $$e^A e^B = e^{A+B}.$$

Proof.   From the equation $AB = BA$ we find that

$$A^2 B = A(BA) = (AB)A = (BA)A = BA^2,$$

so B commutes with A^2. By induction, B commutes with every positive integer power of A. By writing e^{tA} as a power series we find that B also commutes with e^{tA} for every real t.

Now let F be the matrix function defined by the equation

$$F(t) = e^{t(A+B)} - e^{tA} e^{tB}.$$

We will show that $F(t)$ is O for all t. Differentiating $F(t)$ and using the fact that B commutes with e^{tA} we find

$$F'(t) = (A + B)e^{t(A+B)} - Ae^{tA} e^{tB} - e^{tA} B e^{tB}$$
$$= (A + B)e^{t(A+B)} - (A + B)e^{tA} e^{tB} = (A + B)F(t).$$

By the uniqueness theorem we have

$$F(t) = e^{t(A+B)} F(0).$$

But $F(0) = O$, so $F(t) = O$ for all t. Hence

$$e^{t(A+B)} = e^{tA} e^{tB}.$$

When $t = 1$ we obtain (9.16).

EXAMPLE.   The matrices sA and tA commute for all scalars s and t. Hence we have

$$e^{sA} e^{tA} = e^{(s+t)A}.$$

9.9   Existence and uniqueness theorems for homogeneous linear systems with constant coefficients

The matrix differential equation $Y'(t) = AY(t)$, where A is an $n \times n$ constant matrix and Y is an $n \times m$ matrix function, is called a *homogeneous linear system with constant coefficients*. We shall use the exponential matrix to give an explicit formula for the solution of such a system.

THEOREM 9.7.   *Let A be a given n × n constant matrix and let B be a given n × m constant matrix. Then the initial-value problem*

$$(9.17) \qquad\qquad Y'(t) = AY(t), \qquad Y(0) = B$$

has a unique n × m matrix solution on the interval $-\infty < t < +\infty$. This solution is given by the formula

$$(9.18) \qquad\qquad Y(t) = e^{tA}B.$$

More generally, the solution of the intial-value problem

$$Y'(t) = AY(t), \qquad Y(a) = B,$$

is $Y(t) = e^{(t-a)A}B$.

   Proof.   Differentiation of (9.18) gives us $Y'(t) = Ae^{tA}B = AY(t)$. Since $Y(0) = B$, this is a solution of the initial-value problem (9.17).

   To prove that this is the only solution, we argue as in the proof of Theorem 9.5. Let $Z(t)$ be another matrix function satisfying $Z'(t) = AZ(t)$ with $Z(0) = B$, and let $G(t) = e^{-tA}Z(t)$. Then we can easily verify that $G'(t) = O$, so $G(t) = G(0) = Z(0) = B$. In other words, $e^{-tA}Z(t)$, so $Z(t) = e^{tA}B = Y(t)$. The more general case with $Y(a) = B$ is treated in exactly the same way.

9.10   Calculating e^{tA} in special cases

   Although Theorem 9.7 gives an explicit formula for the solution of a homogeneous system with constant coefficients, there still remains the problem of actually computing the exponential matrix e^{tA}. If we were to calculate e^{tA} directly from the series definition we would have to compute all powers A^k for $k = 0, 1, 2, \ldots$, and then compute the sum of each series $\sum_{k=0}^{\infty} t^k c_{ij}^{(k)}/k!$, where $c_{ij}^{(k)}$ is the ij-entry of A^k. In general this is a hopeless task unless A is a matrix whose powers can be readily calculated. For example, if A is a diagonal matrix, say

$$A = \text{diag}(\lambda_1, \ldots, \lambda_n),$$

then every power of A is also a diagonal matrix, in fact, $A^k = \text{diag}(\lambda_1^k, \ldots, \lambda_n^k)$. Therefore in this case e^{tA} is a diagonal matrix given by

$$e^{tA} = \text{diag}\left(\sum_{k=0}^{\infty} \frac{t^k}{k!}\lambda_1^k, \ldots, \sum_{k=0}^{\infty} \frac{t^k}{k!}\lambda_n^k\right) = \text{diag}(e^{t\lambda_1}, \ldots, e^{t\lambda_n}).$$

Another easy case to handle is when A is a matrix that can be diagonalized. For example, if there is a nonsingular matrix C such that $C^{-1}AC$ is a diagonal matrix, say $C^{-1}AC = D$, then we have $A = CDC^{-1}$, from which we find

$$A^2 = (CDC^{-1})(CDC^{-1}) = CD^2C^{-1},$$

and, more generally,

$$A^k = CD^k C^{-1}.$$

Therefore in this case we have

$$e^{tA} = \sum_{k=0}^{\infty} \frac{t^k A^k}{k!} = \sum_{k=0}^{\infty} \frac{t^k}{k!} CD^k C^{-1} = C \left(\sum_{k=0}^{\infty} \frac{t^k D^k}{k!} \right) C^{-1} = Ce^{tD}C^{-1}.$$

The diagonal matrix D has the same eigenvalues as A. If these are $\lambda_1, \ldots, \lambda_n$ then

$$e^{tD} = \text{diag}(e^{t\lambda_1}, \ldots, e^{t\lambda_n}),$$

so $Ce^{tD}C^{-1}$ is a linear combination of the form $e^{t\lambda_1} L_1(A) + e^{t\lambda_2} L_2(A) + \cdots + e^{t\lambda_n} L_n(A)$ where the multipliers $L_1(A), L_2(A), \ldots, L_n(A)$ are $n \times n$ matrices depending on C and hence on A. In other words we have proved Equation (9.19) in the following theorem.

THEOREM 9.8. *If an $n \times n$ matrix A with eigenvalues $\lambda_1, \ldots, \lambda_n$ can be diagonalized, then we have*

(9.19)
$$e^{tA} = \sum_{k=1}^{n} e^{t\lambda_k} L_k(A),$$

where each multiplier $L_k(A)$ is an $n \times n$ matrix depending only on A. This implies that

(9.20)
$$A^r = \sum_{k=1}^{n} \lambda_k^r L_k(A) \qquad \text{for every integer } r \geq 0,$$

and

(9.21)
$$p(A) = \sum_{k=1}^{n} p(\lambda_k) L_k(A) \qquad \text{for any polynomial } p(\lambda).$$

Proof. Take $t = 0$ in (9.19) to get

$$I = \sum_{k=1}^{n} L_k(A).$$

Now differentiate each member of (9.19) and then take $t = 0$. Differentiation gives us

$$Ae^{tA} = \sum_{k=1}^{n} \lambda_k e^{t\lambda_k} L_k(A).$$

and when $t = 0$ this becomes

$$A = \sum_{k=1}^{n} \lambda_k L_k(A).$$

Continue the process. Each differentiation of the left member of (9.19) introduces a new factor A on the left and a new factor λ_k in the kth term of the sum on the right. After taking r derivatives and putting $t = 0$ in the result we get (9.20).

Now let $p(\lambda) = \sum_{r=0}^{m} c_r \lambda^r$ be any polynomial in λ. Multiply each member of (9.20) by c_r and sum on r from 0 to m to obtain (9.21).

By making special choices of the polynomial $p(\lambda)$ we can obtain explicit formulas for the matrix multipliers $L_k(A)$ in Equation (9.19) as shown in the following theorem.

THEOREM 9.9.   *If an $n \times n$ matrix A has n distinct eigenvalues $\lambda_1, \ldots, \lambda_n$, then we have*

$$(9.22) \qquad\qquad e^{tA} = \sum_{k=1}^{n} e^{t\lambda_k} L_k(A),$$

where $L_k(A)$ is a polynomial in A of degree $n - 1$ given by the formula

$$(9.23) \qquad\qquad L_k(A) = \prod_{\substack{j=1 \\ j \neq k}}^{n} \frac{A - \lambda_j i}{\lambda_k - \lambda_j} \qquad for \qquad k = 1, 2, \ldots, n.$$

Proof.   We know that A can be diagonalized because it has n distinct eigenvalues. Therefore, by Theorem 9.8, we have Equation (9.19) as well as (9.21) for any polynomial $p(\lambda)$. If we choose $p(\lambda)$ to be the characteristic polynomial of A, say

$$(9.24) \qquad\qquad p(\lambda) = (\lambda - \lambda_1)(\lambda - \lambda_2) \cdots (\lambda - \lambda_n),$$

then $p(\lambda_k) = 0$ for each k and Equation (9.21) gives us

$$(9.25) \qquad\qquad p(A) = O.$$

This, of course, is the Cayley-Hamilton theorem, discussed in Section 6.13. The foregoing argument gives another proof of the Cayley-Hamilton theorem for matrices with distinct eigenvalues.

By making slightly different choices of the polynomial in (9.21) we can determine the multipliers $L_k(A)$ explicitly. Instead of the characteristic polynomial in (9.24) we choose a polynomial of degree $n - 1$ obtained by suppressing one factor of the characteristic polynomial. For example, let $p_1(\lambda)$ denote the polynomial obtained by omitting the first factor $(\lambda - \lambda_1)$ in (9.24); that is, let

$$p_1(\lambda) = (\lambda - \lambda_2) \cdots (\lambda - \lambda_n).$$

Then $p_1(\lambda_k) = 0$ if $k \geq 2$ while

$$p_1(\lambda_1) = (\lambda_1 - \lambda_2) \cdots (\lambda_1 - \lambda_n).$$

Note that $p_1(\lambda_1) \neq 0$ because the λ_k are distinct. Using this polynomial in Equation (9.21) we find

$$p_1(A) = \sum_{k=1}^{n} p_1(\lambda_k) L_k(A) = p_1(\lambda_1) L_1(A).$$

Since $p_1(\lambda_1) \neq 0$ we can solve this equation for $L_1(A)$ to obtain

$$L_1(A) = \frac{p_1(A)}{p_1(\lambda_1)} = \prod_{j=2}^{n} \frac{A - \lambda_j i}{\lambda_1 - \lambda_j}.$$

Similarly, if we let $p_2(\lambda)$ denote the polynomial obtained by omitting the second factor $(\lambda - \lambda_2)$ in (9.24), that is, if

$$p_2(\lambda) = \prod_{\substack{j=1 \\ j \neq 2}}^{n} (\lambda - \lambda_j),$$

we find $p_2(\lambda_k) = 0$ if $k \neq 2$ and

$$p_2(\lambda_2) = \prod_{\substack{j=1 \\ j \neq 2}}^{n} (\lambda_2 - \lambda_j) \neq 0.$$

Using polynomial $p_2(\lambda)$ in (9.21) we find

$$p_2(A) = \sum_{k=1}^{n} p_2(\lambda_k) L_k(A) = p_2(\lambda_2) L_2(A).$$

Solving for $L_2(A)$ we obtain

$$L_2(A) = \frac{p_2(A)}{p_2(\lambda_2)} = \prod_{\substack{j=1 \\ j \neq 2}}^{n} \frac{A - \lambda_j i}{\lambda_2 - \lambda_j}.$$

The same type of argument gives the general formula for $L_k(A)$ in (9.23).

EXAMPLE 1.   If A is a 3×3 matrix with distinct eigenvalues λ, μ, ν, Theorem 9.9 gives us

$$e^{tA} = e^{\lambda t} \frac{(A - \mu I)(A - \nu I)}{(\lambda - \mu)(\lambda - \nu)} + e^{\mu t} \frac{(A - \lambda I)(A - \nu I)}{(\mu - \lambda)(\mu - \nu)} + e^{\nu t} \frac{(A - \lambda I)(A - \mu I)}{(\nu - \lambda)(\nu - \mu)}.$$

EXAMPLE 2.   Calculate e^{tA} for the 2×2 matrix $A = \begin{bmatrix} 5 & 4 \\ 1 & 2 \end{bmatrix}$.

Solution. This matrix has distinct eigenvalues $\lambda_1 = 6$, $\lambda_2 = 1$, so by Theorem 9.9 we have

$$e^{tA} = e^{6t} L_1(A) + e^{t} L_2(A),$$

where

$$L_1(A) = \frac{A - \lambda_2 I}{\lambda_1 - \lambda_2} = \frac{1}{5}(A - I) = \frac{1}{5}\begin{bmatrix} 4 & 4 \\ 1 & 1 \end{bmatrix}$$

and

$$L_2(A) = \frac{A - \lambda_1 I}{\lambda_2 - \lambda_1} = \frac{1}{-5}(A - 6I) = \frac{1}{-5}\begin{bmatrix} -1 & 4 \\ 1 & -4 \end{bmatrix}.$$

Therefore

$$e^{tA} = \frac{e^{6t}}{5}\begin{bmatrix} 4 & 4 \\ 1 & 1 \end{bmatrix} + \frac{e^t}{-5}\begin{bmatrix} -1 & 4 \\ 1 & -4 \end{bmatrix} = \frac{1}{5}\begin{bmatrix} 4e^{6t} + e^t & 4e^{6t} - 4e^t \\ e^{6t} - e^t & e^{6t} + 4et \end{bmatrix}.$$

EXAMPLE 3.   Solve the linear system

$$y_1' = 5y_1 + 4y_2$$
$$y_2' = y_1 + 2y_2$$

subject to the initial conditions $y_1(0) = 2$, $y_2(0) = 3$.

Solution. In matrix form this system can be written as

$$Y'(t) = AY(t), \qquad Y(0) = \begin{bmatrix} 2 \\ 3 \end{bmatrix}, \qquad \text{where } A = \begin{bmatrix} 5 & 4 \\ 1 & 2 \end{bmatrix}.$$

By Theorem 9.7 the solution is $Y(t) = e^{tA}Y(0)$. Using the matrix e^{tA} calculated in Example 2 we find

$$\begin{bmatrix} y_1 \\ y_2 \end{bmatrix} = \frac{1}{5}\begin{bmatrix} 4e^{6t} + e^t & 4e^{6t} - 4e^t \\ e^{6t} - e^t & e^{6t} + 4et \end{bmatrix}\begin{bmatrix} 2 \\ 3 \end{bmatrix}$$

from which we obtain

(9.26) $y_1 = 4e^{6t} - 2e^t, \qquad y_2 = e^{6t} + 2e^t.$

Alternate solution of Example 3. There is an alternate procedure that does not require calculating e^{tA} explicitly. Because A can be diagonalized, we know that $e^{tA} = Ce^{tD}C^{-1}$ where

$$e^{tD} = \begin{bmatrix} e^{6t} & 0 \\ 0 & e^t \end{bmatrix}.$$

Consequently, without actually calculating e^{tA} we know that each of its entries is a linear combination of e^{6t} and e^t with constant coefficients; and because $Y(t) = e^{tA}Y(0)$, the same is true for each entry of the solution vector $Y(t)$. Therefore the solutions themselves must have the form

$$y_1 = ae^{6t} + be^t, \qquad y_2 = ce^{6t} + de^t,$$

with four undetermined constant coefficients a, b, c, d. We can find these by using the two differential equations and the two initial conditions. In fact, the initial conditions imply $a + b = 2$, $c + d = 3$, and the differential equations imply $a = 4c$, $b = -d$. This gives $a = 4$, $b = -2$, $c = 1$, $d = 2$, which is the same result obtained in (9.26).

Note. The second method used in the foregoing example suggests an alternate procedure that could be used to solve a homogeneous linear system

$$Y'(t) = AY(t), \qquad Y(0) = B,$$

if it is known that matrix A can be diagonalized. If the eigenvalues of A are $\lambda_1, \ldots, \lambda_n$ the entries of the exponential matrix e^{tA} are linear combinations of the exponentials $e^{t\lambda_1}, \ldots, e^{t\lambda_n}$ with constant coefficients, hence so are the entries of the solution matrix $Y(t)$. Therefore we can write each solution function $y_{ij}(t)$ as a linear combination of the exponentials $e^{t\lambda_1}, \ldots, e^{t\lambda_n}$ with undetermined constant coefficients and then use the differential equation $Y'(t) = AY(t)$ and the initial condition $Y(0) = B$ to determine these coefficients.

Many methods are known for calculating e^{tA} when A cannot be diagonalized. Most of them are complicated and require preliminary transformations, the nature of which depends on the multiplicities of the eigenvalues of A. In a later section we shall discuss a practical and straightforward method for calculating e^{tA} that can be used whether or not A can be diagonalized. It is valid for *all* square matrices A and requires no preliminary transformations of any kind. This method was developed by E. J. Putzer in a paper in the *American Mathematical Monthly*, Vol. 73 (1966), pp. 2–7, and is based on the Cayley-Hamilton theorem.

9.11 Exercises

In each of Exercises 1 through 4, (a) express A^{-1}, A^2 and all higher powers of A as a linear combination of I and A. (The Cayley-Hamilton theorem can be of help.) (b) Calculate e^{tA}.

1. $A = \begin{bmatrix} 1 & 0 \\ 1 & 1 \end{bmatrix}.$ 2. $A = \begin{bmatrix} 1 & 0 \\ 1 & 2 \end{bmatrix}.$ 3. $A = \begin{bmatrix} 0 & 1 \\ 1 & 0 \end{bmatrix}.$ 4. $A = \begin{bmatrix} -1 & 0 \\ 0 & 1 \end{bmatrix}.$

In each of Exercises 5, 6, 7, (a) calculate A^n and express A^3 in terms of I, A, and A^2. (b) Calculate e^{tA}.

5. $A = \begin{bmatrix} 0 & 1 & 1 \\ 0 & 0 & 1 \\ 0 & 0 & 0 \end{bmatrix}.$ 6. $A = \begin{bmatrix} 0 & 1 & 1 \\ 0 & 1 & 1 \\ 0 & 0 & 0 \end{bmatrix}.$ 7. $A = \begin{bmatrix} 2 & 0 & 0 \\ 0 & 1 & 0 \\ 0 & 1 & 1 \end{bmatrix}.$

8. (a) If $A = \begin{bmatrix} 0 & 1 \\ -1 & 0 \end{bmatrix}$, prove that $e^{tA} = \begin{bmatrix} \cos t & \sin t \\ -\sin t & \cos t \end{bmatrix}.$

   (b) Find a corresponding formula for e^{tA} when $A = \begin{bmatrix} a & b \\ -b & a \end{bmatrix}$, with a, b real.

9. If $F(t) = \begin{bmatrix} t & t-1 \\ 0 & 1 \end{bmatrix}$, prove that $e^{F(t)} = eF(e^{t-1}).$

10. If $A(t)$ is a scalar function of t, the derivative of $e^{A(t)}$ is $e^{A(t)}A'(t)$. Compute the derivative of $e^{A(t)}$ when $A(t) = \begin{bmatrix} 1 & t \\ 0 & 0 \end{bmatrix}$ and show that the result is not equal to either of the two products $e^{A(t)}A'(t)$ or $A'(t)e^{A(t)}.$

11. If $A = \begin{bmatrix} 0 & -1 & 0 \\ 1 & 0 & 1 \\ 0 & 1 & 0 \end{bmatrix}$, express e^{tA} as a linear combination of I, A, A^2.

12. If $A = \begin{bmatrix} 0 & 1 & 0 \\ 2 & 0 & 1 \\ 0 & 1 & 0 \end{bmatrix}$, prove that $e^A = \begin{bmatrix} x^2 & xy & y^2 \\ 2xy & x^2+y^2 & 2xy \\ y^2 & xy & x^2 \end{bmatrix}$, where $x = \cosh 1$ and $y = \sinh 1$.

    Here $\cosh t = \frac{1}{2}(e^t + e^{-t})$ and $\sinh t = \frac{1}{2}(e^t - e^{-t})$ are hyperbolic functions.

13. This example shows that the equation $e^{A+B} = e^A e^B$ is not always true for matrix exponentials. Compute each of the matrices $e^A e^B$, $e^B e^A$, e^{A+B} when $A = \begin{bmatrix} 1 & 1 \\ 0 & 0 \end{bmatrix}$ and $B = \begin{bmatrix} 1 & -1 \\ 0 & 0 \end{bmatrix}$, and verify that the three results are distinct.

14. Given two $n \times n$ constant matrices A and B.
    (a) If $e^{tA}e^{tB} = e^{tB}e^{tA}$ for all real t, prove that $AB = BA$. [*Suggestion:* Differentiate.]
    (b) Now assume that $e^{t(A+B)} = e^{tA}e^{tB}$ for all real t. Prove or disprove: $AB = BA$.
15. Given an $n \times n$ matrix (a_{ij}), where $a_{ij} = n$ if $i + j = n + 1$, and $a_{ij} = 0$ otherwise.
    (a) Show that $A^{2k} = n^{2k}A$ and that $A^{2k+1} = n^{2k}A$ for every integer $k \geq 1$.
    (b) Calculate e^{tA} in terms of n, A, and I. When $n = 2$ you should get

$$e^{tA} = (\cosh 2t)I + \frac{1}{2}(\sinh 2t)A.$$

16. Given a constant $n \times n$ matrix A such that $A^2 = aI + bA$ for some choice of scalars a and b. All higher powers of A are also linear combinations of I and A, so the series for the exponential matrix e^{tA} simplifies to the form

$$e^{tA} = f(t)I + g(t)A$$

where $f(t)$ and $g(t)$ are scalar functions of t.
    (a) Assume the matrices I and A are linearly independent, and show that both $f(t)$ and $g(t)$ satisfy the second-order differential equation

$$y'' = ay + by'.$$

    (b) For the special case in which $A^2 = 4I + 3A$, determine explicitly $f(t)$ and $g(t)$ such that

$$e^{tA} = f(t)I + g(t)A.$$

9.12 Putzer's method for calculating e^{tA}

The Cayley-Hamilton theorem shows that the nth power of any $n \times n$ matrix A can be expressed as a linear combination of the lower powers I, A, $A^2, \ldots, A^{n-1}$. Therefore each of the higher powers A^{n+1}, $A^{n+2}, \ldots$, can also be expressed as a linear combination of I, A, $A^2, \ldots, A^{n-1}$. Consequently, in the infinite series defining e^{tA}, each term $t^k A^k / k!$ with $k \geq n$ is a linear combination of $t^k I, t^k A, t^k A^2, \ldots, t^k A^{n-1}$. It follows that e^{tA} is expresssible as a polynomial in A of the form

$$(9.27) \qquad e^{tA} = \sum_{k=0}^{n-1} q_k(t)A^k,$$

where the scalar coefficients $q_k(t)$ depend on t. E. J. Putzer developed two useful methods for expressing e^{tA} as a polynomial in A. The next theorem describes the simpler of the two methods.

THEOREM 9.10. *Let $\lambda_1, \ldots, \lambda_n$ be the eigenvalues of an $n \times n$ matrix A, and define a sequence of polynomials in A as follows:*

$$(9.28) \qquad P_0(A) = I, \qquad P_k(A) = \prod_{m=1}^{k}(A - \lambda_m I), \qquad for\ k = 1, 2, \ldots, n.$$

Then we have

$$(9.29) \qquad e^{tA} = \sum_{k=0}^{n-1} r_{k+1}(t)P_k(A),$$

where the scalar coefficients $r_1(t), \ldots, r_n(t)$ are determined recursively from the following triangular system of linear first-order differential equations:

(9.30)
$$r_1'(t) = \lambda_1 r_1(t), \qquad r_1(0) = 1,$$
$$r_{k+1}'(t) = \lambda_{k+1} r_{k+1}(t) + r_k(t), \qquad r_{k+1}(0) = 0, \qquad (k = 1, 2, \ldots, n-1).$$

Note. Equation (9.29) does not express e^{tA} directly in powers of A as indicated in (9.27), but as a linear combination of the polynomials $P_0(A), P_1(A), \ldots, P_{n-1}(A)$. These polynomials are easily calculated once the eigenvalues are known. Also, the multipliers $r_1(t), \ldots, r_n(t)$ in (9.29) are easily determined because the system of differential equations is triangular. When the solutions are determined in succession, each differential equation involves only one unknown function.

Proof. Let $r_1(t), \ldots, r_n(t)$ be the scalar functions determined by (9.30) and define a matrix function F by the equation

(9.31)
$$F(t) = \sum_{k=0}^{n-1} r_{k+1}(t) P_k(A).$$

We will prove that $F(t) = e^{tA}$ by showing that F satisfies the same differential equation as e^{tA}, namely $F'(t) = AF(t)$, and the same initial condition at 0.

First we note that $F(0) = r_1(0)P_0(A) = I$. Next, we differentiate (9.31) and use the recursion formulas in (9.30) to obtain

$$F'(t) = \sum_{k=0}^{n-1} r_{k+1}'(t) P_k(A) = \sum_{k=0}^{n-1} \{r_k(t) + \lambda_{k+1} r_{k+1}(t)\} P_k(A),$$

where $r_0(t)$ is defined to be 0. Rewrite this in the form

$$F'(t) = \sum_{k=0}^{n-2} r_{k+1}(t) P_{k+1}(A) + \sum_{k=0}^{n-1} \lambda_{k+1} r_{k+1}(t) P_k(A),$$

then subtract $\lambda_n F(t) = \sum_{k=0}^{n-1} \lambda_n r_{k+1}(t) P_k(A)$ to obtain the relation

(9.32)
$$F'(t) - \lambda_n F(t) = \sum_{k=0}^{n-2} r_{k+1}(t) \{P_{k+1}(A) + (\lambda_{k+1} - \lambda_n)P_k(A)\}.$$

But from (9.28) we see that $P_{k+1}(A) = (A - \lambda_{k+1}I)P_k(A)$, so

$$P_{k+1}(A) + (\lambda_{k+1} - \lambda_n)P_k(A) = (A - \lambda_{k+1}I)P_k(A) + (\lambda_{k+1} - \lambda_n)P_k(A)$$
$$= (A - \lambda_n I)P_k(A).$$

Therefore Equation (9.32) becomes

$$F'(t) - \lambda_n F(t) = (A - \lambda_n I) \sum_{k=0}^{n-2} r_{k+1}(t) P_k(A) = (A - \lambda_n I)\{F(t) - r_n(t)P_{n-1}(A)\}$$
$$= (A - \lambda_n I)F(t) - r_n(t)P_n(A).$$

The Cayley-Hamilton theorem implies that $P_n(A) = O$, so the last equation becomes

$$F'(t) - \lambda_n F(t) = (A - \lambda_n I)F(t) = AF(t) - \lambda_n F(t),$$

from which we find $F'(t) = AF(t)$. Since $F(0) = I$, the uniqueness theorem shows that $F(t) = e^{tA}$.

EXAMPLE 1.   If A is a 2×2 matrix with both its eigenvalues equal to λ, express e^{tA} as a linear combination of I and A.

Solution. We write $\lambda_1 = \lambda_2 = \lambda$ and consider the system of first-order differential equations

$$r_1'(t) = \lambda r_1(t), \qquad\qquad r_1(0) = 1,$$
$$r_2'(t) = \lambda r_2(t) + r_1(t), \qquad r_2(0) = 0.$$

Solving these equations in succession we find

$$r_1(t) = e^{\lambda t}, \qquad r_2(t) = te^{\lambda t}.$$

Since $P_0(A) = I$ and $P_1(A) = A - \lambda I$, the required formula for e^{tA} is

(9.33) $$e^{tA} = e^{\lambda t}I + te^{\lambda t}(A - \lambda I) = e^{\lambda t}(1 - \lambda t)I + te^{\lambda t}A.$$

EXAMPLE 2.   Solve Example 1 if the eigenvalues of A are λ and μ, where $\lambda \neq \mu$.

Solution. In this case the system of differential equations is

$$r_1'(t) = \lambda r_1(t), \qquad\qquad r_1(0) = 1,$$
$$r_2'(t) = \mu r_2(t) + r_1(t), \qquad r_2(0) = 0.$$

Its solutions are given by

$$r_1(t) = e^{\lambda t}, \qquad r_2(t) = \frac{e^{\lambda t} - e^{\mu t}}{\lambda - \mu}.$$

Since $P_0(A) = I$ and $P_1(A) = A - \lambda I$, the required formula for e^{tA} is

(9.34) $$e^{tA} = e^{\lambda t}I + \frac{e^{\lambda t} - e^{\mu t}}{\lambda - \mu}(A - \lambda I) = \frac{\lambda e^{\mu t} - \mu e^{\lambda t}}{\lambda - \mu}I + \frac{e^{\lambda t} - e^{\mu t}}{\lambda - \mu}A.$$

The reader should verify that this agrees with the result given by Theorem 9.9 when $n = 2$.

It should be noted that the formula for e^{tA} in (9.33) can be regarded as a limiting case of (9.34) obtained by letting $\mu \to \lambda$. (See Exercise 8 in Section 9.14.)

If the eigenvalues λ and μ are complex numbers, the exponentials $e^{\lambda t}$ and $e^{\mu t}$ will also be complex numbers. But if λ and μ are complex conjugates, the scalars multiplying I and A in (9.34) will be real. For example, suppose that

$$\lambda = \alpha + i\beta, \qquad \mu = \alpha - i\beta, \qquad \text{where} \qquad \beta \neq 0.$$

Then $\lambda - \mu = 2i\beta$ so equation (9.34) gives us

$$e^{tA} = e^{(\alpha+i\beta)t}I + \frac{e^{(\alpha+i\beta)t} - e^{(\alpha-i\beta)t}}{2i\beta}\{A - (\alpha + i\beta)I\}$$

$$= e^{\alpha t}\left\{e^{i\beta t}I + \frac{e^{i\beta t} - e^{-i\beta t}}{2i\beta}(A - \alpha I - i\beta I)\right\}$$

$$= e^{\alpha t}\left\{(\cos \beta t + i \sin \beta t)I + \frac{\sin \beta t}{\beta}(A - \alpha I - i\beta I)\right\}.$$

The terms involving i cancel and we find

(9.35)
$$e^{tA} = \frac{e^{\alpha t}}{\beta}\{(\beta \cos \beta t - \alpha \sin \beta t)I + \sin \beta t A\}.$$

EXAMPLE 3. Calculate e^{tA} as a linear combination of I, A, A^2 and A^3 if A is a 4×4 matrix with eigenvalues $\lambda_1 = \lambda_2 = 0$, $\lambda_3 = \lambda_4 = 1$.

Solution. Solving the triangular system in (9.30) we find

$$r_1(t) = 1, \qquad r_2(t) = t, \qquad r_3(t) = e^t - t - 1, \qquad r_4(t) = (t - 2)e^t + t + 2.$$

Therefore by Theorem 9.10 we have

$$e^{tA} = I + tA + (e^t - t - 1)A^2 + \{(t - 2)e^t + t + 2\}A^2(A - I)$$

$$= I + tA + \{(3 - t)e^t - 2t - 3\}A^2 + \{(t - 2)e^t + t + 2\}A^3.$$

9.13 Alternate methods for calculating e^{tA} in special cases

Putzer's method for expressing e^{tA} as a polynomial in A is completely general because it works for all square matrices A. A general method is not necessarily the best method to use in certain special cases. For example, Theorem 9.9 describes a straightforward method for computing e^{tA} when all the eigenvalues of A are distinct. This section gives simpler methods for computing e^{tA} in two more special cases: (a) When all the eigenvalues of A are equal, and (b) when A has two distinct eigenvalues, exactly one of which has multiplicity 1.

THEOREM 9.11. *If A is an $n \times n$ matrix with all its eigenvalues equal to λ, then we have*

(9.36)
$$e^{tA} = e^{\lambda t} \sum_{k=0}^{n-1} \frac{t^k}{k!}(A - \lambda I)^k.$$

Proof. Because the matrices $\lambda t I$ and $t(A - \lambda I)$ commute we have

$$e^{tA} = e^{\lambda t I}e^{t(A-\lambda I)} = (e^{\lambda t}I) \sum_{k=0}^{\infty} \frac{t^k}{k!}(A - \lambda I)^k.$$

The Cayley-Hamilton theorem implies that $(A - \lambda I)^k = O$ for every $k \geq n$, so the series terminates when $k = n - 1$.

THEOREM 9.12.   *Let A be an $n \times n$ matrix ($n \geq 3$) with two distinct eigenvalues λ and μ, where λ has multiplicity $n - 1$ and μ has multiplicity 1. Then we have*

$$e^{tA} = e^{\lambda t} \sum_{k=0}^{n-2} \frac{t^k}{k!}(A - \lambda I)^k + \left\{ \frac{e^{\mu t}}{(\mu - \lambda)^{n-1}} - \frac{e^{\lambda t}}{(\mu - \lambda)^{n-1}} \sum_{k=0}^{n-2} \frac{t^k}{k!}(\mu - \lambda)^k \right\}(A - \lambda I)^{n-1}.$$

Proof.   As in the proof of Theorem 9.11 we begin by writing

$$e^{tA} = e^{\lambda t} \sum_{k=0}^{\infty} \frac{t^k}{k!}(A - \lambda I)^k = e^{\lambda t} \sum_{k=0}^{n-2} \frac{t^k}{k!}(A - \lambda I)^k + e^{\lambda t} \sum_{k=n-1}^{\infty} \frac{t^k}{k!}(A - \lambda I)^k$$

$$= e^{\lambda t} \sum_{k=0}^{n-2} \frac{t^k}{k!}(A - \lambda I)^k + e^{\lambda t} \sum_{r=0}^{\infty} \frac{t^{n-1+r}}{(n-1+r)!}(A - \lambda I)^{n-1+r}.$$

Now we evaluate the series over r in closed form by using the Cayley-Hamilton theorem. Since

$$A - \mu I = A - \lambda I - (\mu - \lambda)I$$

we have

$$(A - \lambda I)^{n-1}(A - \mu I) = (A - \lambda I)^n - (\mu - \lambda)(A - \lambda I)^{n-1}.$$

The left member is O by the Cayley-Hamilton theorem, so

$$(A - \lambda I)^n = (\mu - \lambda)(A - \lambda I)^{n-1}.$$

Using this relation repeatedly we find

$$(A - \lambda I)^{n-1+r} = (\mu - \lambda)^r(A - \lambda I)^{n-1}.$$

Therefore the series over r becomes

$$\sum_{r=0}^{\infty} \frac{t^{n-1+r}}{(n-1+r)!}(A - \lambda I)^{n-1+r}$$

$$= \frac{1}{(\mu - \lambda)^{n-1}} \sum_{k=n-1}^{\infty} \frac{t^k}{k!}(\mu - \lambda)^k(A - \lambda I)^{n-1}$$

$$= \frac{1}{(\mu - \lambda)^{n-1}} \left\{ e^{t(\mu - \lambda)} - \sum_{k=0}^{n-2} \frac{t^k}{k!}(\mu - \lambda)^k \right\}(A - \lambda I)^{n-1}.$$

This completes the proof. Theorem 9.12 can also be deduced by applying Putzer's method, but the details are more complicated.

***Matrices of Order Three*.** The explicit formulas in Theorems 9.11, 9.12 and 9.9 cover all matrices of order $n \leq 3$. Since the 3×3 case often arises in practice, the formulas in this case are listed below for easy reference.

CASE 1. If A is a 3×3 matrix with eigenvalues λ, λ, λ, then by Theorem 9.11 we have

$$e^{tA} = e^{\lambda t}\left\{I + t(A - \lambda I) + \frac{1}{2}t^2(A - \lambda I)^2\right\}.$$

CASE 2. If A is a 3×3 matrix with eigenvalues λ, λ, μ, with $\lambda \neq \mu$, then by Theorem 9.12 we have

$$e^{tA} = e^{\lambda t}\left\{I + t(A - \lambda I) + \frac{e^{\mu t} - e^{\lambda t}}{(\mu - \lambda)^2}(A - \lambda I)^2 - \frac{te^{\lambda t}}{\mu - \lambda}(A - \lambda I)^2\right\}.$$

CASE 3. If A is a 3×3 matrix with distinct eigenvalues λ, μ, ν, then by Theorem 9.9 we have

$$e^{tA} = e^{\lambda t}\frac{(A - \mu I)(A - \nu I)}{(\lambda - \mu)(\lambda - \nu)} + e^{\mu t}\frac{(A - \lambda I)(A - \nu I)}{(\mu - \lambda)(\mu - \nu)} + e^{\nu t}\frac{(A - \lambda I)(A - \mu I)}{(\nu - \lambda)(\nu - \mu)}.$$

EXAMPLE. Compute e^{tA} when $A = \begin{bmatrix} 0 & 1 & 0 \\ 0 & 0 & 1 \\ 2 & -5 & 4 \end{bmatrix}$.

Solution. The eigenvalues of A are 1, 1, 2, so the formula of Case 2 gives us

$$(9.37) \qquad e^{tA} = e^t\{I + t(A - I)\} + (e^{2t} - e^t)(A - I)^2 - te^t(A - I)^2.$$

By collecting like powers of A we can also write this as follows:

$$(9.38) \qquad e^{tA} = (-2te^t + e^{2t})I + \{(3t + 2)e^t - 2e^{2t}\}A - \{(t + 1)e^t - e^{2t}\}A^2.$$

At this stage we can calculate $(A - I)^2$ or A^2 and perform the indicated operations in (9.37) or (9.38) to write the result as a 3×3 matrix

$$e^{tA} = \begin{bmatrix} -2te^t + e^{2t} & (3t + 2)e^t - 2e^{2t} & -(t + 1)e^t + e^{2t} \\ -2(t + 1)e^t + 2e^{2t} & (3t + 5)e^t - 4e^{2t} & -(t + 2)e^t + 2e^{2t} \\ -2(t + 2)e^t + 4e^{2t} & (3t + 8)e^t - 8e^{2t} & -(t + 4)e^t + 4e^{2t} \end{bmatrix}.$$

9.14 Exercises

For each of the matrices in Exercises 1 through 6, express e^{tA} as a polynomial in A.

1. $A = \begin{bmatrix} 5 & -2 \\ 4 & -1 \end{bmatrix}$.

2. $A = \begin{bmatrix} 1 & 2 \\ 2 & -1 \end{bmatrix}$

3. $A = \begin{bmatrix} 1 & 0 & 2 \\ 0 & 1 & 3 \\ 0 & 0 & 1 \end{bmatrix}$.

4. $A = \begin{bmatrix} 0 & 1 & 0 \\ 0 & 0 & 1 \\ -6 & -11 & -6 \end{bmatrix}$.

5. $A = \begin{bmatrix} 3 & -1 & 1 \\ 2 & 0 & 1 \\ 1 & -1 & 2 \end{bmatrix}$.

6. $A = \begin{bmatrix} 1 & 1 & 0 & 0 \\ 0 & 2 & 1 & 0 \\ 0 & 0 & 3 & 0 \\ 0 & 0 & 0 & 4 \end{bmatrix}$.

7. (a) A 3×3 matrix A is known to have all its eigenvalues equal to λ. Prove that

$$e^{tA} = \frac{1}{2}e^{\lambda t}\left\{(\lambda^2 t^2 - 2\lambda t + 2)I + (-2\lambda t^2 + 2t)A + t^2 A^2\right\}.$$

(b) Find a corresponding formula if A is a 4×4 matrix with all its eigenvalues equal to λ.

8. Show that the formula for e^{tA} in (9.33) can be regarded as a limiting case of (9.34) obtained by letting $\mu \to \lambda$. [*Hint:* Use L'Hôpital's rule for the indeterminate form $0/0$.]

In each of Exercises 9 through 16, solve the system $Y' = AY$ subject to the given initial condition.

9. $A = \begin{bmatrix} 1 & 2 \\ 2 & -1 \end{bmatrix}$, $Y(0) = \begin{bmatrix} c_1 \\ c_2 \end{bmatrix}$.

10. $A = \begin{bmatrix} -5 & 3 \\ -15 & 7 \end{bmatrix}$, $Y(0) = \begin{bmatrix} 1 \\ 1 \end{bmatrix}$.

11. $A = \begin{bmatrix} 3 & -1 & 1 \\ 2 & 0 & 1 \\ 1 & -1 & 2 \end{bmatrix}$, $Y(0) = \begin{bmatrix} 1 \\ -1 \\ 2 \end{bmatrix}$.

12. $A = \begin{bmatrix} 2 & 0 & 0 \\ 0 & 1 & 0 \\ 0 & 1 & 1 \end{bmatrix}$, $Y(0) = \begin{bmatrix} c_1 \\ c_2 \\ c_3 \end{bmatrix}$.

13. $A = \begin{bmatrix} 0 & 1 & 0 \\ 0 & 0 & 1 \\ -6 & -11 & -6 \end{bmatrix}$, $Y(0) = \begin{bmatrix} 1 \\ 0 \\ 0 \end{bmatrix}$.

14. $A = \begin{bmatrix} -2 & 2 & -3 \\ 2 & 1 & -6 \\ -1 & -2 & 0 \end{bmatrix}$, $Y(0) = \begin{bmatrix} 8 \\ 0 \\ 0 \end{bmatrix}$.

15. $A = \begin{bmatrix} 1 & 1 & 0 & 0 \\ 0 & 2 & 1 & 0 \\ 0 & 0 & 3 & 0 \\ 0 & 0 & 0 & 4 \end{bmatrix}$, $Y(0) = \begin{bmatrix} 0 \\ 0 \\ 1 \\ 1 \end{bmatrix}$.

16. $A = \begin{bmatrix} 0 & 0 & 2 & 0 \\ 1 & 0 & 0 & 2 \\ 0 & 0 & 0 & 4 \\ 0 & 0 & 1 & 0 \end{bmatrix}$, $Y(0) = \begin{bmatrix} 1 \\ 0 \\ 2 \\ 1 \end{bmatrix}$.

9.15 Nonhomogeneous linear systems with constant coefficients

We consider next the nonhomogeneous initial-value problem

$$(9.39) \qquad Y'(t) = AY(t) + Q(t), \qquad Y(a) = B,$$

on an interval J. Here A is an $n \times n$ constant matrix, $Q(t)$ is an $n \times m$ matrix function continuous on J, and a is a given point in J. The initial-value matrix B is $n \times m$. We can obtain an explicit formula for the $n \times m$ matrix solution of this problem by the same process used to treat the scalar case.

First we multiply both members of (9.39) by the exponential matrix e^{-tA} and rewrite the differential equation in the form

$$(9.40) \qquad e^{-tA}\left\{Y'(t) - AY(t)\right\} = e^{-tA}Q(t).$$

The left member of (9.40) is the derivative of the product $e^{-tA}Y(t)$. Therefore, if we integrate both members of (9.40) from a to x, where $x \in J$, we obtain

$$e^{-xA}Y(x) - e^{-aA}Y(a) = \int_a^x e^{-tA}Q(t)\,dt.$$

Multiplying by e^{xA} we obtain the explicit formula (9.41) that appears in the following theorem.

THEOREM 9.13. *Let A be an $n \times n$ constant matrix and let Q be an $n \times m$ matrix function continuous on an interval J. Then the initial-value problem*

$$Y'(t) = AY(t) + Q(t), \qquad Y(a) = B,$$

has a unique solution on J given by the explicit formula

$$(9.41) \qquad Y(x) = e^{(x-a)A}B + e^{xA} \int_a^x e^{-tA}Q(t)\, dt.$$

As in the homogeneous case, the difficulty in applying this formula in practice lies in the calculation of the exponential matrices.

Note that the first term, $e^{(x-a)A}B$, is the solution of the homogeneous problem $Y'(t) = AY(t)$, $Y(a) = B$. The second term is the solution of the nonhomogeneous problem

$$Y'(t) = AY(t) + Q(t), \qquad Y(a) = O.$$

We illustrate Theorem 9.13 with an example.

EXAMPLE. Solve the initial-value problem

$$Y'(t) = AY(t) + Q(t), \qquad Y(0) = B,$$

on the interval $(-\infty, +\infty)$, where

$$A = \begin{bmatrix} 2 & -1 & 1 \\ 0 & 3 & -1 \\ 2 & 1 & 3 \end{bmatrix}, \qquad Q(t) = \begin{bmatrix} e^{2t} \\ 0 \\ te^{2t} \end{bmatrix}, \qquad B = \begin{bmatrix} 0 \\ 0 \\ 0 \end{bmatrix}.$$

Solution. According to Theorem 9.13, the solution is given by

$$(9.42) \qquad Y(x) = e^{xA} \int_0^x e^{-tA}Q(t)\, dt = \int_0^x e^{(x-t)A}Q(t)\, dt.$$

The eigenvalues of A are 2, 2, and 4. To calculate e^{xA} we use the formula from Case 2, Section 9.12 to obtain

$$e^{xA} = e^{2x}\{I + x(A - 2I)\} + \frac{1}{4}(e^{4x} - e^{2x})(A - 2I)^2 - \frac{1}{2}xe^{2x}(A - 2I)^2$$

$$= e^{2x}\left\{I + x(A - 2I) + \frac{1}{4}(e^{2x} - 2x - 1)(A - 2I)^2\right\}.$$

We can replace x by $x - t$ in this formula to obtain $e^{(x-t)A}$. Therefore the integrand in (9.42) is

$$e^{(x-t)A}Q(t)$$

$$= e^{2(x-t)}\left\{I + (x - t)(A - 2I) + \frac{1}{4}[e^{2(x-t)} - 2(x - t) - 1](A - 2I)^2\right\}Q(t)$$

$$= e^{2x}\begin{bmatrix} 1 \\ 0 \\ t \end{bmatrix} + (A - 2I)e^{2x}\begin{bmatrix} x - t \\ 0 \\ t(x - t) \end{bmatrix} + \frac{1}{4}(A - 2I)^2 e^{2x}\begin{bmatrix} e^{2x}e^{-2t} - 2(x - t) - 1 \\ 0 \\ e^{2x}te^{-2t} - 2t(x - t) - t \end{bmatrix}.$$

Integrating, we find

$$Y(x) = \int_0^x e^{(x-t)A} Q(t)\, dt$$

$$= e^{2x} \begin{bmatrix} x \\ 0 \\ \frac{1}{2}x^2 \end{bmatrix} + (A - 2I)e^{2x} \begin{bmatrix} \frac{1}{2}x^2 \\ 0 \\ \frac{1}{6}x^3 \end{bmatrix}$$

$$+ \frac{1}{4}(A - 2I)^2 e^{2x} \begin{bmatrix} \frac{1}{2}e^{2x} - \frac{1}{2} - x - x^2 \\ 0 \\ \frac{1}{4}e^{2x} - \frac{1}{4} - \frac{1}{2}x - \frac{1}{2}x^2 - \frac{1}{3}x^3 \end{bmatrix}.$$

Since we have

$$A - 2I = \begin{bmatrix} 0 & -1 & 1 \\ 0 & 1 & -1 \\ 2 & 1 & 1 \end{bmatrix} \quad \text{and} \quad (A - 2I)^2 = \begin{bmatrix} 2 & 0 & 2 \\ -2 & 0 & -2 \\ 2 & 0 & 2 \end{bmatrix},$$

we obtain

$$Y(x) = e^{2x} \begin{bmatrix} x \\ 0 \\ \frac{1}{2}x^2 \end{bmatrix} + e^{2x} \begin{bmatrix} \frac{1}{6}x^3 \\ -\frac{1}{6}x^3 \\ x^2 + \frac{1}{6}x^3 \end{bmatrix} + e^{2x} \begin{bmatrix} \frac{3}{8}e^{2x} - \frac{3}{8} - \frac{3}{4}x - \frac{3}{4}x^2 - \frac{1}{6}x^3 \\ -\frac{3}{8}e^{2x} + \frac{3}{8} + \frac{3}{4}x + \frac{3}{4}x^2 + \frac{1}{6}x^3 \\ \frac{3}{8}e^{2x} - \frac{3}{8} - \frac{3}{4}x - \frac{3}{4}x^2 - \frac{1}{6}x^3 \end{bmatrix}$$

$$= e^{2x} \begin{bmatrix} \frac{3}{8}e^{2x} - \frac{3}{8} + \frac{1}{4}x - \frac{3}{4}x^2 \\ -\frac{3}{8}e^{2x} + \frac{3}{8} + \frac{3}{4}x + \frac{3}{4}x^2 \\ \frac{3}{8}e^{2x} - \frac{3}{8} - \frac{3}{4}x + \frac{3}{4}x^2 \end{bmatrix}.$$

The components of $Y(x)$ are the required functions y_1, y_2, y_3.

9.16 Exercises

1. Let Z be the solution of the nonhomogeneous system

$$Z'(t) = AZ(t) + Q(t),$$

on an interval J with initial value $Z(a)$. Prove that there is only one solution of the nonhomogeneous system

$$Y'(t) = AY(t) + Q(t)$$

on J with another initial value $Y(a)$ and that it is given by the formula

$$Y(t) = Z(t) + e^{(t-a)A}\{Y(a) - Z(a)\}.$$

Special methods are often available for determining a particular solution $Z(t)$ that resembles the given function $Q(t)$. Exercises 2, 3, 5, and 7 indicate such methods for $Q(t) = C$, $Q(t) = e^{\alpha t}C$, $Q(t) = t^m C$, and $Q(t) = (\cos \alpha t)C + (\sin \alpha t)D$, where C and D are constant vectors. If the particular solution $Z(t)$ so obtained does not have the required initial value, we modify $Z(t)$ as indicated in Exercise 1 to obtain another solution $Y(t)$ with the required initial value.

2. (a) Let A be a constant $n \times n$ matrix, and let B, C be constant vectors in $\mathbf{R}^n$. Prove that the solution of the system

$$Y'(t) = AY(t) + C, \qquad Y(a) = B,$$

on $(-\infty, +\infty)$ is given by the formula

$$Y(x) = e^{(x-a)A}B + \left(\int_0^{x-a} e^{uA}\, du \right) C.$$

(b) If A is nonsingular, show that the integral in part (a) has the value $\{e^{(x-a)A} - I\}A^{-1}$.

(c) Compute $Y(x)$ explicitly when

$$A = \begin{bmatrix} -1 & 2 \\ -2 & 3 \end{bmatrix}, \qquad C = \begin{bmatrix} 1 \\ 2 \end{bmatrix}, \qquad B = \begin{bmatrix} b \\ c \end{bmatrix}, \qquad a = 0.$$

3. Given an $n \times n$ constant matrix A, constant vectors B, C in $\mathbf{R}^n$, and a scalar α.

(a) Prove that the nonhomogeneous system $Z'(t) = AZ(t) + e^{\alpha t}C$ has a solution of the form $Z(t) = e^{\alpha t}B$ if and only if $(\alpha I - A)B = C$.

(b) If α is not an eigenvalue of A, prove that the vector B can always be chosen so that the system in (a) has a solution of the form $Z(t) = e^{\alpha t}B$.

(c) If α is not an eigenvalue of A, prove that every solution of the system $Y'(t) = AY(t) + e^{\alpha t}C$ has the form $Y(t) = e^{tA}(Y(0) - B) + e^{\alpha t}B$, where $B = (\alpha I - A)^{-1}C$.

4. Use the method suggested by Exercise 3 to find a solution of the nonhomogeneous system $Y'(t) = AY(t) + e^{2t}C$ with

$$A = \begin{bmatrix} 3 & 1 \\ 2 & 2 \end{bmatrix}, \qquad C = \begin{bmatrix} -1 \\ -1 \end{bmatrix}, \qquad Y(0) = \begin{bmatrix} 0 \\ 1 \end{bmatrix}.$$

5. Given an $n \times n$ constant matrix A, constant vectors B, C in $\mathbf{R}^n$, and a positive integer m.

(a) Prove that the nonhomogeneous system $Y'(t) = AY(t) + t^m C$, $Y(0) = B$, has a particular solution of the form

$$Y(t) = B_0 + tB_1 + t^2 B_2 + \cdots + t^m B_m,$$

where $B_0, B_1, \ldots, B_m$ are constant vectors, if and only if $C = -\frac{1}{m!}A^{m+1}B$. Determine the coefficients $B_0, B_1, \ldots, B_m$ for such a solution.

(b) If A is nonsingular, prove that the initial vector B can always be chosen so that the system in (a) has a solution of the specified form.

6. Consider the nonhomogeneous system

$$y_1' = 3y_1 + y_2 + t^2$$

$$y_2' = 2y_1 + 2y_2 + t^3.$$

(a) Find a particular solution of the form $Y(t) = B_0 + tB_1 + t^2 B_2 + t^3 B_3$.

(b) Find a solution of the system with $y_1(0) = y_2(0) = 1$.

7. Given an $n \times n$ constant matrix A, constant vectors B, C, D in $\mathbf{R}^n$, and a nonzero real α. Prove that the nonhomogeneous system

$$Y'(t) = AY(t) + (\cos \alpha t)C + (\sin \alpha t)D, \qquad Y(0) = B,$$

has a particular solution of the form $Y(t) = (\cos \alpha t)E + (\sin \alpha t)F$, where E and F are constant vectors, if and only if $(A^2 + \alpha^2 I)B = -(AC + \alpha D)$. Determine E and F in terms of A, B, C for such a solution. Note that if $A^2 + \alpha^2 I$ is nonsingular, the initial vector B can always be chosen so that the system has a solution of the specified form.

8. (a) Find a particular solution of the nonhomogeneous system

$$y_1' = y_1 + 3y_2 + 4 \sin 2t$$

$$y_2' = y_1 - y_2.$$

(b) Find a solution of the system with $y_1(0) = y_2(0) = 1$.

In each of Exercises 9 and 10, solve the nonhomogeneous system $Y'(t) = AY(t) + Q(t)$ subject to the given initial condition.

9. $A = \begin{bmatrix} 4 & 1 \\ -2 & 1 \end{bmatrix}$, $Q(t) = \begin{bmatrix} 0 \\ -2e^t \end{bmatrix}$, $Y(0) = \begin{bmatrix} 1 \\ 0 \end{bmatrix}$.

10. $A = \begin{bmatrix} -1 & -1 & 0 \\ 0 & -1 & -1 \\ 0 & 0 & -1 \end{bmatrix}$, $Q(t) = \begin{bmatrix} t^2 \\ 2t \\ t \end{bmatrix}$, $Y(0) = \begin{bmatrix} 6 \\ -2 \\ 1 \end{bmatrix}$.

9.17   The general linear system $Y'(t) = P(t)Y(t) + Q(t)$

Theorem 9.13 gives an explicit formula for the solution of the linear system

$$Y'(t) = AY(t) + Q(t), \qquad Y(a) = B,$$

where A is a constant $n \times n$ matrix and $Q(t)$, $Y(t)$ are $n \times m$ matrix functions. We turn now to the more general case

(9.43) $$Y'(t) = P(t)Y(t) + Q(t), \qquad Y(a) = B,$$

where the $n \times n$ matrix $P(t)$ is not necessarily constant.

If P and Q are continuous on an open interval J, a general existence-uniqueness theorem that we shall prove in Chapter 10 tells us that for each a in J and each initial matrix B there is exactly one solution to the initial-value problem (9.43). In this section we use this result to obtain an explicit formula for the solution, generalizing Theorem 9.13.

In the scalar case ($m = n = 1$) the differential equation (9.43) can be solved as follows. Introduce the function $A(x) = \int_a^x P(t) \, dt$, then multiply both members of (9.43) by $e^{-A(t)}$ to rewrite the differential equation in the form

(9.44) $$e^{-A(t)}\{Y'(t) - P(t)Y(t)\} = e^{-A(t)}Q(t).$$

The left member of (9.44) is the derivative of the product $e^{-A(t)}Y(t)$. Therefore we can integrate both members from a to x, where a and x are points in J, to obtain

$$e^{-A(x)}Y(x) - e^{-A(a)}Y(a) = \int_a^x e^{-A(t)}Q(t) \, dt.$$

Multiplying by $e^{A(x)}$ we obtain the explicit formula

(9.45) $$Y(x) = e^{A(x)}e^{-A(a)}Y(a) + e^{A(x)}\int_a^x e^{-A(t)}Q(t) \, dt.$$

The only part of this argument that does not apply immediately to matrix functions is the statement that the left-hand member of (9.44) is the derivative of the product $e^{-A(t)}Y(t)$. At this stage we used the fact that the derivative of $e^{-A(t)}$ is $-P(t)e^{-A(t)}$. In the scalar case this is a consequence of the following differentiation property of exponential functions:

$$\textit{If } E(t) = e^{A(t)} \qquad \textit{then} \qquad E'(t) = A'(t)e^{A(t)}.$$

Unfortunately, this differentiation formula is not always true when A is a matrix function. For example, it is false for the 2×2 matrix function $A(t) = \begin{bmatrix} 1 & t \\ 0 & 0 \end{bmatrix}$. (See Exercise 10 of Section 9.11.) Therefore a modified argument is needed to extend Equation (9.45) to the matrix case.

Suppose we multiply each member of (9.43) by an unspecified $n \times n$ matrix $F(t)$. This gives us the relation

$$F(t)Y'(t) = F(t)P(t)Y(t) + F(t)Q(t).$$

Now we add $F'(t)Y(t)$ to both members to transform the left member into the derivative of the product $F(t)Y(t)$. If we do this the last equation gives us

$$\{F(t)Y(t)\}' = \{F'(t) + F(t)P(t)\}Y(t) + F(t)Q(t).$$

If we can choose the matrix $F(t)$ so that the sum $\{F'(t) + F(t)P(t)\}$ on the right is the zero matrix, the last equation simplifies to

$$\{F(t)Y(t)\}' = F(t)Q(t).$$

Integrating this from a to x we obtain

$$F(x)Y(x) - F(a)Y(a) = \int_a^x F(t)Q(t)\, dt.$$

If, in addition, the matrix $F(x)$ is nonsingular, we obtain the explicit formula

(9.46) $$Y(x) = F(x)^{-1}F(a)Y(a) + F(x)^{-1} \int_a^x F(t)Q(t)\, dt.$$

This is a generalization of the scalar formula (9.45). The process will work if we can find a $n \times n$ nonsingular matrix function $F(t)$ that satisfies the matrix differential equation

$$F'(t) = -F(t)P(t).$$

Note that this differential equation is very much like the original differential equation (9.43) with $Q(t) = 0$, except that the unknown function $F(t)$ is a square matrix instead of an $n \times m$ matrix. Also, the unknown function is multiplied on the right by $-P(t)$ instead of on the left by $P(t)$.

We shall prove next that the differential equation for F always has a nonsingular solution. The proof will depend on the following existence theorem for homogeneous linear systems.

THEOREM 9.14.   *Assume $A(t)$ is an $n \times n$ matrix function continuous on an open interval J. If $a \in J$ and if B is a given $n \times m$ constant matrix, the homogeneous linear system*

$$Y'(t) = A(t)Y(t), \qquad Y(a) = B,$$

has an $n \times m$ matrix solution Y on J.

A proof of Theorem 9.14 appears in Chapter 10 (Theorem 10.3). With the help of this result we can prove the following theorem.

THEOREM 9.15.   *Given an n × n matrix function P, continuous on an open interval J, and given any point a in J, there exists an n × n matrix function F that satisfies the matrix differential equation*

(9.47) $$F'(x) = -F(x)P(x)$$

on J with initial value F(a) = I. Moreover, F(x) is nonsingular for each x in J.

Proof.   Let $Y_k(x)$ be an $n \times 1$ column matrix solution of the differential equation

$$Y_k'(x) = -P(x)^t Y_k(x)$$

on *J* with initial vector $Y_k(a) = I_k$, where I_k is the *k*th column of the $n \times n$ identity matrix *I*. Here $P(x)^t$ denotes the transpose of $P(x)$. Let $G(x)$ be the $n \times n$ matrix whose *k*th column is $Y_k(x)$. Then *G* satisfies the matrix differential equation

(9.48) $$G'(x) = -P(x)^t G(x)$$

on *J* with initial value $G(a) = I$. Now take the transpose of each member of (9.48). Since the transpose of a product is the product of transposes in reverse order, we obtain

$$\{G'(x)\}^t = -G(x)^t P(x).$$

Also, the transpose of the derivative G' is the derivative of the transpose G^t. Therefore the matrix $F(x) = G(x)^t$ satisfies the differential equation (9.47) with initial value $F(a) = I$.

Now we prove that $F(x)$ is nonsingular by exhibiting its inverse. Let *H* be the $n \times n$ matrix function whose *k*th column is the solution of the differential equation

$$H'(x) = P(x)H(x), \qquad H(a) = I,$$

on *J*. The product $F(x)H(x)$ has derivative

$$F(x)H'(x) + F'(x)H(x) = F(x)P(x)H(x) - F(x)P(x)H(x) = O$$

for each *x* in *J*. Therefore the product $F(x)H(x)$ is constant, $F(x)H(x) = F(a)H(a) = I$, so $H(x)$ is the inverse of $F(x)$. This completes the proof.

The results of this section are summarized in the following theorem.

THEOREM 9.16.   *Given an n × n matrix function P and an n × m matrix function Q, both continuous on an open interval J, the solution of the initial-value problem*

(9.49) $$Y'(x) = P(x)Y(x) + Q(x), \qquad Y(a) = B,$$

on J is given by the formula

(9.50) $$Y(x) = F(x)^{-1}(a) + F(x)^{-1} \int_a^x F(t)Q(t)\,dt.$$

The n × n matrix F(x) is the transpose of the matrix whose kth column is the solution of the initial-value problem

(9.51) $$Y'(x) = -P(x)^t Y(x), \qquad Y(a) = I_k,$$

where I_k is the kth column of the identity matrix I.

We remind the reader once more that the proof of Theorem 9.16 was based on Theorem 9.14, the existence theorem for homogeneous linear systems, which we have not yet proved.

Although Theorem 9.16 provides an explicit formula for the solution of the general linear system (9.49), the formula is not always a useful one for calculating the solution because of the difficulty involved in determining the matrix function F.

9.18   A power series method for solving homogeneous linear systems

Consider a homogeneous linear system

$$(9.52) \qquad Y'(x) = A(x)Y(x), \qquad Y(0) = B,$$

in which the given $n \times n$ matrix $A(x)$ has a power-series expansion in x convergent in some open interval containing the origin, say

$$A(x) = A_0 + xA_1 + x^2A_2 + \cdots + x^kA_k + \cdots \qquad \text{for} \qquad |x| < r_1,$$

where the coefficients $A_0, A_1, A_2, \ldots$ are given $n \times n$ matrices. Let us try to find a power-series solution of the form

$$Y(x) = B_0 + xB_1 + x^2B_2 + \cdots + x^kB_k + \cdots,$$

with $n \times m$ constant matrix coefficients $B_0, B_1, B_2, \ldots$. Since $Y(0) = B_0$, the initial condition will be satisfied by taking $B_0 = B$, the prescribed initial $n \times m$ matrix. To determine the remaining coefficients we substitute the power series for $Y(x)$ in the differential equation and equate coefficients of like powers of x to obtain the following system of equations:

$$(9.53) \qquad B_1 = A_0B, \qquad (k+1)B_{k+1} = \sum_{r=0}^{k} A_rB_{k-r} \qquad \text{for} \qquad k = 1, 2, \ldots.$$

These equations can be solved in succession for the matrices $B_1, B_2, \ldots$. If the resulting power series for $Y(x)$ converges in some interval $|x| < r_2$, then $Y(x)$ will be a solution of the initial-value problem (9.52) in the interval $|x| < r$, where $r = \min\{r_1, r_2\}$.

For example, if $A(x)$ is a constant matrix A, then $A_0 = A$ and $A_k = O$ for $k \geq 1$, so the system of equations in (9.53) becomes

$$B_1 = AB, \qquad (k+1)B_{k+1} = AB_k \qquad \text{for} \qquad k \geq 1.$$

Solving these equations in succession we find

$$B_k = \frac{1}{k!}A^kB \qquad \text{for} \qquad k \geq 1.$$

Therefore the series solution in this case becomes

$$Y(x) = B + \sum_{k=1}^{\infty} \frac{x^k}{k!}A^kB = e^{xA}B.$$

This agrees with the result obtained in Theorem 9.5 for homogeneous linear systems with constant coefficients.

9.19 Exercises

1. Let p be a real-valued function and Q an $n \times 1$ matrix function, both continuous on an open interval J. Let A be an $n \times n$ constant matrix. Prove that the initial-value problem

$$Y'(x) = p(x)AY(x) + Q(x), \qquad Y(a) = B,$$

has the solution

$$Y(x) = e^{q(x)A}B + e^{q(x)A}\int_a^x e^{-q(t)A}Q(t)\,dt$$

on J, where $q(x) = \int_a^x p(t)\,dt$.

2. Consider the special case of Exercise 1 in which A is nonsingular, $a = 0$, $p(x) = 2x$, and $Q(x) = xC$, where C is a constant vector. Show that the solution becomes

$$Y(x) = e^{x^2 A}\left(B + \frac{1}{2}A^{-1}C\right) - \frac{1}{2}A^{-1}C.$$

3. Let $A(t)$ be an $n \times n$ matrix function and let $E(t) = e^{A(t)}$. Let $Q(t)$, $Y(t)$, and B be $n \times 1$ column matrices. Assume that $E'(t) = A'(t)E(t)$ on an open interval J. If $a \in J$ and if A' and Q are continuous on J, prove that the initial-value problem $Y'(t) = A'(t)Y(t) + Q(t)$, $Y(a) = B$, has the following solution on J:

$$Y(x) = e^{A(x)}e^{-A(a)}B + e^{A(x)}\int_a^x e^{-A(t)}Q(t)\,dt.$$

4. Let $E(t) = e^{A(t)}$. This exercise describes examples of matrix functions $A(t)$ for which $E'(t) = A'(t)E(t)$.

    (a) Let $A(t) = t^r A$, where A is an $n \times n$ constant matrix and r is a positive integer. Prove that $E'(t) = A'(t)E(t)$ for all real t.

    (b) Let $A(t)$ be a polynomial in t with matrix coefficients that commute, say

$$A(t) = \sum_{r=0}^m t^r A_r,$$

   where $A_r A_s = A_s A_r$ for all r and s. Prove that $E'(t) = A'(t)E(t)$ for all real t.

    (c) Solve the homogeneous linear system $Y'(t) = (I + tA)Y(t)$, $Y(0) = B$, on $(-\infty, +\infty)$, where A is an $n \times n$ constant matrix.

5. Assume that an $n \times n$ matrix function $A(x)$ has a power-series expansion convergent for $|x| < r$. Develop a power-series procedure for solving the following homogeneous linear system of second order:

$$Y(x) = A(x)Y(x), \qquad \text{with} \qquad Y(0) = B \qquad \text{and} \qquad Y'(0) = C.$$

6. Consider the second-order system $Y(x) + AY(x) = O$, with $Y(0) = B$ and $Y'(0) = C$, where A is a constant $n \times n$ matrix. Prove that the system has the power-series solution

$$Y(x) = \left(I + \sum_{k=1}^\infty \frac{(-1)^k x^{2k} A^k}{(2k)!}\right)B + \left(xI + \sum_{k=1}^\infty \frac{(-1)^k x^{2k+1} A^k}{(2k+1)!}\right)C,$$

convergent for all real x.

10

THE METHOD OF SUCCESSIVE APPROXIMATIONS

10.1 Introduction

This chapter proves the existence and uniqueness theorems for differential equations referred to in Chapters 8 and 9. Proofs are given by the *method of successive approximations*, a powerful iterative procedure that also has applications to many other problems. The method was first published by Joseph Liouville in 1838 in connection with the study of linear differential equations of second order. It was later extended by J. Caqué in 1864, L. Fuchs in 1870, and G. Peano in 1888 to the study of linear equations of order n. In 1890 Émile Picard (1856–1941) extended the method to encompass nonlinear differential equations as well. In recognition of his fundamental contributions some writers refer to the process as *Picard's method*. The method is not only of theoretical interest but can also be used to obtain numerical approximations to solutions in some cases.

10.2 Application to the homogeneous linear system $Y'(t) = A(t)Y(t)$

To introduce the method of successive approximations we use it to prove the existence and uniqueness of a solution for the initial-value problem

$$(10.1) \qquad Y'(t) = A(t)Y(t), \qquad Y(a) = B.$$

Here we assume that $A(t)$ is a known $n \times n$ matrix function, continuous on an open interval J, and a is any point in J. The unknown function Y is an $n \times m$ matrix function, and B is any given $n \times m$ constant matrix. In many applications we are interested in the case $m = 1$, where B and Y are $n \times 1$ column matrices, but the method works equally well for the general case of $n \times m$ matrices.

The method of successive approximations is based on a ridiculously simple idea. We begin with an initial guess at a solution, which we call $Y_0(t)$, substitute this for $Y(t)$ in the right-hand side of the differential equation, to get a new equation

$$(10.2) \qquad Y'(t) = A(t)Y_0(t)$$

that no longer contains the unknown function in the right-hand member. It is common practice to choose $Y_0(t)$ to be the initial vector B, although that is not essential. Equation (10.2) has exactly one solution Y_1 on J satisfying the initial condition $Y_1(a) = B$, namely,

$$Y_1(x) = B + \int_a^x A(t)Y_0(t)\, dt.$$

Now we replace $Y(t)$ by $Y_1(t)$ in the right member of (10.1) and integrate again from a to x, calling the new solution Y_2, with $Y_2(a) = B$. In other words, we let

$$Y_2(x) = B + \int_a^x A(t)Y_1(t)\, dt.$$

When this procedure is repeated indefinitely, it generates a sequence of functions Y_0, Y_1, $Y_2,\ldots$, where Y_{k+1} is determined from Y_k by the recursion formula

(10.3) $$Y_{k+1}(x) = B + \int_a^x A(t)Y_k(t)\, dt \qquad \text{for} \qquad k = 0, 1, 2,\ldots.$$

It is a remarkable fact that the sequence of functions so defined always converges to a limit function Y that satisfies the initial-value problem (10.1) and, moreover, it is the *only* solution of (10.1). Before investigating the convergence of the process, we illustrate the method with an example whose solution we have already found by other means.

EXAMPLE.   Consider the initial-value problem $Y'(t) = AY(t)$, $Y(0) = B$, where A is a constant $n \times n$ matrix. We know from Theorem 9.7 that the solution is given by the formula $Y(x) = e^{xA}B$ for all real x. To obtain this solution by successive approximations we make initial guess $Y_0(x) = B$. Repeated use of the recursion formula (10.3) gives us

$$Y_1(x) = B + \int_0^x AB\, dt = B + xAB,$$

$$Y_2(x) = B + \int_0^x AY_1(t)\, dt = B + \int_0^x (AB + tA^2B)\, dt = B + xAB + \frac{1}{2}x^2A^2B.$$

$$\vdots$$

$$Y_k(x) = B + xAB + \frac{1}{2}x^2A^2B + \cdots + \frac{1}{k!}x^kA^kB = \left(\sum_{r=0}^k \frac{(xA)^r}{r!}\right)B.$$

The last sum on the right is a partial sum of the series for e^{xA}. Therefore when $k \to \infty$ we find

$$\lim_{k \to \infty} Y_k(x) = e^{xA}B.$$

10.3   Convergence of the sequence of successive approximations

THEOREM 10.1.   EXISTENCE THEOREM FOR HOMOGENEOUS LINEAR SYSTEMS.   *Given an $n \times n$ matrix function $A(t)$ continuous on an open interval J, and given any $n \times m$ constant matrix B. If $a \in J$, define a sequence of $n \times m$ matrix functions by the recursion* *formula*

(10.4) $Y_0(x) = B,$ $Y_{k+1}(x) = B + \int_a^x A(t)Y_k(t)\,dt$ *for* $k = 0, 1, 2, \ldots.$

Then for each x in J the sequence $\{Y_k(x)\}$ converges to an $n \times m$ matrix function $Y(x)$ that satisfies the initial-value problem

(10.5) $Y'(x) = A(x)Y(x),$ $Y(a) = B.$

   Proof.   First we write each term $Y_k(x)$ as a telescoping sum,

(10.6) $$Y_k(x) = Y_0(x) + \sum_{r=0}^{k-1}\{Y_{r+1}(x) - Y_r(x)\}.$$

To prove that $Y_k(x)$ tends to a limit as $k \to \infty$ we shall prove that the infinite series

(10.7) $$\sum_{r=0}^{\infty}\{Y_{r+1}(x) - Y_r(x)\}$$

converges for each x in J. For this purpose it suffices to prove that the series of matrix norms

(10.8) $$\sum_{r=0}^{\infty}\|Y_{r+1}(x) - Y_r(x)\|$$

converges. In this series we use the matrix norm introduced in Section 9.3; the norm of a matrix is the sum of the absolute values of all its entries.

   Consider a closed and bounded subinterval J_1 of J containing a. We shall prove that for every x in J_1 the series in (10.8) is dominated by a convergent series of constants independent of x. This implies not only that the series in (10.7) converges, but also that it converges *uniformly* on J_1.

   To estimate the size of the terms in (10.8) we use the recursion formula repeatedly. Initially, we have

$$Y_1(x) - Y_0(x) = \int_a^x A(t)B\,dt.$$

For simplicity, we assume that $a < x$. Then we can write

(10.9) $$\|Y_1(x) - Y_0(x)\| = \left\|\int_a^x A(t)B\,dt\right\| \le \int_a^x \|A(t)\|\|B\|\,dt.$$

   Since each entry of $A(t)$ is continuous on J, each entry is bounded on the closed bounded subinterval J_1. Therefore $\|A(t)\| \le M$, where M is the sum of the bounds of all the entries of $A(t)$ on the interval J_1. (The number M depends on J_1.) Therefore the integrand in (10.9) is bounded by $\|B\|M$, so we have

$$\|Y_1(x) - Y_0(x)\| \le \int_a^x \|B\|M\,dt = \|B\|M(x - a)$$

for all $x > a$ in J_1.

Now we use the recursion formula once more to express the difference $Y_2 - Y_1$ in terms of $Y_1 - Y_0$, and then use the estimate just obtained for $Y_1 - Y_0$ to obtain

$$\|Y_2(x) - Y_1(x)\| = \left\| \int_a^x A(t)\{Y_1(t) - Y_0(t)\}\, dt \right\| \leq \int_a^x \|A(t)\|\|B\|M(t - a)\, dt$$

$$\leq \|B\|M^2 \int_a^x (t - a)\, dt = \|B\| \frac{M^2(x - a)^2}{2!}$$

for all $x > a$ in J_1. By induction we find

$$\|Y_{r+1}(x) - Y_r(x)\| \leq \|B\| \frac{M^{r+1}(x - a)^{r+1}}{(r + 1)!} \qquad \text{for} \qquad r = 0, 1, 2, \ldots,$$

for all $x > a$ in J_1. If $x < a$ a similar argument gives the same inequality with $|x - a|$ appearing instead of $(x - a)$. If we denote by L the length of the interval J_1, then we have $|x - a| \leq L$ for all x in J_1, so we obtain the estimate

$$\|Y_{r+1}(x) - Y_r(x)\| \leq \|B\| \frac{M^{r+1}L^{r+1}}{(r + 1)!} \qquad \text{for} \qquad r = 0, 1, 2, \ldots,$$

and for all x in J_1. Therefore the series in (10.8) is dominated by the convergent series

$$\|B\| \sum_{r=0}^{\infty} \frac{(ML)^{r+1}}{(r + 1)!} = \|B\|(e^{ML} - 1).$$

This proves that the series in (10.8) converges uniformly on J_1.

The foregoing argument shows that the sequence of successive approximations always converges and that the convergence is uniform on J_1. Let Y denote the limit function. That is, define $Y(x)$ for each x in J_1 by the equation

$$Y(x) = \lim_{k \to \infty} Y_k(x).$$

We shall prove that Y has the following properties:
   (a) Y is continuous on J_1.
   (b) $Y(x) = B + \int_a^x A(t)Y(t)\, dt$ for all x in J_1.
   (c) $Y(a) = B$ and $Y'(x) = A(x)Y(x)$ for all x in J_1.
   Part (c) shows that Y is a solution of the initial-value problem on J_1.

Proof of (a).   Each function Y_k is an $n \times m$ matrix whose entries are scalar functions, continuous on J_1. Each entry of the limit function Y is the limit of a uniformly convergent sequence of continuous functions, so each entry of Y is also continuous on J_1. Therefore Y itself is continuous on J_1.

Proof of (b).   The recursion formula in (10.4) states that

$$Y_{k+1}(x) = B + \int_a^x A(t)Y_k(t)\, dt.$$

Therefore

$$Y(x) = \lim_{k \to \infty} Y_{k+1}(x) = B + \lim_{k \to \infty} \int_a^x A(t)Y_k(t)\,dt = B + \int_a^x A(t)\lim_{k \to \infty} Y_k(t)\,dt$$

$$= B + \int_a^x A(t)Y(t)\,dt.$$

The interchange of the limit symbol with the integral sign is valid because of the uniform convergence of the sequence $\{Y_k\}$ on J_1.

Proof of (c).   The equation $Y(a) = B$ follows at once from (b). Because of (a), the integrand in (b) is continuous on J_1, so the first fundamental theorem of calculus tells us that $Y'(x)$ exists and equals $A(x)Y(x)$ on J_1.

The interval J_1 was any closed and bounded subinterval of J containing a. If J_1 is enlarged, the process for obtaining $Y(x)$ doesn't change because it only involves integration from a to x. For every x in J there is a closed bounded subinterval of J containing both a and x. Consequently, a solution exists over the full interval J.

THEOREM 10.2.   UNIQUENESS THEOREM FOR HOMOGENEOUS LINEAR SYSTEMS.   *If $A(t)$ is continuous on an open interval J, the differential equation*

$$Y'(t) = A(t)Y(t)$$

has at most one solution on J satisfying a given initial condition $Y(a) = B$.

Proof.   Let Y and Z be two solutions on J. Let J_1 be any closed and bounded subinterval of J containing a. We will prove that $Z(x) = Y(x)$ for every x in J_1. This implies that $Z = Y$ on the full interval J.

Because both Y and Z are solutions we have

$$Z'(t) - Y'(t) = A(t)\{Z(t) - Y(t)\}.$$

Choose x in J_1 and integrate this inequality from a to x, using the fact that $Z(a) = Y(a)$, to obtain

$$Z(x) - Y(x) = \int_a^x A(t)\{Z(t) - Y(t)\}\,dt.$$

This implies the inequality

(10.10) $$\|Z(x) - Y(x)\| \leq M \left| \int_a^x \|Z(t) - Y(t)\|\,dt \right|,$$

where M is an upper bound for $\|A(t)\|$ on J_1. Let M_1 be an upper bound for the continuous function $\|Z(t) - Y(t)\|$ on J_1. Then inequality (10.10) gives us

(10.11) $$\|Z(x) - Y(x)\| \leq MM_1|x - a|.$$

Using (10.11) in the right-hand member of (10.10) we obtain

$$\|Z(x) - Y(x)\| \le M^2 M_1 \left| \int_a^x |t - a| \, dt \right| = M^2 M_1 \frac{|x - a|^2}{2}.$$

By induction we find

(10.12)
$$\|Z(x) - Y(x)\| \le M^r M_1 \frac{|x - a|^r}{r!}$$

for every $r \ge 1$. When $r \to \infty$ the right member approaches 0, so $\|Z(x) - Y(x)\| = 0$, hence $Z(x) = Y(x)$. This completes the proof.

Theorems 10.1 and 10.2 can be summarized in the following existence-uniqueness theorem.

THEOREM 10.3.   *Let $A(t)$ be an $n \times n$ matrix function continuous on an open interval J. If $a \in J$ and if B is any $n \times m$ constant matrix, the homogeneous linear system*

$$Y'(t) = A(t)Y(t), \qquad Y(a) = B,$$

has one and only one $n \times m$ matrix solution Y on J.

10.4   The method of successive approximations applied to first-order nonlinear systems

The method of successive approximations can also be applied to some nonlinear systems. Consider a first-order system of the form

(10.13)
$$Y' = F(t, Y),$$

where F is a given $n \times m$ matrix function, and Y is an unknown $n \times m$ matrix function to be determined. We seek a solution Y that satisfies the equation

$$Y'(t) = F[t, Y(t)]$$

for each t in some interval J and that also satisfies a given initial condition $Y(a) = B$, where $a \in J$ and B is a given $n \times m$ constant matrix.

In a manner parallel to the linear case, we construct a sequence of successive approximations $Y_0, Y_1, Y_2, \ldots$, by taking $Y_0 = B$ and defining Y_{k+1} in terms of Y_k by the recursion formula

(10.14)
$$Y_{k+1}(x) = B + \int_a^x F[t, Y_k(t)] \, dt \qquad \text{for} \qquad k = 0, 1, 2, \ldots.$$

With certain restrictions on F, this sequence will converge to a limit function Y that will satisfy the given differential equation and the given initial condition.

Before we investigate restrictions on F that guarantee convergence of the process we discuss two scalar examples chosen to illustrate some of the difficulties that can arise in practice.

EXAMPLE 1.   Consider the nonlinear initial-value problem $y' = x^2 + y^2$ with $y = 0$ when $x = 0$. We shall compute a few approximations to the solution. We choose $Y_0(x) = 0$ and determine the next three approximations by using the recursion formula in (10.14):

$$Y_1(x) = \int_0^x t^2 \, dt = \frac{x^3}{3},$$

$$Y_2(x) = \int_0^x [t^2 + Y_1^2(t)] \, dt = \int_0^x \left(t^2 + \frac{t^6}{9} \right) dt = \frac{x^3}{3} + \frac{x^7}{63},$$

$$Y_3(x) = \int_0^x \left[t^2 + \left(\frac{t^3}{3} + \frac{t^7}{63} \right)^2 \right] dt = \frac{x^3}{3} + \frac{x^7}{63} + \frac{2x^{11}}{2079} + \frac{x^{15}}{59535}.$$

It is now apparent that a great deal of labor will be needed to compute further approximations. For example, the next two approximations Y_4 and Y_5 will be polynomials of degrees 31 and 63, respectively.

The next example exhibits a further difficulty that can arise in the computation of the successive approximations.

EXAMPLE 2.   Consider the nonlinear initial-value problem $y' = 2x + e^x$, with $y = 0$ when $x = 0$. We begin with the initial guess $Y_0(x) = 0$ and we find

$$Y_1(x) = \int_0^x (2t + 1) \, dt = x^2 + x,$$

$$Y_2(x) = \int_0^x (2t + e^{t^2 + t}) \, dt = x^2 + \int_0^x e^{t^2 + t} \, dt.$$

Here further progress is impeded by the fact that the last integral cannot be evaluated in terms of elementary functions. However, for a given x it is possible to calculate a numerical approximation to the integral and thereby obtain an approximation to $Y_2(x)$.

Because of the difficulties displayed in the last two examples, the method of successive approximations is sometimes not very useful for the explicit determination of solutions in practice. The real value of the method is its use in establishing existence theorems.

10.5   Proof of an existence-uniqueness theorem for first-order nonlinear systems

We turn now to an existence-uniqueness theorem for first-order nonlinear systems. By placing suitable restrictions on the function F that appears in the right-hand member of the differential equation

$$Y' = F(x, Y),$$

we can extend the method of proof used for the linear case in Section 10.3.

Let J denote the open interval over which we seek a solution. Assume $a \in J$ and let B be a given $n \times m$ matrix of constants. Let S denote the set of ordered pairs (x, Y) given by

$$S = \{(x, Y) : |x - a| \leq h, \|Y - B\| \leq b\},$$

where $h > 0$ and $b > 0$. [If $n = m = 1$ this is a rectangle with center at (a, B) and with base $2h$ and altitude $2b$.] We assume that the domain of F includes a set S of this type and that F is bounded on S, say

$$(10.15) \qquad \|F(x, Y)\| \leq M$$

for all (x, Y) in S, where M is a positive constant.

Next, we assume that the composite matrix function $G(x) = F[x, Y(x)]$ is continuous on the interval $(a - h, a + h)$ for every function Y that is continuous on $(a - h, a + h)$ and that has the property that $(x, Y(x)) \in S$ for all x in $(a - h, a + h)$. This assumption guarantees the existence of the integrals that occur in the method of successive approximations, and it also implies continuity of the functions so constructed.

Finally, we assume that F satisfies a condition of the form

$$\|F(x, Y) - F(x, Z)\| \leq C\|Y - Z\|$$

whenever (x, Y) and (x, Z) are in S, where C is a positive constant. This restriction is called a *Lipschitz condition*, in honor of Rudolph Lipschitz who first introduced it in 1876. A Lipschitz condition does not restrict a function very seriously and it enables us to extend the proof of the existence theorem from the linear to the nonlinear case.

THEOREM 10.4.   *Assume F satisfies the boundedness, continuity, and Lipschitz conditions specified above on a set S. Let I denote the open interval $(a - L, a + L)$, where L is the smaller of h and b/M. Then there is one and only one $n \times m$ matrix function Y defined on I with $Y(a) = B$ such that $(x, Y(x)) \in S$ and*

$$Y'(x) = F[x, Y(x)] \qquad \text{for each } x \text{ in } I.$$

Proof.   Since the proof is analogous to that for the linear case we sketch only the principal steps. We let $Y_0(x) = B$ and define $n \times m$ matrix functions $Y_1, Y_2, \ldots$ on I by the recursion formula

$$(10.16) \qquad Y_{k+1}(x) = B + \int_a^x F[t, Y_k(t)] \, dt \qquad \text{for} \qquad k = 0, 1, 2, \ldots.$$

For the recursion formula to be meaningful we need to know that $(x, Y_k(x)) \in S$ for each x in I. This is is easily proved by induction on k. When $k = 0$ we have $(x, Y_0(x)) = (x, B)$, which is in S. Assume then that $(x, Y_k(x)) \in S$ for some k and each x in I. Using (10.16) and (10.15) we obtain

$$\|Y_{k+1}(x) - B\| \leq \left| \int_a^x \|F[t, Y_k(t)]\| \, dt \right| \leq M \left| \int_a^x dt \right| = M|x - a|.$$

Since $|x - a| \leq L$ for x in I, this implies that

$$\|Y_{k+1}(x) - B\| \leq ML \leq b,$$

which shows that $(x, Y_{k+1}(x)) \in S$ for each x in I. Therefore the recursion formula is meaningful for every $k \geq 0$ and every x in I.

The convergence of the sequence $\{Y_k(x)\}$ is now established exactly as in Section 10.3. We write

$$Y_k(x) = Y_0(x) + \sum_{r=0}^{k-1}\{Y_{r+1}(x) - Y_r(x)\},$$

and prove that $Y_k(x)$ tends to a limit as $k \to \infty$ by proving that the infinite series

$$\sum_{r=0}^{\infty} \|Y_{r+1}(x) - Y_r(x)\|$$

converges on I. This is deduced from the inequality

$$\|Y_{r+1}(x) - Y_r(x)\| \le \frac{MC^r|x - a|^{r+1}}{(r + 1)!} \le \frac{MC^rL^{r+1}}{(r + 1)!}$$

which is proved by induction, using the recursion formula and the Lipschitz condition. We then define the limit function Y by the equation

$$Y(x) = \lim_{k \to \infty} Y_k(x)$$

for each x in I and verify that it satisfies the integral equation

$$Y(x) = B + \int_a^x F[t, Y(t)]\,dt,$$

exactly as in the linear case. This proves the existence of a solution. The uniqueness can then be proved by the same method used to prove Theorem 10.2.

10.6 Exercises

1. Consider the linear initial-value problem $y' + y = 2e^x$, with $y = 1$ when $x = 0$.
   (a) Find the exact solution Y of this linear first-order differential equation.
   (b) Apply the method of successive approximations, starting with the initial guess $Y_0(x) = 1$. Determine $Y_n(x)$ explicitly and show that $\lim_{n\to\infty} Y_n(x) = Y(x)$ for all real x.
2. Apply the method of successive approximations to the nonlinear initial-value problem

$$y' = x + y^2, \qquad \text{with} \qquad y = 0 \text{ when } x = 0.$$

Take $Y_0(x) = 0$ as initial guess and compute $Y_3(x)$.
3. Apply the method of successive approximations to the nonlinear initial-value problem

$$y' = 1 + xy^2, \qquad \text{with} \qquad y = 0 \text{ when } x = 0.$$

Take $Y_0(x) = 0$ as initial guess and compute $Y_3(x)$.
4. Apply the method of successive approximations to the nonlinear initial-value problem

$$y' = x^2 + y^2, \qquad \text{with} \qquad y = 0 \text{ when } x = 0.$$

Start with the "bad" initial guess $Y_0(x) = 1$, compute $Y_3(x)$, and compare with the results of Example 1 of Section 10.4.

5. If a and b are real, let $M(a, b)$ denote the larger of a and b. Consider the following initial-value problem on the interval $[-1, 1]$: $y' = M(x, y)$, with $y = 1$ when $x = 0$.
   (a) Start with $Y_0(x) = 1$ as initial guess and use the method of successive approximations to determine $Y_n(x)$ for $n = 1, 2, 3,$ and 4.
   (b) Determine the solution of the initial-value problem.

6. Consider the nonlinear initial-value problem

   $$y' = x^2 + y^2, \qquad \text{with} \qquad y = 1 \text{ when } x = 0.$$

   (a) Apply the method of successive approximations, starting with initial guess $Y_0(x) = 1$, and compute $Y_2(x)$.
   (b) Let $f(x, y) = x^2 + y^2$. Consider the rectangle $R = [-1, 1] \times [-1, 1]$, and find the smallest M such that $|f(x, y)| \le M$ on R. Find an interval $I = (-c, c)$ such that the graph of every approximating function Y_n over I will lie in R.
   (c) Assume the solution $y = Y(x)$ has a power-series expansion in a neighborhood of the origin. Determine the first six nonzero terms of this power-series expansion and compare with the result of part (a).

7. Consider the nonlinear initial-value problem

   $$y' = 1 + y^2, \qquad \text{with} \qquad y = 0 \text{ when } x = 0.$$

   (a) Apply the method of successive approximations, starting with initial guess $Y_0(x) = 0$, and compute $Y_4(x)$.
   (b) Prove that every approximating function Y_n is defined on the entire real axis.
   (c) Use Theorem 10.4 to show that the initial-value problem has at most one solution in any interval of the form $(-h, h)$.
   (d) Verify that the differential equation is satisfied on the interval $(-\pi/2, \pi/2)$ by $y = \tan x$. In this example, the successive approximations are defined on the entire real axis, but they converge to a limit function only on the interval $(-\pi/2, \pi/2)$.

8. We seek two functions $y = Y(x)$ and $z = Z(x)$ that simultaneously satisfy the system of equations $y' = z$, $z' = x^3(y + z)$ with initial conditions $y = 1$ and $z = 1/2$ when $x = 0$. Start with the initial guesses $Y_0(x) = 1$, $Z_0(x) = 1/2$, and use the method of successive approximations to obtain the approximating functions

   $$Y_3(x) = 1 + \frac{x}{2} + \frac{3x^5}{40} + \frac{x^6}{60} + \frac{x^9}{192},$$

   $$Z_3(x) = \frac{1}{2} + \frac{3x^4}{8} + \frac{x^5}{10} + \frac{3x^8}{64} + \frac{7x^9}{360} + \frac{x^{12}}{256}.$$

9. Consider the system of equations $y' = 2x + z$, $z' = 3xy + x^2 z$, with initial conditions $y = 2$ and $z = 0$ when $x = 0$. Start with initial guesses $Y_0(x) = 2$ and $Z_0(x) = 0$, and use the method of successive approximations to determine $Y_3(x)$ and $Z_3(x)$.

10. Consider the initial-value problem $y'' = x^2 y' + x^4 y$, with $y = 5$ and $y' = 1$ when $x = 0$. Change this to an equivalent problem involving a system of two equations for two unknown functions $y = Y(x)$ and $z = Z(x)$, where $z = y'$. Then use the method of successive approximations, starting with initial guesses $Y_0(x) = 5$ and $Z_0(x) = 1$, and determine $Y_3(x)$ and $Z_3(x)$.

★ 10.7* Successive approximations and fixed points of operators

The basic idea underlying the method of successive approximations can be used not only to establish existence theorems for differential equations but also for many other important problems in analysis. The rest of this chapter reformulates the method of successive approximations in a setting that greatly increases the scope of its applications.

---

*Starred sections in this chapter contain optional material that illustrate further applications of linear algebra concepts to topics in analysis.

In the proof of Theorem 10.1 we constructed a sequence of functions $\{Y_k\}$ according to the recursion formula

$$Y_{k+1}(x) = B + \int_a^x A(t)Y_k(t)\,dt.$$

The right-hand member of this formula can be regarded as an operator T that converts certain functions Y into new functions $T(Y)$ defined by the equation

$$T(Y) = B + \int_a^x A(t)Y(t)\,dt.$$

In the proof of Theorem 10.1 we found that the solution Y of the initial-value problem $Y'(x) = A(x)Y(x)$, $Y(a) = B$, satisfies the integral equation.

$$Y = B + \int_a^x A(t)Y(t)\,dt.$$

In operator notation this states that $T(Y) = Y$. In other words, the solution Y remains unaltered by the operator T. Such a function Y is called a *fixed point* of the operator T.

Many important problems in analysis can be formulated so their solution depends on the existence of a fixed point for some operator. Therefore it is worthwhile to try to discover properties of operators that guarantee the existence of a fixed point. We turn now to a systematic treatment of this problem.

★ 10.8 Normed linear spaces

To formulate the method of successive approximations in a general form it is convenient to work within the framework of linear spaces. Let S be an arbitrary linear space. When we speak of approximating one element x in S by another element y in S, we consider the difference $x - y$, which we call the *error* of the approximation. To measure the size of this error we introduce a norm in the space. A Euclidean space always has a norm that it inherits from the inner product, namely, $\|x\| = (x, x)^{1/2}$. But we are also interested in norms that do not arise from an inner product, so we define a more general concept of norm as follows.

DEFINITION. NORM. *Let S be any linear space. A real-valued function N defined on S is called a norm if it has the following properties:*
  (a) $N(x) \geq 0$ *for each x in S.*
  (b) $N(cx) = |c|N(x)$ *for each x in S and each scalar c.*
  (c) $N(x + y) \leq N(x) + N(y)$ *for all x and y in S.*
  (d) $N(x) = 0$ *implies $x = O$.*

A linear space with a norm assigned to it is called a *normed linear space*.

The norm of x is sometimes written $\|x\|$ instead of $N(x)$. In this notation the defining properties become:
  (a) $\|x\| \geq 0$ for all x in S.
  (b) $\|cx\| = |c|\|x\|$ for all x in S and all scalars c.
  (c) $\|x + y\| \leq \|x\| + \|y\|$ for all x and y in S.
  (d) $\|x\| = 0$ implies $x = O$.
If x and y are in S, we refer to $\|x - y\|$ as the *distance* from x to y.

EXAMPLE. *The max norm.* Let $C(J)$ denote the linear space of real-valued functions continuous on a closed and bounded interval J. If $\varphi \in C(J)$, define

$$\|\varphi\| = \max_{x \in J} |\varphi(x)|,$$

where the symbol on the right stands for the maximum absolute value of φ on J. The reader can verify that this norm has the four fundamental properties listed in the definition.

The max norm is not derived from an inner product. To prove this we show that the max norm violates some property possessed by all inner-product norms. For example, if a norm is derived from an inner product, then the *parallelogram law*

$$\|x + y\|^2 + \|x - y\|^2 = 2\|x\|^2 + 2\|y\|^2$$

holds for all x and y in S. (See Exercise 8 in Section 3.13.) The parallelogram law is not always satisfied by the max norm. For example, let x and y be the functions on the interval $[0, 1]$ given by

$$x(t) = t, \qquad y(t) = 1 - t.$$

Then we have $\|x\| = \|y\| = \|x + y\| = \|x - y\| = 1$, so the parallelogram law is violated.

★ 10.9   Contraction operators

In this section we consider the normed linear space $C(J)$ of all real functions continuous on a closed bounded interval J in which $\|\varphi\|$ is the max norm. Consider an operator

$$T : C(J) \to C(J)$$

whose domain is $C(J)$ and whose range is a subset of $C(J)$. That is, if φ is continuous on J, then $T(\varphi)$ is also continuous on J. The following formulas illustrate a few simple examples of such operators. In each case φ is an arbitrary function in $C(J)$ and $T(\varphi)(x)$ is defined for each x in J by the formula given here:

$$T(\varphi)(x) = \lambda \varphi(x), \qquad \text{where } \lambda \text{ is a fixed real number,}$$

$$T(\varphi)(x) = \int_c^x \varphi(t)\, dt, \qquad \text{where } c \text{ is a given point in } J,$$

$$T(\varphi)(x) = b + \int_c^x f[t, \varphi(t)]\, dt,$$

where b is constant and the composition $f[t, \varphi(t)]$ is continuous on J.

We are interested in those operators T for which the distance $\|T(\varphi) - T(\psi)\|$ is less than a fixed constant multiple of $\|\varphi - \psi\|$, the multiplier being a constant $\alpha < 1$. These are called *contraction operators*, and are defined as follows:

DEFINITION. CONTRACTION OPERATOR. *An operator* $T : C(J) \to C(J)$ *is called a contraction operator if there is a constant* α *satisfying* $0 \leq \alpha < 1$ *such that for every pair of functions* φ *and* ψ *in* $C(J)$ *we have*

(10.17) $$\|T(\varphi) - T(\psi)\| \leq \alpha \|\varphi - \psi\|.$$

The constant α *is called a contraction constant for* T.

Note. Inequality (10.17) holds if and only if we have $|T(\varphi)(x) - T(\psi)(x)| \leq \alpha\|\varphi - \psi\|$ for every x in J.

EXAMPLE 1. Let T be the operator defined by $T(\varphi)(x) = \lambda\varphi(x)$, where λ is constant. Since

$$|T(\varphi)(x) - T(\psi)(x)| = |\lambda||\varphi(x) - \psi(x)|,$$

we have $\|T(\varphi) - T(\psi)\| = |\lambda|\|\varphi - \psi\|$. Therefore this operator is a contraction operator if and only if $|\lambda| < 1$, in which case $|\lambda|$ may be used as a contraction constant.

EXAMPLE 2. Let $T(\varphi)(x) = b + \int_c^x f[t, \varphi(t)]\,dt$, where f satisfies a Lipschitz condition of the form

$$|f(x, y) - f(x, z)| \leq K|y - z|$$

for all x in J and all real y and z; here K is a positive constant. Let $L(J)$ denote the length of the interval J. If $KL(J) < 1$ we can easily show that T is a contraction operator with contraction constant $KL(J)$. In fact, for every x in J we have

$$|T(\varphi)(x) - T(\psi)(x)| = \left|\int_c^x \{f[t, \varphi(t)] - f[t, \psi(t)]\}\,dt\right| \leq K\left|\int_c^x |\varphi(t) - \psi(t)|\,dt\right|$$

$$\leq K\|\varphi - \psi\|\left|\int_c^x dt\right| \leq KL(J)\|\varphi - \psi\|.$$

If $KL(J) < 1$, then T is a contraction operator with contraction constant $\alpha = KL(J)$.

★ 10.10 Fixed-point theorem for contraction operators

The next theorem shows that every contraction operator has a unique fixed point.

THEOREM 10.5. *Let $T : C(J) \to C(J)$ be a contraction operator. Then there exists one and only one function φ in $C(J)$ such that*

$$(10.18) \qquad\qquad\qquad T(\varphi) = \varphi.$$

Proof. Let φ_0 be any function in $C(J)$ and define a sequence of functions $\{\varphi_n\}$ by the recursion formula

$$\varphi_{n+1} = T(\varphi_n) \qquad \text{for} \qquad n = 0, 1, 2, \ldots.$$

Note that $\varphi_{n+1} \in C(J)$ for each n. We shall prove that the sequence $\{\varphi_n\}$ converges to a limit function φ in $C(J)$. The method is similar to that used in the proof of Theorem 10.3. We write each φ_n as a telescoping sum,

$$(10.19) \qquad\qquad \varphi_n(x) = \varphi_0(x) + \sum_{k=0}^{n-1}\{\varphi_{k+1}(x) - \varphi_k(x)\}$$

and prove convergence of $\{\varphi_n\}$ by showing that the infinite series

$$(10.20) \qquad \varphi_0(x) + \sum_{k=0}^{\infty} \{\varphi_{k+1}(x) - \varphi_k(x)\}$$

converges uniformly on J. Then we show that the sum of this series is the required fixed point.

The uniform convergence of the series will be established by comparing it with the convergent geometric series

$$M \sum_{k=0}^{\infty} \alpha^k,$$

where $M = \|\varphi_0\| + \|\varphi_1\|$, and α is a contraction constant for T. The comparison is provided by the inequality

$$(10.21) \qquad |\varphi_{k+1}(x) - \varphi_k(x)| \leq M\alpha^k,$$

which holds for every x in J and every $k \geq 1$. To prove (10.21) we note that

$$(10.22) \qquad |\varphi_{k+1}(x) - \varphi_k(x)| = |T(\varphi_k)(x) - T(\varphi_{k-1})(x)| \leq \alpha \|\varphi_k - \varphi_{k-1}\|.$$

Therefore inequality (10.21) will be proved if we show that

$$(10.23) \qquad \|\varphi_k - \varphi_{k-1}\| \leq M\alpha^{k-1}$$

for every $k \geq 1$. We prove (10.23) by induction. First we have

$$\|\varphi_1 - \varphi_0\| \leq \|\varphi_1\| + \|\varphi_0\| = M,$$

which is the same as (10.23) when $k = 1$. To prove that (10.23) holds for $k + 1$ if it holds for k we note that (10.22) gives us

$$|\varphi_{k+1}(x) - \varphi_k(x)| \leq \alpha \|\varphi_k - \varphi_{k-1}\| \leq M\alpha^k.$$

Since this is valid for each x in J we must also have

$$\|\varphi_{k+1} - \varphi_k\| \leq M\alpha^k.$$

This proves (10.23) by induction. Therefore the series in (10.20) converges uniformly on J. If we let φ denote its sum we have

$$(10.24) \qquad \varphi(x) = \lim_{n \to \infty} \varphi_n(x) = \varphi_0(x) + \sum_{k=0}^{\infty} \{\varphi_{k+1}(x) - \varphi_k(x)\}.$$

The function φ is continuous on J because it is a sum of a uniformly convergent series of continuous functions. To prove that φ is a fixed point of T we compare $T(\varphi)$ with

$\varphi_{n+1} = T(\varphi_n)$. Using the contraction property of T we have

(10.25) $\qquad |T(\varphi)(x) - \varphi_{n+1}(x)| = |T(\varphi)(x) - T(\varphi_n)(x)| \leq \alpha |\varphi(x) - \varphi_n(x)|.$

But from (10.19) and (10.24) we find

$$|\varphi(x) - \varphi_n(x)| = \left| \sum_{k=n}^{\infty} \{\varphi_{k+1}(x) - \varphi_k(x)\} \right| \leq \sum_{k=n}^{\infty} |\varphi_{k+1}(x) - \varphi_k(x)| \leq M \sum_{k=n}^{\infty} \alpha^k,$$

where in the last step we used (10.21). Therefore (10.25) implies

$$|T(\varphi)(x) - \varphi_{n+1}(x)| \leq M \sum_{k=n}^{\infty} \alpha^{k+1}.$$

When $n \to \infty$ the series on the right tends to 0, so $\varphi_{n+1}(x) \to T(\varphi)(x)$. But since $\varphi_{n+1}(x) \to \varphi(x)$ as $n \to \infty$, this proves that $\varphi(x) = T(\varphi)(x)$ for each x in J. Therefore $\varphi = T(\varphi)$, so φ is a fixed point.

Finally, we prove that the fixed point φ is unique. Let ψ be another function in $C(J)$ such that $T(\psi) = \psi$. Then we have

$$\|\varphi - \psi\| = \|T(\varphi) - T(\psi)\| \leq \alpha \|\varphi - \psi\|.$$

This gives us $(1 - \alpha)\|\varphi - \psi\| \leq 0$. Since $\alpha < 1$ we may divide by $1 - \alpha$ to obtain the inequality $\|\varphi - \psi\| \leq 0$. But since we also have $\|\varphi - \psi\| \geq 0$ this means that $\|\varphi - \psi\| = 0$, and hence $\varphi = \psi$. The proof of the fixed-point theorem is now complete.

★ **10.11 Applications of the fixed-point theorem**

As already noted, the fixed-point theorem gives a general existence-uniqueness theorem for systems of linear differential equations. To illustrate the broad scope of its applications, we use the fixed-point theorem to prove two further important results in analysis. The first gives a sufficient condition for an equation of the form $f(x, y) = 0$ to define y as a function of x.

THEOREM 10.6. AN IMPLICIT-FUNCTION THEOREM. *Let f be defined on a rectangular strip R of the form*

$$R = \{(x, y) : a \leq x \leq b, -\infty < y < +\infty\}.$$

Assume that the partial derivative $D_2 f(x, y)$ *exists*[†] *and satisfies an inequality of the form*

(10.26) $\qquad\qquad\qquad 0 < m \leq D_2 f(x, y) \leq M$

for all (x, y) *in R, where m and M are positive constants with* $m \leq M$. *Assume also that for each function* φ *continuous on* $[a, b]$ *the composite function* $g(x) = f[x, \varphi(x)]$ *is continuous on* $[a, b]$. *Then there exists one and only one function* $y = Y(x)$, *continuous on* $[a, b]$, *such*

[†]$D_2 f(x, y)$ is the derivative of $f(x, y)$ with respect to y, holding x fixed.

that

(10.27) $$f[x, Y(x)] = 0$$

for all x in $[a, b]$.

Note. We describe this result by saying that the equation $f(x, y) = 0$ serves to define y implicitly as a function of x in $[a, b]$.

Proof. Let C be the linear space of continuous functions on $[a, b]$, and define an operator $T : C \to C$ by the equation

$$T(\varphi)(x) = \varphi(x) - \frac{1}{M} f[x, \varphi(x)]$$

for each x in $[a, b]$. Here M is the positive constant in (10.26). The function $T(\varphi) \in C$ whenever $\varphi \in C$. We shall prove that T is a contraction operator. Once we know this it follows that T has a unique fixed point Y in C. For this function we have $Y = T(Y)$, which means

$$Y(x) = Y(x) - \frac{1}{M} f[x, Y(x)]$$

for each x in $[a, b]$. This gives us (10.27), as required.

To show that T is a contraction operator we consider the difference

(10.28) $$T(\varphi)(x) - T(\psi)(x) = \varphi(x) - \psi(x) - \frac{f[x, \varphi(x)] - f[x, \psi(x)]}{M}.$$

The mean-value theorem for derivatives tells us that

$$f[x, \varphi(x)] - f[x, \psi(x)] = D_2 f[x, z(x)][\varphi(x) - \psi(x)],$$

where $z(x)$ lies between $\varphi(x)$ and $\psi(x)$. Therefore (10.28) gives us

(10.29) $$T(\varphi)(x) - T(\psi)(x) = [\varphi(x) - \psi(x)] \left(1 - \frac{D_2 f[x, z(x)]}{M} \right).$$

The hypothesis (10.26) implies that

$$0 \le \left(1 - \frac{D_2 f[x, z(x)]}{M} \right) \le 1 - \frac{m}{M}.$$

Therefore (10.29) gives us the inequality

(10.30) $$|T(\varphi)(x) - T(\psi)(x)| \le |\varphi(x) - \psi(x)| \left(1 - \frac{m}{M} \right) \le \alpha \|\varphi - \psi\|,$$

where $\alpha = 1 - \frac{m}{M}$. Since $0 < m \le M$, we have $0 \le \alpha < 1$. Inequality (10.30) is valid for every x in $[a, b]$. Hence T is a contraction operator. This completes the proof.

The next application of the fixed-point theorem establishes an existence theorem for the integral equation

(10.31)
$$\varphi(x) = \psi(x) + \lambda \int_a^b K(x,t)\varphi(t)\,dt.$$

Here ψ is a given function, continuous on $[a,b]$, λ is a given constant, and K is a given function defined and bounded on the square

$$S = \{(x,y) : a \le x \le b, a \le y \le b\}.$$

The function K is called the *kernel* of the integral equation. Let C be the linear space of continuous functions on $[a,b]$. We assume that the kernel K is such that the operator T given by

$$T(\varphi)(x) = \psi(x) + \lambda \int_a^b K(x,t)\varphi(t)\,dt.$$

maps C into C. In other words, we assume that $T(\varphi) \in C$ whenever $\varphi \in C$. A solution of the integral equation is any function φ in C that satisfies (10.31).

THEOREM 10.7.   AN EXISTENCE-UNIQUENESS THEOREM FOR INTEGRAL EQUATIONS.   *If, under the foregoing conditions, we have*

(10.32)
$$|K(x,y)| \le M$$

for all (x,y) in S, where $M > 0$, then for each λ such that

(10.33)
$$|\lambda| < \frac{1}{M(b-a)}$$

there is one and only one function φ in C that satisfies the integral equation (10.31).

  Proof. We shall prove that T is a contraction operator. Take any two function φ_1 and φ_2 in C and consider the difference

$$T(\varphi_1)(x) - T(\varphi_2)(x) = \lambda \int_a^b K(x,t)[\varphi_1(t) - \varphi_2(t)]\,dt.$$

Using the upper bound for the kernel $K(x,t)$ in (10.32) we find

$$|T(\varphi_1)(x) - T(\varphi_2)(x)| = |\lambda| M(b-a) \|\varphi_1 - \varphi_2\| = \alpha \|\varphi_1 - \varphi_2\|,$$

where $\alpha = |\lambda| M(b-a)$. Because of (10.33) we have $0 \le \alpha < 1$, so T is a contraction operator with contraction constant α. Therefore T has a unique fixed point φ in C. This function φ satisfies the integral equation (10.31).

ANSWERS TO EXERCISES

0.8   Exercises (page 8)

3. (b) $(a, b)(a, b) = (a^2, b^2)$, so the square of a complex number can never be $(-1, 0)$. The product of two complex numbers can be zero without either factor being zero. For example, $(1, 0)(0, 1) = (0, 0)$. Therefore, the common method of solving an algebraic equation $f(x) = 0$ by factoring $f(x)$ into linear factors and setting each factor equal to 0 would not be available with this definition of multiplication.

0.11   Exercises (pages 10–11)

1. (a) $2i$ (b) $-i$ (c) $\frac{1}{2} - \frac{1}{2}i$ (d) $18 + i$ (e) $-\frac{1}{5} + \frac{3}{5}i$ (f) $1 + i$ (g) 0
   (h) $1 + i$
2. (a) $\sqrt{2}$ (b) 5 (c) 1 (d) 1 (e) $\sqrt{2}$ (f) $\sqrt{65}$
3. (a) $r = 2, \theta = \frac{1}{2}\pi$ (b) $r = 3, \theta = -\frac{1}{2}\pi$ (c) $r = 1, \theta = \pi$ (d) $r = 1, \theta = 0$
   (e) $r = 2\sqrt{3}, \theta = 5\pi/6$ (f) $r = 1, \theta = \frac{1}{4}\pi$ (g) $r = 2\sqrt{2}, \theta = \frac{1}{4}\pi$
   (h) $r = 2\sqrt{2}, \theta = -\frac{1}{4}\pi$ (i) $r = \frac{1}{2}\sqrt{2}, \theta = -\frac{1}{4}\pi$ (j) $r = \frac{1}{2}, \theta = -\frac{1}{2}\pi$
4. (a) $y = 0, x$ arbitrary   (b) $x \geq 0, y = 0$ (c) All x and y (d) $x = 0, y$ arbitrary;
   or $y = 0, x$ arbitrary   (e) $x = 1, y = 0$ (f) $x = 1, y = 0$

0.26   Exercises (page 24)

1. $(1, \frac{7}{3})$ and $(3, 1)$.
2. $a = -2, b = 4$.
3. $a = 1; b = 0, c = -1$.
4. (b) $f'(t) = 144 - 32t$.
   (c) Velocity is 0 at the midpoint of the motion, when $t = 9/2$.
   (d) Acceleration is $f''(t) = -32$.
5. (a) $f'(x) = \dfrac{1 - \log x}{x^2} > 0$ if $0 < x < e$; $f''(x) = \dfrac{-3 - 2\log x}{x^3} > 0$ if $x > e^{3/2}$.
   (b) The derivative of $\frac{1}{2}(\log x)^2$ is $\dfrac{\log x}{x}$ so $\displaystyle\int_1^t \dfrac{\log x}{x}\,dx = \frac{1}{2}(\log t)^2$; the constant is $\frac{1}{2}$.
   (c) $m = \frac{1}{2e}$, point of contact $(\sqrt{e}, \frac{1}{2e})$
   (d) $b = \sqrt{e}$
6. $p(x) = 3 + x + \frac{1}{6}x^3$
8. (b) $a = 1, b = 0$.

1.4 Exercises (pages 30–31)

1. (a) $(5, 0, 9)$ (b) $(-3, 6, 3)$ (c) $(3, -1, 4)$ (d) $(-7, 24, 21)$ (e) $(0, 0, 0)$
5. $x = \frac{1}{5}(3c_1 - c_2), y = \frac{1}{5}(2c_2 - c_1)$
6. (a) $(x + z, x + y + z, x + y)$ (c) $x = 2, y = 1, z = -1$
7. (a) $(x + 2z, x + y + z, x + y + z)$ (b) One example: $x = -2, y = z = 1$
8. (a) $(x + z, x + y + z, x + y, y)$ (c) $x = -1, y = 4, z = 2$
12. The diagonals of a parallelogram bisect each other

1.8 Exercises (pages 36–37)

1. (a) -6 (b) 2 (c) 6 (d) 0 (e) 4
2. (a) $(A \cdot B)C = (21, 28, -35)$ (b) $A \cdot (B + C) = 64$ (c) $(A + B) \cdot C = 72$
   (d) $A(B \cdot C) = (30, 60, -105)$ (e) $A/(B \cdot C) = (\frac{2}{15}, \frac{4}{15}, \frac{-7}{15})$
5. One example: $(1, -5, -3)$
6. One example: $x = -2, y = 1$
7. $C = \frac{4}{9}(-1, -2, 2), D = \frac{1}{9}(22, -1, 10)$
8. $C = \frac{1}{11}(1, 2, 3, 4, 5), D = (\frac{5}{11}, \frac{7}{44}, \frac{1}{33}, \frac{-5}{88}, \frac{-7}{55})$
9. (a) $\sqrt{74}$ (b) $\sqrt{14}$ (c) $\sqrt{53}$ (d) 5
10. (a) $(1, -1)$ or $(-1, 1)$ (b) $(1, 1)$ or $(-1, -1)$ (c) $(3, 2)$ or $(-3, -2)$
    (d) $(b, -a)$ or $(-b, a)$
11. (a) $\dfrac{1}{\sqrt{42}}(4, -1, 5)$ (b) $\dfrac{1}{\sqrt{14}}(-2, -3, 1)$ (c) $\dfrac{1}{\sqrt{2}}(1, 0, 1)$
    (d) $\dfrac{1}{\sqrt{42}}(-5, -4, -1)$ (e) $\dfrac{1}{\sqrt{42}}(-1, -5, 4)$
12. A and B, C and D, C and E, D and E
13. (a) $(2, -1)$ and $(-2, 1)$ (b) $(2, 1)$ and $(-2, -1)$ (c) $(1, 2)$ and $(-1, -2)$
    (d) $(1, 2)$ and $(-1, -2)$
14. One example: $C = (8, 1, 1)$
15. One example: $C = (1, -5, -3)$
16. $P = \frac{11}{25}(3, 4), Q = \frac{2}{25}(-4, 3)$
17. $P = \frac{5}{2}(1, 1, 1, 1), Q = \frac{1}{2}(-3, -1, 1, 3)$
18. $\pm\dfrac{1}{\sqrt{2}}(0, 1, 1)$
20. The sum of the squares of the sides of any parallelogram is equal to the sum of the squares of the diagonals.
22. $4; 12\sqrt{2}$
23. $C = \frac{1}{11}(1, 2, 3, 4, 5), D = \frac{1}{11}(10, \frac{7}{2}, \frac{2}{3}, \frac{-5}{4}, \frac{-14}{5})$
24. $C = tA, D = B - tA$, where $t = (A \cdot B)/(A \cdot A)$

1.11 Exercises (pages 41–42)

1. $\frac{11}{9}B$
2. $\frac{5}{2}B$
3. (a) $\frac{6}{7}, \frac{3}{7}, \frac{-2}{7}$ (b) $(\frac{6}{7}, \frac{3}{7}, \frac{-2}{7})$ and $(\frac{-6}{7}, \frac{-3}{7}, \frac{2}{7})$
5. $0, \sqrt{\frac{35}{41}}, \sqrt{\frac{6}{41}}$
6. $7\pi/8$
8. $\pi/6$
9. 0
10. (b) If $\cos\theta = 1$, equation holds for all x, y; otherwise the only solution is $x = y = 0$
14. All except (b)
17. (c) All except Theorem 1.4(a)
18. (a) All

1.15 Exercises (pages 47–48)

1. (a) $x = y = \frac{1}{2}$ (b) $x = -\frac{1}{2}, y = \frac{1}{2}$ (c) $x = 4, y = -1$ (d) $x = 1, y = 6$
2. $x = \frac{1}{4}, y = \frac{7}{8}$
3. $x = 3, y = -4$
7. All $t \neq 0$
9. (c) $7\mathbf{i} - 4(\mathbf{i} + \mathbf{j})$
10. (b) $\mathbf{j} = B - A, \mathbf{k} = \frac{1}{3}(C - B)$ (c) $\frac{1}{3}(15A - 14B + 5C)$
11. $\{A\}, \{B\}, \{C\}, \{D\}, \{A, B\}, \{A, C\}, \{A, D\}, \{B, C\}, \{C, D\}$
12. (a) Independent   (b) One example $D = A$ (c) One example: $E = (0, 0, 0, 1)$
    (d) For the choice $E = (0, 0, 0, 1)$ we have $X = 2A + B - C + 3E$
13. (c) $t = 0, \sqrt{2}, -\sqrt{2}$
14. (a) $\{(1, 0, 1, 0), (0, 1, 0, 1), (2, 0, -1, 0)\}$ (b) The set given   (c) The set given
17. $\{(0, 1, 1), (1, 1, 1), (0, 1, 0)\}, \{(0, 1, 1), (1, 1, 1), (0, 0, 1)\}$
18. $\{(1, 1, 1, 1), (0, 1, 1, 1), (0, 0, 1, 1), (0, 0, 0, 1)\}, \{(1, 1, 1, 1), (0, 1, 1, 1), (0, 1, 0, 0), (0, 0, 1, 0)\}$
19. $L(U) \subseteq L(T) \subseteq L(S)$
20. One example: $A = \{I_1, \ldots, I_n\}, B = \{I_1 + I_2, I_2 + I_3, \ldots, I_{n-1} + I_n, I_n + I_1\}$

1.17 Exercises (page 50)

1. (a) $-1 - i$ (b) $-1 + i$ (c) $1 - i$ (d) $-1 + i$ (e) $-1 - i$ (f) $2 - i$
   (g) $-i$ (h) $-1 + 2i$ (i) $-3 - 2i$ (j) $2i$
2. One example: $(1 + i, -5 - 3i, 1 - 3i)$
8. $\pi/3$
9. $3A - B + 2C$

2.6 Exercises (page 57)

1. (b), (d) and (e)
2. (a) and (e)
3. (c), (d) and (e)
4. (b), (e) and (f)
5. (a) No   (b) No   (c) No
6. A, B, C, D, F are collinear
7. Intersect at $(5, 9, 13)$
8. (b) No
9. (a) $9t^2 + 8t + 9$ (b) $\frac{1}{3}\sqrt{65}$

2.9 Exercises (pages 62–63)

1. (c) and (e)
2. (a), (b) and (c)
3. (a) $x = 1 + t, y = 2 + s + t, z = 1 + 4t$ (b) $x = s + t, y = 1 + s, z = s + 4t$
4. (a) $(1, 2, 0)$ and $(2, -3, -3)$ (b) $M = \{(1, 2, 0) + s(1, 1, 2) + t(-2, 4, 1)\}$
6. (a), (b) and (c) $x - 2y + z = -3$
7. (a) $(0, -2, -1)$ and $(-1, -2, 2)$ (b) $M = \{(0, -2, -1) + s(-1, 0, 3) + t(3, 3, 6)\}$
8. Two examples: $(-5, 2, 6)$ and $(-14, 3, 17)$
9. (a) Yes   (b) Two examples: $(1, 0, -1)$ and $(-1, 0, 1)$
10. $(-2, \frac{5}{2}, -\frac{7}{2})$
11. (a), (b) and (c) No
13. $x - y = -1$

2.12 Exercises (pages 67–68)

1. (a) $(-2, 3, -1)$ (b) $(4, -5, 3)$ (c) $(4, -4, 2)$ (d) $(8, 10, 4)$ (e) $(8, 3, -7)$
    (f) $(10, 11, 5)$ (g) $(-2, -8, -12)$ (h) $(2, -2, 0)$ (i) $(-2, 0, 4)$
2. (a) $\pm \dfrac{1}{\sqrt{26}}(-4, 3, 1)$ (b) $\pm \dfrac{1}{\sqrt{2054}}(-41, -18, 7)$ (c) $\pm \dfrac{1}{\sqrt{6}}(1, 2, 1)$
3. (a) $\frac{15}{2}$ (b) $\frac{3}{2}\sqrt{35}$ (c) $\frac{1}{2}\sqrt{3}$
4. $8i + j - 2k$
6. (b) $\cos\theta$ is negative   (c) $\sqrt{5}$
9. (a) One solution is $B = -i - 3k$ (b) $i - j - k$ is the only solution
11. (a) Three possibilities: $D = B + C - A = (0, 0, 2)$, $D = A + C - B = (4, -2, 2)$,
    $D = A + B - C = (-2, 2, 0)$ (b) $\frac{1}{2}\sqrt{6}$
12. -4; $8\sqrt{3}$; $-\frac{1}{2}\sqrt{3}$

2.15 Exercises (pages 71–73)

1. (a) 96   (b) 27   (c) -84
2. $0, \sqrt{2}, -\sqrt{2}$
3. 2
6. (a) $(2b - 1)i + bj + ck$, where b and c are arbitrary   (b) $-\frac{1}{5}i + \frac{2}{5}j$
11. $-3i + 2j + 5k$
14. (b) $1/3$
15. (b) $\sqrt{2005/41}$
17. $x = 1, y = -1, z = 2$
18. $x = 1, y = -1, z = 2$
19. $x = 1, y = 4, z = 1$

2.18 Exercises (pages 76–77)

1. (a) $(-7, 2, -2)$ (b) $-7x + 2y - 2z = 0$ (c) $-7x + 2y - 2z = -9$
2. (a) $(\frac{1}{3}, \frac{2}{3}, -\frac{2}{3})$ (b) $-7, -\frac{7}{2}, \frac{7}{2}$ (c) $\frac{7}{3}$ (d) $(-\frac{7}{9}, -\frac{14}{9}, \frac{14}{9})$
3. $3x - y + 2z = -5$; $9/\sqrt{14}$
4. (b) $\frac{19}{18}\sqrt{6}$
5. (a) $(1, 2, -2)$ (b) $x + 2y - 2z = 5$ (c) $\frac{5}{3}$
6. $10x - 3y - 7z + 17 = 0$
7. Two angles: $\pi/3$ and $2\pi/3$
8. $x + 2y + 9z + 55 = 0$
9. $X(t) = (2, 1, -3) + t(4, -3, 1)$
10. (b) $N = (1, 3, -2)$ (c) $t = 1$ (d) $2x + 3y + 2z + 15 = 0$
    (e) $x + 3y - 2z + 19 = 0$
11. $x + \sqrt{2}y + z = 2 + \sqrt{2}$
12. 6
13. $\dfrac{1}{\sqrt{122}}(7, -8, -3)$
14. $x - y + z = 2$
15. $(\frac{3}{2}, 0, \frac{1}{2})$
17. $X(t) = (1, 2, 3) + t(1, -2, 1)$
19. (b) $P = -\frac{1}{25}(5, -14, 2)$
20. $2x - y + 2z - 8 = 0$
21. (b) 1
23. $3x - z = 0$
24. $2x - 3z = 2$

2.22 Exercises (pages 82–83)

3. $r = ed/(1 - e\sin\theta)$; $r = -ed/(1 + e\sin\theta)$
4. $e = 1, d = 2$
5. $e = \frac{1}{2}, d = 6$
6. $e = \frac{1}{3}, d = 6$
7. $e = 2, d = 1$
8. $e = 2, d = 2$
9. $e = 1, d = 4$
10. $d = 5, r = 25/(10 + 3\cos\theta + 4\sin\theta)$
11. $d = 5, r = 25/(5 + 4\cos\theta + 3\sin\theta)$
12. $d = \frac{1}{2}\sqrt{2}, r = 1/(\cos\theta + \sin\theta + \frac{1}{2}\sqrt{2}), r = 1/(\cos\theta + \sin\theta - \frac{1}{2}\sqrt{2})$
13. (a) $r = 1.5 \times 10^8/(1 + \cos\theta)$; 7.5×10^7 miles
    (b) $r = 5 \times 10^7/(1 - \cos\theta)$; 2.5×10^7 miles

2.27 Exercises (pages 88–89)

1. Center at $(0,0)$; foci at $(\pm 8, 0)$; vertices at $(\pm 10, 0)$; $e = \frac{4}{5}$
2. Center at $(0,0)$; foci at $(0, \pm 8)$; vertices at $(0, \pm 10)$; $e = \frac{4}{5}$
3. Center at $(2, -3)$; foci at $(2 \pm \sqrt{7}, -3)$; vertices at $(6, -3), (-2, -3)$; $e = \dfrac{\sqrt{7}}{4}$
4. Center at $(0,0)$; foci at $(\pm\frac{4}{3}, 0)$; vertices at $(\pm\frac{5}{3}, 0)$; $e = \frac{4}{5}$
5. Center at $(0,0)$; foci at $(\pm\sqrt{3}/6, 0)$; vertices at $(\pm\sqrt{3}/3, 0)$; $e = \frac{1}{2}$
6. Center at $(-1, -2)$; foci at $(-1, 1), (-1, -5)$; vertices at $(-1, 3), (-1, -7)$; $e = \frac{3}{5}$
7. $7x^2 + 16y^2 = 7$
8. $\dfrac{(x + 3)^2}{16} + \dfrac{(y - 4)^2}{9} = 1$
9. $\dfrac{(x + 3)^2}{9} + \dfrac{(y - 4)^2}{16} = 1$
10. $\dfrac{(x + 4)^2}{9} + \dfrac{(y - 2)^2}{1} = 1$
11. $\dfrac{(x - 8)^2}{25} + \dfrac{(y + 2)^2}{9} = 1$
12. $\dfrac{(x - 2)^2}{16} + \dfrac{(y - 1)^2}{4} = 1$
13. Center at $(0,0)$; foci at $(\pm 2\sqrt{41}, 0)$; vertices at $(\pm 10, 0)$; $e = \dfrac{\sqrt{41}}{5}$
14. Center at $(0,0)$; foci at $(0, \pm 2\sqrt{41})$; vertices at $(0, \pm 10)$; $e = \dfrac{\sqrt{41}}{5}$
15. Center at $(-3, 3)$; foci at $(-3 \pm \sqrt{5}, 3)$; vertices at $(-1, 3), (-5, 3)$; $e = \dfrac{\sqrt{5}}{2}$
16. Center at $(0,0)$; foci at $(\pm 5, 0)$; vertices at $(\pm 4, 0)$; $e = \frac{5}{4}$
17. Center at $(0,0)$; foci at $(0, \pm 3)$; vertices at $(0, \pm 2)$; $e = \frac{3}{2}$
18. Center at $(1, -2)$; foci at $(1 \pm \sqrt{13}, -2)$; vertices at $(3, -2), (-1, -2)$; $e = \dfrac{\sqrt{13}}{2}$
19. $\dfrac{x^2}{4} - \dfrac{y^2}{12} = 1$
20. $y^2 - x^2 = 1$
21. $\dfrac{x^2}{4} - \dfrac{y^2}{16} = 1$
22. $\dfrac{(y - 4)^2}{1} - \dfrac{(x + 1)^2}{3} = 1$

23. $\dfrac{8(y+3)^2}{27} - \dfrac{5(x-2)^2}{27} = 1$

24. $4x^2 - y^2 = 11$

25. Vertex at $(0,0)$; directrix $x = 2$; axis $y = 0$

26. Vertex at $(0,0)$; directrix $x = -\frac{3}{4}$; axis $y = 0$

27. Vertex at $(\frac{1}{2}, 1)$; directrix $x = -\frac{5}{2}$; axis $y = 1$

28. Vertex at $(0,0)$; directrix $y = -\frac{3}{2}$; axis $x = 0$

29. Vertex at $(0,0)$; directrix $y = 2$; axis $x = 0$

30. Vertex at $(-2, -\frac{9}{4})$; directrix $y = -\frac{13}{4}$; axis $x = -2$

31. $x^2 = -y$

32. $y^2 = 8x$

33. $(x+4)^2 = -8(y-3)$

34. $(y+1)^2 = 5(x - \frac{7}{4})$

35. $(x - \frac{3}{2})^2 = 2(y + \frac{1}{8})$

36. $(y-3)^2 = -8(x-1)$

37. $x^2 - 4xy + 4y^2 + 40x + 20y - 100 = 0$

38. $x^2 - 2xy + y^2 - 2x - 2y = 1$

39. $\dfrac{x^2}{12} + \dfrac{y^2}{16} = 1$

40. $y^2 - 4x^2 - 4y + 4x = 0$

2.28 Miscellaneous exercises on conic sections (page 90)

1. (a) $e = \sqrt{2/(p+2)}$; foci at $(\sqrt{2}, 0)$ and $(-\sqrt{2}, 0)$ (b) $6x^2 - 3y^2 = 4$

5. (b) $y = Cx^2, C \neq 0$

6. $(x - \frac{2}{5})^2 + (y - \frac{4}{5})^2 = \frac{4}{5}$

11. $\pm\sqrt{\dfrac{23}{3}}$

12. $\frac{2}{3}bh$

13. 16π

14. (a) $\frac{8}{3}$ (b) 2π (c) $48\pi/5$

16. $A = \frac{1}{2}(1 + \sqrt{5})B$

17. $(4, 8)$

3.5 Exercises (pages 95–96)

1. Yes	6. Yes	11. Yes	16. Yes
2. No	7. Yes	12. Yes	17. Yes
3. Yes	8. Yes	13. No	18. No
4. No	9. Yes	14. Yes	19. Yes
5. Yes	10. Yes	15. Yes	20. Yes

23. (a) No   (b) No   (c) No   (d) No

25. (a) Yes   (b) No   (c) Yes   (d) Yes

3.10 Exercises (pages 102–103)

1. Yes; 2	6. No	10. Yes; 1
2. Yes; 2	7. No	11. Yes; n
3. Yes; 2	8. No	12. Yes; n
4. Yes; 2	9. Yes; 1	13. Yes; n
5. Yes; 1		

14. Yes; dim $= 1 + \frac{1}{2}n$ if n is even, $\frac{1}{2}(n + 1)$ if n is odd
15. Yes; dim $= \frac{1}{2}n$ if n is even, $\frac{1}{2}(n + 1)$ if n is odd
16. No
17. Yes; $k + 1$
18. No
19. Yes; n
20. Yes; $n - 1$
21. Yes; n
22. Yes; n
23. (a) dim $= 3$ (b) dim $= 3$ (c) dim $= 2$ (d) dim $= 2$
24. (a) If $a \neq 0$ and $b \neq 0$, set is independent; dim $= 3$; if one of a or b is zero, set is dependent; dim $= 2$.   (b) Independent, dim $= 2$ (c) If $a \neq 0$, independent, dim $= 3$; if $a = 0$, dependent, dim $= 2$ (d) Independent; dim $= 3$ (e) Dependent; dim $= 2$
    (f) Independent; dim $= 2$ (g) Independent; dim $= 2$ (h) Dependent; dim $= 2$
    (i) Independent; dim $= 2$ (j) Independent; dim $= 2$

3.13   Exercises (pages 109–111)

1. (a) No   (b) No   (c) No   (d) No   (e) Yes
10. (b) $\dfrac{(n + 1)(2n + 1)}{6n}a + \dfrac{n + 1}{2}b$ (c) $g(t) = a\left(t - \dfrac{2n + 1}{3n}\right)$, a arbitrary
11. (a) $\frac{1}{2}\sqrt{e^2 + 1}$ (b) $g(x) = b(x - \frac{1}{4}(e^2 + 1))$, b arbitrary
13. (c) 43 (d) $g(t) = a(1 - \frac{2}{3}t)$, a arbitrary
14. (a) No   (b) No   (c) No   (d) No
15. (c) $n!/2^{n+1}$
16. (c) 1 (d) $e^{1/2} - 1$

3.17   Exercises (page 118)

1. (a) and (b) $\frac{1}{3}\sqrt{3}(1, 1, 1)$, $\frac{1}{6}\sqrt{6}(1, -2, 1)$
2. (a) $\frac{1}{2}\sqrt{2}(1, 1, 0, 0)$, $\frac{1}{6}\sqrt{6}(-1, 1, 2, 0)$, $\frac{1}{6}\sqrt{3}(1, -1, 1, 3)$
   (b) $\frac{1}{3}\sqrt{3}(1, 1, 0, 1)$, $\dfrac{1}{\sqrt{42}}(1, -2, 6, 1)$
6. $\frac{2}{3} - \frac{1}{2}\log^2 3$
7. $e^2 - 1$
8. $\frac{1}{2}(e - e^{-1}) + \frac{3}{e}x$; $1 - 7e^{-2}$
9. $\pi - 2\sin x$
10. $\frac{3}{4} - \frac{1}{4}x$

4.4   Exercises (pages 123–124)

1. Linear; nullity 0, rank 2
2. Linear; nullity 0, rank 2
3. Linear; nullity 1, rank 1
4. Linear; nullity 1, rank 1
5. Nonlinear
6. Nonlinear
7. Nonlinear
8. Nonlinear
9. Linear; nullity 0, rank 2
10. Linear; nullity 0, rank 2
11. Linear; nullity 0, rank 2
12. Linear; nullity 0, rank 2
13. Nonlinear
14. Linear; nullity 0, rank 2
15. Nonlinear
16. Linear; nullity 0, rank 3
17. Linear; nullity 1, rank 2
18. Linear; nullity 1, rank 3
19. Nonlinear
20. Nonlinear

21. Nonlinear
22. Nonlinear
23. Linear; nullity 1, rank 2
25. Linear; nullity 0, rank $n + 1$
26. Linear; nullity 1, rank infinite
27. Linear; nullity 2, rank infinite
28. Linear; nullity infinite, rank 2
29. $N(T)$ is the set of constant sequences; $T(V)$ is the set of sequences with limit 0
30. (d) $\{1, \cos x, \sin x\}$ is a basis for $T(V)$; $\dim T(V) = 3$
    (e) $N(T) = S$
    (f) If $T(f) = cf$ with $c \neq 0$, then $cf \in T(V)$ hence $f \in T(V)$ so we have
    $f(x) = c_1 + c_2 \cos x + c_3 \sin x$. The equation $T(f) = cf$ implies

$$2\pi c_1 + \pi c_2 \cos x + \pi c_3 \sin x = c(c_1 + c_2 \cos x + c_3 \sin x).$$

If $c_1 = 0$, then $c = \pi$ and $f(x) = c_2 \cos x + c_3 \sin x$ where c_2, c_3 are not both 0 but otherwise arbitrary; if $c_1 \neq 0$, then $c = 2\pi$ and $f(x) = a$, where $a \neq 0$ but otherwise arbitrary.

4.8   Exercises (pages 130–131)

3. Yes; $x = v, y = u$
4. Yes; $x = u, y = -v$
5. No
6. No
7. No
8. Yes; $x = \log u, y = \log v$
9. No
10. Yes; $x = u - 1, y = v - 1$
11. Yes; $x = \frac{1}{2}(v + u), y = \frac{1}{2}(v - u)$
12. Yes; $x = \frac{1}{3}(v + u), y = \frac{1}{3}(2v - u)$
13. Yes; $x = w, y = v, z = u$
14. No
15. Yes; $x = u, y = \frac{1}{2}v, z = \frac{1}{3}w$
16. Yes; $x = u, y = v, z = w - u - v$
17. Yes; $x = u - 1, y = v - 1, z = w + 1$
18. Yes; $x = u - 1, y = v - 2, z = w - 3$
19. Yes; $x = u, y = v - u, z = w - v$
20. Yes; $x = \frac{1}{2}(u - v + w), y = \frac{1}{2}(v - w + u), z = \frac{1}{2}(w - u + v)$
25. $(S + T)^2 = S^2 + ST + TS + T^2$;
    $(S + T)^3 = S^3 + TS^2 + STS + S^2T + ST^2 + TST + T^2S + T^3$
27. (a) $(ST)(x, y, z) = (x + y + z, x + y, x)$; $(TS)(x, y, z) = (z, z + y, z + y + x)$;
    $(ST - TS)(x, y, z) = (x + y, x - z, -y - z)$; $S^2(x, y, z) = (x, y, z)$;
    $T^2(x, y, z) = (x, 2x + y, 3x + 2y + z)$;
    $(ST)^2(x, y, z) = (3x + 2y + z, 2x + 2y + z, x + y + z)$;
    $(TS)^2(x, y, z) = (x + y + z, x + 2y + 2z, x + 2y + 3z)$;
    $(ST - TS)^2(x, y, z) = (2x + y - z, x + 2y + z, -x + y + 2z)$;
    (b) $S^{-1}(u, v, w) = (w, v, u)$; $T^{-1}(u, v, w) = (u, v - u, w - v)$;
    $(ST)^{-1}(u, v, w) = (w, v - w, u - v)$; $(TS)^{-1}(u, v, w) = (w - v, v - u, u)$;
    (c) $(T - I)(x, y, z) = (0, x, x + y)$; $(T - I)^2(x, y, z) = (0, 0, x)$;
    $(T - I)^n(x, y, z) = (0, 0, 0)$ if $n \geq 3$
28. (a) $Rp(x) = 2$; $Sp(x) = 3 - x + x^2$; $Tp(x) = 2x + 3x^2 - x^3 + x^4$;
    $(ST)p(x) = 2 + 3x - x^2 + x^3$; $(TS)p(x) = 3x - x^2 + x^3$; $(TS)^2 p(x) = 3x - x^2 + x^3$;
    $(T^2S^2)p(x) = -x^2 + x^3$; $(S^2T^2)p(x) = 2 + 3x - x^2 + x^3$; $(TRS)p(x) = 3x$;
    $(RST)p(x) = 2$
    (b) $N(R) = \{p : p(0) = 0\}$; $R(V) = \{p : p \text{ is constant}\}$; $N(S) = \{p : p \text{ is constant}\}$;
    $S(V) = V$; $N(T) = \{O\}$; $T(V) = \{p : p(0) = 0\}$
    (c) $T^{-1} = S$ (d) $(TS)^n = I - R$; $S^nT^n = I$

29. (a) $Dp(x) = 3 - 2x + 12x^2$; $Tp(x) = 3x - 2x^2 + 12x^3$; $(DT)p(x) = 3 - 4x + 36x^2$;
$(TD)p(x) = -2x + 24x^2$; $(DT - TD)p(x) = 3 - 2x + 12x^2$;
$(T^2D^2 - D^2T^2)p(x) = 8 - 192x$

    (b) $p(x) = ax$, where a is an arbitrary scalar

    (c) $p(x) = ax^2 + b$, where a, b are arbitrary scalars

    (d) All p in V

32. T is not one-to-one on V because it maps all constant sequences onto the same sequence.

4.12 Exercises (pages 139–140)

1. (a) The identity matrix $I = (\delta_{jk})$, where $\delta_{jk} = 1$ if $j = k$, and $\delta_{jk} = 0$ if $j \neq k$

    (b) The zero matrix O, each of whose entries is 0

    (c) The matrix $(c\delta_{jk})$, where (δ_{jk}) is the identity matrix of part (a)

2. (a) $\begin{bmatrix} 1 & 0 & 0 \\ 0 & 1 & 0 \end{bmatrix}$
  (b) $\begin{bmatrix} 0 & 1 & 0 \\ 0 & 0 & 1 \end{bmatrix}$
  (c) $\begin{bmatrix} 0 & 1 & 0 & 0 & 0 \\ 0 & 0 & 1 & 0 & 0 \\ 0 & 0 & 0 & 1 & 0 \end{bmatrix}$

3. (a) $-5i + 7j$, $9i - 12j$

    (b) $\begin{bmatrix} 1 & 2 \\ 1 & -1 \end{bmatrix}, \begin{bmatrix} 3 & 0 \\ 0 & 3 \end{bmatrix}$
  (c) $\begin{bmatrix} -\frac{7}{4} & -\frac{1}{4} \\ \frac{1}{4} & \frac{7}{4} \end{bmatrix}, \begin{bmatrix} 3 & 0 \\ 0 & 3 \end{bmatrix}$

4. $\begin{bmatrix} -2 & 0 \\ 0 & 2 \end{bmatrix}, \begin{bmatrix} 4 & 0 \\ 0 & 4 \end{bmatrix}$

5. (a) $3i + 4j + 4k$; nullity 0, rank 3
    (b) $\begin{bmatrix} -1 & -1 & 2 \\ 1 & -3 & 3 \\ -1 & -5 & 5 \end{bmatrix}$

6. $\begin{bmatrix} 2 & 0 & -2 \\ 1 & -1 & 1 \\ 2 & 1 & 0 \end{bmatrix}$

7. (a) $T(4i - j + k) = (0, -2)$; nullity 1, rank 2
    (b) $\begin{bmatrix} 0 & 1 & 1 \\ 0 & 1 & -1 \end{bmatrix}$

    (c) $\begin{bmatrix} 0 & 1 & 3 \\ 0 & 0 & -2 \end{bmatrix}$
  (d) $e_1 = j, e_2 = k, e_3 = i, w_1 = (1, 1), w_2 = (1, -1)$

8. (a) $(5, 0, -1)$; nullity 0, rank 2
    (b) $\begin{bmatrix} 1 & -1 \\ 0 & 0 \\ 1 & 1 \end{bmatrix}$

    (c) $e_1 = i, e_2 = i + j, w_1 = (1, 0, 1), w_2 = (0, 0, 2), w_3 = (0, 1, 0)$

9. (a) $(-1, -3, -1)$; nullity 0, rank 2
    (b) $\begin{bmatrix} 1 & 1 \\ 0 & 1 \\ 1 & 1 \end{bmatrix}$

    (c) $e_1 = i, e_2 = j - i, w_1 = (1, 0, 1), w_2 = (0, 1, 0), w_3 = (0, 0, 1)$

10. (a) $e_1 - e_2$; nullity 0, rank 2
    (b) $\begin{bmatrix} 1 & 2 \\ 5 & 4 \end{bmatrix}$
  (c) $a = 5, b = 4$

11. $\begin{bmatrix} 0 & -1 \\ 1 & 0 \end{bmatrix}, \begin{bmatrix} -1 & 0 \\ 0 & -1 \end{bmatrix}$

12. $\begin{bmatrix} 0 & 1 & 0 \\ 0 & 0 & 0 \\ 0 & 0 & 1 \end{bmatrix}, \begin{bmatrix} 0 & 0 & 0 \\ 0 & 0 & 0 \\ 0 & 0 & 1 \end{bmatrix}$

13. $\begin{bmatrix} 0 & 1 & 1 \\ 0 & 0 & -1 \\ 0 & 0 & 1 \end{bmatrix}, \begin{bmatrix} 0 & 0 & 0 \\ 0 & 0 & -1 \\ 0 & 0 & 1 \end{bmatrix}$

14. $\begin{bmatrix} 1 & 1 \\ 0 & 1 \end{bmatrix}, \begin{bmatrix} 1 & 2 \\ 0 & 1 \end{bmatrix}$

15. $\begin{bmatrix} 0 & -1 \\ 1 & 0 \end{bmatrix}, \begin{bmatrix} -1 & 0 \\ 0 & -1 \end{bmatrix}$

16. $\begin{bmatrix} 0 & -1 & 1 & 0 \\ 1 & 0 & 0 & 1 \\ 0 & 0 & 0 & -1 \\ 0 & 0 & 1 & 0 \end{bmatrix}, \begin{bmatrix} -1 & 0 & 0 & -2 \\ 0 & -1 & 2 & 0 \\ 0 & 0 & -1 & 0 \\ 0 & 0 & 0 & -1 \end{bmatrix}$

17. $\begin{bmatrix} 1 & -1 \\ 1 & 1 \end{bmatrix}, \begin{bmatrix} 0 & -2 \\ 2 & 0 \end{bmatrix}$

18. $\begin{bmatrix} 2 & -3 \\ 3 & 2 \end{bmatrix}, \begin{bmatrix} -5 & -12 \\ 12 & -5 \end{bmatrix}$

19. (a) $\begin{bmatrix} 0 & 0 & 0 & 0 \\ 0 & 1 & 0 & 0 \\ 0 & 0 & 2 & 0 \\ 0 & 0 & 0 & 3 \end{bmatrix}$ (b) $\begin{bmatrix} 0 & 1 & 0 & 0 \\ 0 & 0 & 4 & 0 \\ 0 & 0 & 0 & 9 \\ 0 & 0 & 0 & 0 \end{bmatrix}$ (c) $\begin{bmatrix} 0 & 0 & 0 & 0 \\ 0 & 0 & 2 & 0 \\ 0 & 0 & 0 & 6 \\ 0 & 0 & 0 & 0 \end{bmatrix}$

     (d) $\begin{bmatrix} 0 & -1 & 0 & 0 \\ 0 & 0 & -2 & 0 \\ 0 & 0 & 0 & -3 \\ 0 & 0 & 0 & 0 \end{bmatrix}$ (e) $\begin{bmatrix} 0 & 0 & 0 & 0 \\ 0 & 1 & 0 & 0 \\ 0 & 0 & 4 & 0 \\ 0 & 0 & 0 & 9 \end{bmatrix}$ (f) $\begin{bmatrix} 0 & 0 & -8 & 0 \\ 0 & 0 & 0 & -48 \\ 0 & 0 & 0 & 0 \\ 0 & 0 & 0 & 0 \end{bmatrix}$

20. Choose $(x^3, x^2, x, 1)$ as a basis for V, and (x^2, x) as a basis for W. Then the matrix of TD is
$\begin{bmatrix} 6 & 0 & 0 & 0 \\ 0 & 2 & 0 & 0 \end{bmatrix}$

4.16   Exercises (pages 146–147)

1. $B + C = \begin{bmatrix} 3 & 4 \\ 0 & 2 \\ 6 & -5 \end{bmatrix}$, $AB = \begin{bmatrix} 15 & -14 \\ -15 & 14 \end{bmatrix}$, $BA = \begin{bmatrix} -1 & 4 & -2 \\ -4 & 16 & -8 \\ 7 & -28 & 14 \end{bmatrix}$,

     $AC = \begin{bmatrix} 0 & 0 \\ 0 & 0 \end{bmatrix}$, $CA = \begin{bmatrix} 0 & 0 & 0 \\ 2 & -8 & 4 \\ 4 & -16 & 8 \end{bmatrix}$, $A(2B - 3C) = \begin{bmatrix} 30 & -28 \\ -30 & 28 \end{bmatrix}$

2. (a) $\begin{bmatrix} a & b \\ 0 & 0 \end{bmatrix}$, a and b arbitrary   (b) $\begin{bmatrix} -2a & a \\ -2b & b \end{bmatrix}$, a and b arbitrary

3. (a) $a = 9, b = 6, c = 1, d = 5$ (b) $a = 1, b = 6, c = 0, d = -2$

4. (a) $\begin{bmatrix} -9 & -2 & -10 \\ 6 & 14 & 8 \\ -7 & 5 & -5 \end{bmatrix}$ (b) $\begin{bmatrix} -3 & 5 & -4 \\ 0 & 3 & 24 \\ 12 & -27 & 0 \end{bmatrix}$

6. $A^n = \begin{bmatrix} 1 & n \\ 0 & 1 \end{bmatrix}$

7. $A^n = \begin{bmatrix} \cos n\theta & -\sin n\theta \\ \sin n\theta & \cos n\theta \end{bmatrix}$

8. $A^n = \begin{bmatrix} 1 & n & \dfrac{n(n+1)}{2} \\ 0 & 1 & n \\ 0 & 0 & 1 \end{bmatrix}$

9. $\begin{bmatrix} 1 & 0 \\ -100 & 1 \end{bmatrix}$

10. $\begin{bmatrix} a & b \\ c & -a \end{bmatrix}$, where b and c are arbitrary, and a is any number satisfying $a^2 = -bc$

11. (b) $\begin{bmatrix} a & 0 \\ 0 & a \end{bmatrix}$, where a is arbitrary

12. $\begin{bmatrix} 1 & 0 \\ 0 & 1 \end{bmatrix}$, $\begin{bmatrix} -1 & 0 \\ 0 & -1 \end{bmatrix}$, and $\begin{bmatrix} a & b \\ c & -a \end{bmatrix}$, where b, c are arbitrary and a is any number satisfying $a^2 = 1 - bc$

13. $C = \begin{bmatrix} \frac{15}{2} & \frac{13}{2} \\ 8 & 7 \end{bmatrix}$, $D = \begin{bmatrix} \frac{33}{4} & \frac{19}{4} \\ \frac{43}{4} & \frac{25}{4} \end{bmatrix}$

14. (b) $(A + B)^2 = A^2 + AB + BA + B^2$; $(A + B)(A - B) = A^2 + BA - AB - B^2$
    (c) For those that commute

4.20 Exercises (pages 158–159)

1. $(x, y, z) = (\frac{8}{5}, -\frac{7}{5}, \frac{8}{5})$
2. No solution
3. $(x, y, z) = (1, -1, 0) + t(-3, 4, 1)$
4. $(x, y, z) = (1, -1, 0) + t(-3, 4, 1)$
5. $(x, y, z, u) = (1, 1, 0, 0) + t(1, 14, 5, 0)$
6. $(x, y, z, u) = (1, 8, 0, -4) + t(2, 7, 3, 0)$
7. $(x, y, z, u, v) = t_1(-1, 1, 0, 0, 0) + t_2(-1, 0, 3, -3, 1)$
8. $(x, y, z, u) = (1, 1, 1, -1) + t_1(-1, 3, 7, 0) + t_2(4, 9, 0, 7)$
9. $(x, y, z) = (\frac{4}{3}, \frac{2}{3}, 0) + t(5, 1, -3)$
10. (a) $(x, y, z, u) = (1, 6, 3, 0) + t_1(4, 11, 7, 0) + t_2(0, 0, 0, 1)$
    (b) $(x, y, z, u) = (\frac{3}{11}, 4, \frac{19}{11}, 0) + t(4, -11, 7, 22)$

12. $\begin{bmatrix} -1 & 2 & 1 \\ 5 & -8 & -6 \\ -3 & 5 & 4 \end{bmatrix}$

13. $\begin{bmatrix} \frac{-5}{3} & \frac{2}{3} & \frac{4}{3} \\ -1 & 0 & 1 \\ \frac{7}{3} & \frac{-1}{3} & \frac{-5}{3} \end{bmatrix}$

14. $\begin{bmatrix} 14 & 8 & 3 \\ 8 & 5 & 2 \\ 3 & 2 & 1 \end{bmatrix}$

15. $\begin{bmatrix} 1 & -2 & 1 & 0 \\ 0 & 1 & -2 & 1 \\ 0 & 0 & 1 & -2 \\ 0 & 0 & 0 & 1 \end{bmatrix}$

16. $\begin{bmatrix} 0 & \frac{1}{2} & 0 & -1 & 0 & 1 \\ 1 & 0 & 0 & 0 & 0 & 0 \\ 0 & 0 & 0 & 1 & 0 & -1 \\ -3 & 0 & 1 & 0 & 0 & 0 \\ 0 & 0 & 0 & 0 & 0 & \frac{1}{2} \\ 9 & 0 & -3 & 0 & 1 & 0 \end{bmatrix}$

4.21 Miscellaneous exercises on matrices (pages 159–160)

3. $P = \begin{bmatrix} 2 & 1 \\ 5 & -1 \end{bmatrix}$

4. $\begin{bmatrix} 0 & 0 \\ 0 & 0 \end{bmatrix}$, $\begin{bmatrix} 1 & 0 \\ 0 & 1 \end{bmatrix}$, and $\begin{bmatrix} a & b \\ c & 1 - a \end{bmatrix}$, where b and c are arbitrary and a is any solution of the quadratic equation $a^2 - a + bc = 0$

10. (a) $\begin{bmatrix} 1 & 1 \\ -1 & 1 \end{bmatrix}, \begin{bmatrix} 1 & 1 \\ 1 & -1 \end{bmatrix}, \begin{bmatrix} -1 & 1 \\ 1 & 1 \end{bmatrix}, \begin{bmatrix} 1 & -1 \\ 1 & 1 \end{bmatrix}, \begin{bmatrix} -1 & -1 \\ 1 & -1 \end{bmatrix},$

$\begin{bmatrix} -1 & -1 \\ -1 & 1 \end{bmatrix}, \begin{bmatrix} 1 & -1 \\ -1 & -1 \end{bmatrix}, \begin{bmatrix} -1 & 1 \\ -1 & -1 \end{bmatrix}$

5.8   Exercises (pages 167–168)

1. (a) 6   (b) 76   (c) $a^3 - 4a$
2. (a) 1   (b) 1   (c) 1
3. (b) $(b - a)(c - a)(c - b)(a + b + c)$ and $(b - a)(c - a)(c - b)(ab + ac + bc)$
5. (a) -8 (b) $(b - a)(c - a)(d - a)(c - b)(d - b)(d - c)$
    (c) $(b - a)(c - a)(d - a)(c - b)(d - b)(d - c)(a + b + c + d)$
    (d) $a(a^2 - 4)(a^2 - 16)$ (e) -160

7. $\det A = 16, \det(A^{-1}) = \frac{1}{16}, A^{-1} = \begin{bmatrix} \frac{1}{2} & -\frac{3}{4} & \frac{1}{8} & \frac{1}{16} \\ 0 & \frac{1}{2} & -\frac{3}{4} & \frac{1}{8} \\ 0 & 0 & \frac{1}{2} & -\frac{3}{4} \\ 0 & 0 & 0 & \frac{1}{2} \end{bmatrix}$

10. $F' = \begin{vmatrix} f_1' & f_2' & f_3' \\ g_1 & g_2 & g_3 \\ h_1 & h_2 & h_3 \end{vmatrix} + \begin{vmatrix} f_1 & f_2 & f_3 \\ g_1' & g_2' & g_3' \\ h_1 & h_2 & h_3 \end{vmatrix} + \begin{vmatrix} f_1 & f_2 & f_3 \\ g_1 & g_2 & g_3 \\ h_1' & h_2' & h_3' \end{vmatrix}$

11. (b) If $F = \begin{vmatrix} f_1 & f_2 & f_3 \\ f_1' & f_2' & f_3' \\ f_1'' & f_2'' & f_3'' \end{vmatrix}$ then $F' = \begin{vmatrix} f_1 & f_2 & f_3 \\ f_1' & f_2' & f_3' \\ f_1''' & f_2''' & f_3''' \end{vmatrix}$

5.15   Exercises (pages 174–175)

6. $\det A = (\det B)(\det D)$
7. (a) Independent   (b) Independent   (c) Dependent

5.20   Exercises (pages 181–182)

1. (a) $\begin{bmatrix} 4 & -3 \\ -2 & 1 \end{bmatrix}$ (b) $\begin{bmatrix} 2 & -1 & 1 \\ -6 & 3 & 5 \\ -4 & -2 & 2 \end{bmatrix}$ (c) $\begin{bmatrix} 109 & 113 & -41 & -13 \\ -40 & -92 & 74 & 16 \\ -41 & -79 & 7 & 47 \\ -50 & 38 & 16 & 20 \end{bmatrix}$

2. (a) $-\frac{1}{2} \begin{bmatrix} 4 & -2 \\ -3 & 1 \end{bmatrix}$ (b) $\frac{1}{8} \begin{bmatrix} 2 & -6 & -4 \\ -1 & 3 & -2 \\ 1 & 5 & 2 \end{bmatrix}$

    (c) $\frac{1}{306} \begin{bmatrix} 109 & -40 & -41 & -50 \\ 113 & -92 & -79 & 38 \\ -41 & 74 & 7 & 16 \\ -13 & 16 & 47 & 20 \end{bmatrix}$

3. (a) $\lambda = 1$ (b) $\lambda = 0, \lambda = \pm 3$ (c) $\lambda = 3, \lambda = \pm i$
5. (a) $x = 0, y = 1, z = 2$ (b) $x = 1, y = 1, z = -1$

5.22   Miscellaneous exercises on determinants (pages 184–186)

1. (b) $\det \begin{bmatrix} x - x_1 & y - y_1 & z - z_1 \\ x_2 - x_1 & y_2 - y_1 & z_2 - z_1 \\ x_3 - x_1 & y_3 - y_1 & z_3 - z_1 \end{bmatrix} = 0$; $\det \begin{bmatrix} x & y & z & 1 \\ x_1 & y_1 & z_1 & 1 \\ x_2 & y_2 & z_2 & 1 \\ x_3 & y_3 & z_3 & 1 \end{bmatrix} = 0$

(c) $\det \begin{bmatrix} (x - x_1)^2 + (y - y_1)^2 & (x - x_1) & (y - y_1) \\ (x_2 - x_1)^2 + (y_2 - y_1)^2 & (x_2 - x_1) & (y_2 - y_1) \\ (x_3 - x_1)^2 + (y_3 - y_1)^2 & (x_3 - x_1) & (y_3 - y_1) \end{bmatrix} = 0;$

$\det \begin{bmatrix} x^2 + y^2 & x & y & 1 \\ x_1^2 + y_1^2 & x_1 & y_1 & 1 \\ x_2^2 + y_2^2 & x_2 & y_2 & 1 \\ x_3^2 + y_3^2 & x_3 & y_3 & 1 \end{bmatrix} = 0.$

2. (a) $\operatorname{cof} A = \begin{bmatrix} 1 & 0 & 0 \\ -x & 1 & 0 \\ x^2 - y & -x & 1 \end{bmatrix}$

5. $6(x - x^2 + e^{3x} - e^{2x})$

6. $a = -7, b = c = 12$

6.4 Exercises (pages 191–192)

1. (b) $a\lambda_1 + b\lambda_2$
7. The nonzero constant polynomials
8. Eigenfunctions: $f(t) = Ct^\lambda$, where $C \neq 0$
9. Eigenfunctions: $f(t) = Ce^{t/\lambda}$, where $C \neq 0$
10. Eigenfunctions: $f(t) = Ce^{t^2/(2\lambda)}$, where $C \neq 0$
11. Eigenvectors belonging to $\lambda = 0$ are all constant sequences with limit $a \neq 0$.
    Eigenvectors belonging to $\lambda = -1$ are all nonconstant sequences with limit $a = 0$

6.10 Exercises (pages 202–203)

	Eigenvalues	Eigenvectors	$\dim E(\lambda)$
1. (a)	1, 1	$(a, b) \neq (0, 0)$	2
(b)	1, 1	$t(1, 0), t \neq 0$	1
(c)	1, 1	$t(0, 1), t \neq 0$	1
(d)	2	$t(1, 1), t \neq 0$	1
	0	$t(1, -1), t \neq 0$	1
2.	$1 + \sqrt{ab}$	$t(\sqrt{a}, \sqrt{b}), t \neq 0$	1
	$1 - \sqrt{ab}$	$t(\sqrt{a}, -\sqrt{b}), t \neq 0$	1

3. If the field of scalars is the set of real numbers **R**, then real eigenvalues exist only when $\sin \theta = 0$, in which case there are two equal eigenvalues $\lambda_1 = \lambda_2 = \cos \theta$, where $\cos \theta = 1$ or -1. In this case every nonzero vector is an eigenvector, so $\dim E(\lambda_1) = \dim E(\lambda_2) = 2$. If the field of scalars is the set of complex numbers **C**, then the eigenvalues are $\lambda_1 = \cos \theta + i \sin \theta$ and its conjugate $\lambda_2 = \cos \theta - i \sin \theta$. If $\sin \theta = 0$ these are real and equal. If $\sin \theta \neq 0$ they are distinct complex conjugates; the eigenvectors belonging to λ_1 are $t(i, 1), t \neq 0$; those belonging to λ_2 are $t(1, i), t \neq 0$; $\dim E(\lambda_1) = \dim E(\lambda_2) = 1$.

4. $\begin{bmatrix} a & b \\ c & -a \end{bmatrix}$, where b and c are arbitrary and a is any number satisfying $a^2 = 1 - bc$

5. Let $A = \begin{bmatrix} a & b \\ c & d \end{bmatrix}$, and let $\Delta = (a - d)^2 + 4bc$. The eigenvalues are real and distinct if $\Delta > 0$, real and equal if $\Delta = 0$, complex conjugates if $\Delta < 0$.

6. $a = b = c = d = e = f = 1$

Eigenvalues	Eigenvectors	dim $E(\lambda)$
7. (a) 1, 1, 1	$t(0, 0, 1), t \neq 0$	1
(b) 1	$t(1, -1, 0), t \neq 0$	1
2	$t(3, 3, -1), t \neq 0$	1
21	$t(1, 1, 6), t \neq 0$	1
(c) 1	$t(3, -1, 3), t \neq 0$	1
2, 2	$t(2, 2, -1), t \neq 0$	1

8. 1, 1, -1, -1 for each matrix

6.12   Exercises (pages 207–208)

2. (a) Eigenvalues 1, 3; $C = \begin{bmatrix} -2c & 0 \\ c & d \end{bmatrix}$, where $cd \neq 0$

(b) Eigenvalues 6, -1; $C = \begin{bmatrix} 2a & b \\ 5a & -b \end{bmatrix}$, where $ab \neq 0$

(c) Eigenvalues 3, 3; if a nonsingular C exists then $C^{-1}AC = 3I$, $AC = 3C$, $A = 3I$

(d) Eigenvalues 1, 1; if a nonsingular C exists then $C^{-1}AC = I$, $AC = C$, $A = I$

3. $C = A^{-1}B$

4. (a) Eigenvalues 1, 1, -1; eigenvectors $(1, 0, 1)$, $(0, 1, 0)$, $(1, 0, -1)$;

$$C = \begin{bmatrix} 1 & 0 & 1 \\ 0 & 1 & 0 \\ 1 & 0 & -1 \end{bmatrix}$$

(b) Eigenvalues 2, 2, 1; eigenvectors $(1, 0, -1)$, $(0, 1, -1)$, $(1, -1, 1)$;

$$C = \begin{bmatrix} 1 & 0 & 1 \\ 0 & 1 & -1 \\ -1 & -1 & 1 \end{bmatrix}$$

5. (a) Eigenvalues 2, 2; eigenvectors $t(1, 0)$, $t \neq 0$. If $C = \begin{bmatrix} a & b \\ -b & 0 \end{bmatrix}$, $b \neq 0$, then

$$C^{-1}AC = \begin{bmatrix} 2 & 0 \\ 1 & 2 \end{bmatrix}$$

(b) Eigenvalues 3, 3; eigenvectors $t(1, 1)$, $t \neq 0$. If $C = \begin{bmatrix} a & b \\ a+b & b \end{bmatrix}$, $b \neq 0$, then

$$C^{-1}AC = \begin{bmatrix} 3 & 0 \\ 1 & 3 \end{bmatrix}$$

6. Eigenvalues 1, 1, 1; eigenvectors $t(1, -1, -1)$, $t \neq 0$

6.14   Exercises (page 211)

1. $A^{-1} = 2I - A$, $A^n = nA - (n-1)I$

2. $A^{-1} = \frac{3}{2}I - \frac{1}{2}A$, $A^n = (2^n - 1)A - (2^n - 2)I$

3. $A^{-1} = A$, $A^n = \dfrac{1 + (-1)^n}{2}I + \dfrac{1 - (-1)^n}{2}A$

4. $A^{-1} = A$, $A^n = \dfrac{1 + (-1)^n}{2}I + \dfrac{1 - (-1)^n}{2}A$

5. $A^n = O$ if $n \geq 3$

6. $A^n = A$ if $n \geq 1$

7. $A^3 = 4A^2 - 5A + 2I$; $A^n = \begin{bmatrix} 2^n & 0 & 0 \\ 0 & 1 & 0 \\ 0 & n & 1 \end{bmatrix}$

6.16 Miscellaneous exercises on eigenvalues and eigenvectors (pages 214–216)

1. (a) Eigenvalue $\lambda_1 = -1$, eigenvectors $t(1, 0, 0)$, $t \neq 0$.
   Eigenvalue $\lambda_2 = 4$, eigenvectors $t(-2, 1, 0)$, $t \neq 0$.
   Eigenvalue $\lambda_3 = 0$, eigenvectors $t(1, -2, 1)$, $t \neq 0$.
   (b) $C = \begin{bmatrix} 1 & -2 & 1 \\ 0 & 1 & -2 \\ 0 & 0 & 1 \end{bmatrix}$

2. $C = \begin{bmatrix} 1 & a & 0 \\ -1 & -a & a \\ -1 & a & -a \end{bmatrix}$, where $a \neq 0$ but otherwise arbitrary

3. (a) Eigenvalues $\lambda_1 = 1$, $\lambda_2 = \lambda_3 = 3$. Eigenvectors belonging to λ_1 are $a(1, -1, -1)$, where $a \neq 0$. Those belonging to λ_2 and λ_3 are $a(-1, 2, 0)$, where $a \neq 0$.
   (b) $C = \begin{bmatrix} 1 & -1 & -1 \\ -1 & 2 & 0 \\ -1 & 0 & 3 \end{bmatrix}$

4. (b) $C = \begin{bmatrix} 1 & 0 & x \\ 1 & 0 & 0 \\ 1 & 2x & y \end{bmatrix}$, where $x \neq 0$ and y is arbitrary.
   (c) $D = \begin{bmatrix} 0 & x & c \\ 0 & 0 & c \\ 2x & y & c \end{bmatrix}$, where $xc \neq 0$ and y is arbitrary.

5. Eigenvalues: $2 + i$, $2 - i$; orthogonal eigenvectors $(1, i)$, $(1, -i)$

6. $\begin{bmatrix} \frac{7}{2} & -\frac{1}{2} & -\frac{1}{2} \\ 1 & 2 & -1 \\ \frac{1}{2} & \frac{1}{2} & \frac{1}{2} \end{bmatrix}$

7. (a) $J^2 = nJ$, $J^3 = n^2 J, \ldots, J^n = n^{n-1} J$ (b) $\det(\lambda I - J) = \lambda^{n-1}(\lambda - n)$
   (c) Orthogonal set of eigenvectors: $(1, 1, 1, 1)$, $(1, -1, 1, -1)$, $(-1, 1, 1, -1)$, $(-1, 1, 1, -1)$.
   The first belongs to $\lambda = 4$, the other three belong to $\lambda = 0$.
   (d) All the eigenvalues are equal to zero except one, which is nt

8. All eigenvalues are equal to $a - t$, except one which is $nt + a - t$

9. $A^2 = n^2 I$; $A^3 = n^2 I$; $A^{-1} = \frac{1}{n^2} A$

10. (a) $a = \frac{1}{2}(1 + (-1)^n)$, $b = \frac{1}{2}(1 - (-1)^n)$ (b) Eigenvalues $1, 1, -1, -1$
    (c) $C = \begin{bmatrix} 1 & 1 & 1 & 1 \\ 1 & 0 & 1 & 0 \\ 1 & 0 & -1 & 0 \\ 1 & 1 & -1 & -1 \end{bmatrix}$

11. (a) $A^4 = 16I$
    (b) $\det(\lambda I - A) = \lambda^4 - 16$; Eigenvalues: $2, 2, -2, -2$

12. (a) $A^2 = 2A + 3I$
    (b) $A^{-1} = \frac{1}{3}(A - 2I) = \frac{1}{3} \begin{bmatrix} -2 & 1 & 1 & 1 \\ 1 & -2 & 1 & 1 \\ 1 & 1 & -2 & 1 \\ 1 & 1 & 1 & -2 \end{bmatrix}$
    (c) Eigenvalues: $-1, -1, -1, 3$
    (d) Eigenvectors: $(1, 0, 0, -1)$, $(0, 1, 0, -1)$, $(0, 0, 1, -1)$, $(1, 1, 1, 1)$

7.4 Exercises (pages 219–220)

3. (b) T^n is Hermitian if n is even, skew-Hermitian if n is odd
6. (d) $Q(x + ty) = Q(x) + t\bar{t}Q(y) + \bar{t}(T(x), y) + t(T(y), x)$

7.10 Exercises (pages 225–227)

1. (a) Symmetric and Hermitian
   (b) None of the four types
   (c) Skew-symmetric
   (d) Skew-symmetric and skew-Hermitian

4. (b) $\begin{bmatrix} \cos\theta & \sin\theta \\ \sin\theta & -\cos\theta \end{bmatrix}$

5. Eigenvalues $\lambda_1 = 0$, $\lambda_2 = 25$; orthonormal eigenvectors $u_1 = \frac{1}{5}(4, -3)$, $u_2 = \frac{1}{5}(3, 4)$;
   $C = \frac{1}{5}\begin{bmatrix} 4 & 3 \\ -3 & 4 \end{bmatrix}$

6. Eigenvalues $\lambda_1 = 2i$, $\lambda_2 = -2i$; orthonormal eigenvectors
   $u_1 = \frac{1}{\sqrt{2}}(1, -i)$, $u_2 = \frac{1}{\sqrt{2}}(1, i)$; $C = \frac{1}{\sqrt{2}}\begin{bmatrix} 1 & 1 \\ -i & i \end{bmatrix}$

7. Eigenvalues $\lambda_1 = 1$, $\lambda_2 = 3$, $\lambda_3 = -4$; orthonormal eigenvectors
   $u_1 = \frac{1}{\sqrt{10}}(1, 0, 3)$, $u_2 = \frac{1}{\sqrt{14}}(3, 2, -1)$, $u_3 = \frac{1}{\sqrt{35}}(3, -5, -1)$;
   $C = \begin{bmatrix} \frac{1}{\sqrt{10}} & \frac{3}{\sqrt{14}} & \frac{3}{\sqrt{35}} \\ 0 & \frac{2}{\sqrt{14}} & \frac{-5}{\sqrt{35}} \\ \frac{3}{\sqrt{10}} & \frac{-1}{\sqrt{14}} & \frac{-1}{\sqrt{35}} \end{bmatrix}$

8. Eigenvalues $\lambda_1 = 1$, $\lambda_2 = 6$, $\lambda_3 = -4$; orthonormal eigenvectors
   $u_1 = \frac{1}{5}(0, 4, -3)$, $u_2 = \frac{1}{\sqrt{50}}(5, 3, 4)$, $u_3 = \frac{1}{\sqrt{50}}(5, -3, -4)$;
   $C = \frac{1}{\sqrt{50}}\begin{bmatrix} 0 & 5 & 5 \\ 4\sqrt{2} & 3 & -3 \\ -3\sqrt{2} & 4 & -4 \end{bmatrix}$

9. (a), (b) and (c) are unitary; (b) and (c) are orthogonal

11. (a) Eigenvalues $\lambda_1 = ia$, $\lambda_2 = -ia$; orthonormal eigenvectors
    $u_1 = \frac{1}{\sqrt{2}}(1, i)$, $u_2 = \frac{1}{\sqrt{2}}(1, -i)$; $C = \frac{1}{\sqrt{2}}\begin{bmatrix} 1 & 1 \\ i & -i \end{bmatrix}$

7.14 Exercises (page 235)

1. (a) $A = \begin{bmatrix} 4 & 2 \\ 2 & 1 \end{bmatrix}$ (b) $\lambda_1 = 0$, $\lambda_2 = 5$
   (c) $u_1 = \frac{1}{\sqrt{5}}(1, -2)$, $u_2 = \frac{1}{\sqrt{5}}(2, 1)$;   (d) $C = \frac{1}{\sqrt{5}}\begin{bmatrix} 1 & 2 \\ -2 & 1 \end{bmatrix}$

2. (a) $A = \begin{bmatrix} 0 & \frac{1}{2} \\ \frac{1}{2} & 0 \end{bmatrix}$ (b) $\lambda_1 = \frac{1}{2}$, $\lambda_2 = -\frac{1}{2}$
   (c) $u_1 = \frac{1}{\sqrt{2}}(1, 1)$, $u_2 = \frac{1}{\sqrt{2}}(1, -1)$;   (d) $C = \frac{1}{\sqrt{2}}\begin{bmatrix} 1 & 1 \\ 1 & -1 \end{bmatrix}$

3. (a) $A = \begin{bmatrix} 1 & 1 \\ 1 & -1 \end{bmatrix}$ (b) $\lambda_1 = \sqrt{2}$, $\lambda_2 = -\sqrt{2}$
   (c) $u_1 = t(1 + \sqrt{2}, 1)$, $u_2 = t(-1, 1 + \sqrt{2})$, where $t = 1/\sqrt{4 + 2\sqrt{2}}$.

(d) $C = t \begin{bmatrix} 1 + \sqrt{2} & -1 \\ 1 & 1 + \sqrt{2} \end{bmatrix}$, where $t = 1/\sqrt{4 + 2\sqrt{2}}$

4. (a) $A = \begin{bmatrix} 34 & -12 \\ -12 & 41 \end{bmatrix}$ (b) $\lambda_1 = 50, \lambda_2 = 25$

    (c) $u_1 = \frac{1}{5}(3, -4), u_2 = \frac{1}{5}(4, 3)$;   (d) $C = \frac{1}{5} \begin{bmatrix} 3 & 4 \\ -4 & 3 \end{bmatrix}$

5. (a) $A = \begin{bmatrix} 1 & \frac{1}{2} & \frac{1}{2} \\ \frac{1}{2} & 0 & \frac{1}{2} \\ \frac{1}{2} & \frac{1}{2} & 0 \end{bmatrix}$ (b) $\lambda_1 = 0, \lambda_2 = \frac{3}{2}, \lambda_3 = -\frac{1}{2}$

    (c) $u_1 = \frac{1}{\sqrt{3}}(1, -1, -1), u_2 = \frac{1}{\sqrt{6}}(2, 1, 1), u_3 = \frac{1}{\sqrt{2}}(0, 1, -1)$

    (d) $C = \frac{1}{\sqrt{6}} \begin{bmatrix} \sqrt{2} & 2 & 0 \\ -\sqrt{2} & 1 & \sqrt{3} \\ -\sqrt{2} & 1 & -\sqrt{3} \end{bmatrix}$

6. (a) $A = \begin{bmatrix} 2 & 0 & 2 \\ 0 & 1 & 0 \\ 2 & 0 & -1 \end{bmatrix}$ (b) $\lambda_1 = 1, \lambda_2 = 3, \lambda_3 = -2$

    (c) $u_1 = (0, 1, 0), u_2 = \frac{1}{\sqrt{5}}(2, 0, 1), u_3 = \frac{1}{\sqrt{5}}(1, 0, -2)$

    (d) $C = \frac{1}{\sqrt{5}} \begin{bmatrix} 0 & 2 & 1 \\ \sqrt{5} & 0 & 0 \\ 0 & 1 & -2 \end{bmatrix}$

7. (a) $A = \begin{bmatrix} 3 & 2 & 4 \\ 2 & 0 & 2 \\ 4 & 2 & 3 \end{bmatrix}$ (b) $\lambda_1 = \lambda_2 = -1, \lambda_3 = 8$

    (c) $u_1 = \frac{1}{\sqrt{2}}(1, 0, -1), u_2 = \frac{1}{3\sqrt{2}}(-1, 4, -1), u_3 = \frac{1}{3}(2, 1, 2)$

    (d) $C = \frac{1}{3\sqrt{2}} \begin{bmatrix} 3 & -1 & 2\sqrt{2} \\ 0 & 4 & \sqrt{2} \\ -3 & -1 & 2\sqrt{2} \end{bmatrix}$

8. Ellipse; center at $(0, 0)$
9. Hyperbola; center at $(-\frac{5}{2}, -\frac{5}{2})$
10. Parabola; vertex at $(\frac{5}{16}, -\frac{15}{16})$
11. Ellipse; center at $(0, 0)$
12. Ellipse; center at $(6, -4)$
13. Parabola; vertex at $(\frac{2}{25}, \frac{11}{25})$
14. Ellipse; center at $(0, 0)$
15. Parabola; vertex at $(\frac{3}{4}, \frac{3}{4})$
16. Ellipse; center at $(-1, \frac{1}{2})$
17. Hyperbola; center at $(0, 0)$
18. Two intersecting lines
19. -14

7.20   Exercises (pages 242–243)

8. $a = \pm \frac{1}{3}\sqrt{3}$
13. (a), (b) and (e)

7.22   Exercises (pages 245–246)

2. (a) Symmetric   (b) Neither   (c) Symmetric   (d) Symmetric

8.3   Exercises (pages 250-251)

1. $y = e^{3x} - e^{2x}$
2. $y = \frac{2}{3}x^2 + \frac{1}{3}x^5$
3. $y = 4\cos x - 2\cos^2 x$
4. Four times the initial amount
5. $f(x) = Cx^n$, or $f(x) = Cx^{1/n}$
6. (b) $y = e^{4x} - e^{-x^3/3}$
7. $y = -1 \pm \sqrt{2 + ce^{-x^2}}$, where $c \geq -2$
8. $y = c_1 e^{2x} + c_2 e^{-2x}$
9. $y = c_1 \cos 2x + c_2 \sin 2x$
10. $y = e^x(c_1 \cos 2x + c_2 \sin 2x)$
11. $y = e^{-x}(c_1 + c_2 x)$
12. (a) $y = 10e^{3x} - 5e^{-x}$ (b) $A = -\frac{1}{2}, B = 2, C = -1$
13. $k = n^2\pi^2$; $f_k(x) = C \sin n\pi x$ $(n = 1, 2, 3, \ldots)$
15. (a) $y'' - y = 0$
    (b) $y'' - 4y' + 4y = 0$
    (c) $y'' + y' + \frac{5}{4}y = 0$
    (d) $y'' + 4y = 0$
    (e) $y'' - y = 0$
16. $y = \frac{1}{3}\sqrt{6}$, $y'' = -12y = -4\sqrt{6}$

8.9   Exercises (pages 259–260)

1. $y = c_1 + c_2 e^{-x} + c_3 e^{3x}$
2. $y = c_1 + c_2 e^x + c_3 e^{-x}$
3. $y = c_1 + (c_2 + c_3 x)e^{-2x}$
4. $y = (c_1 + c_2 x + c_3 x^2)e^x$
5. $y = (c_1 + c_2 x + c_3 x^2 + c_4 x^3)e^{-x}$
6. $y = c_1 e^{2x} + c_2 e^{-2x} + c_3 \cos 2x + c_4 \sin 2x$
7. $y = e^{\sqrt{2}x}(c_1 \cos \sqrt{2}x + c_2 \sin \sqrt{2}x) + e^{-\sqrt{2}x}(c_3 \cos \sqrt{2}x + c_4 \sin \sqrt{2}x)$
8. $y = c_1 e^x + e^{-x/2}(c_2 \cos \frac{1}{2}\sqrt{3}x + c_3 \sin \frac{1}{2}\sqrt{3}x)$
9. $y = e^{-x}\{(c_1 + c_2 x)\cos x + (c_3 + c_4 x)\sin x\}$
10. $y = (c_1 + c_2 x)\cos x + (c_3 + c_4 x)\sin x$
11. $y = c_1 + c_2 x + (c_3 + c_4 x)\cos \sqrt{2}x + (c_5 + c_6 x)\sin \sqrt{2}x$
12. $y = c_1 + c_2 x + (c_3 + c_4 x)\cos 2x + (c_5 + c_6 x)\sin 2x$
13. $f(x) = \dfrac{1}{2m^2}(e^{mx} - \cos mx - \sin mx)$
15. (a) $y^{(4)} - 5y'' + 4y = 0$
    (b) $y''' + 6y'' + 12y' + 8y = 0$
    (c) $y^{(4)} - 2y''' + y'' = 0$
    (d) $y^{(4)} - 2y''' + y'' = 0$
    (e) $y^{(5)} - 2y^{(4)} + y''' = 0$
    (f) $y^{(4)} + 8y''' + 33y'' + 68y' + 52y = 0$
    (g) $y^{(4)} - 2y'' + y = 0$
    (h) $y^{(6)} + 4y'' = 0$

8.15   Exercises (pages 270–272)

1. $y_1 = -2x - x^2 - \frac{1}{3}x^3$
2. $y_1 = \frac{1}{4}xe^{2x}$
3. $y_1 = (x - \frac{4}{3})e^x$
4. $y_1 = \frac{1}{3}\sin x$

5. $y_1 = \frac{1}{2}x^2 e^x + e^{2x}$

6. $y_1 = \frac{1}{2}xe^x$

7. $y_1 = x \cosh x$

8. $y_1 = \frac{1}{24}x^4 e^{-x}$

9. $50y_1 = (11 - 5x)e^x \sin 2x + (2 - 10x)e^x \cos 2x$

10. $y_1 = -(\frac{5}{8}x + \frac{3}{8}x^2 + \frac{1}{12}x^3)e^{-x}$

12. (c) $y_1 = \dfrac{x^m e^{\alpha x}}{p_A^{(m)}(\alpha)}$

15. (b) $2D$ (c) $3D^2$ (d) nD^{n-1}

16. $u(x) = 6(e^{4x} - e^{-x})/5; v(x) = e^x - e^{-5x}$

17. $u(x) = \frac{1}{2}e^{2x-\pi} \sin 5x; v(x) = \frac{5}{6}e^{-2x-\pi} \sin 3x$

18. $u(x) = e^{-x^2}; Q(x) = 4x^2 + 2$

20. $y = (A + Bx^3)e^x + (x^2 - 2x + 2)e^{2x}$

21. $y = Ax^{1/2} + Bx^{-1/2}$

22. $y = A(x^2 - 2) + B/x$

23. $y = x^{-2}[A + B(x - 1)^3 + \frac{1}{9}x^3 + \frac{2}{3}x^2 - \frac{7}{6}x + \frac{1}{2} - (x - 1)^3 \log|x - 1|]$

24. (a) $P(x) = -5/x; Q(x) = 9/x^2$

    (b) $y = c_1 x^3 + c_2 x^3 \log x + x^4$

25. Every $\lambda > 0$ is an eigenvalue; corresponding eigenfunctions $f(t) = Ce^{t/\sqrt{\lambda}}, C \neq 0$

27. Only possible eigenvalues are $\lambda = 0, \lambda = -n^2$, where $n = 1, 2, 3, \ldots$. Eigenfunctions belonging to $\lambda = 0$ are $f(t) = a$, where $a \neq 0$. Those belonging to $\lambda = -n^2$ are $f(t) = a \cos nt$, where $a \neq 0$.

28. Eigenfunctions belonging to $\lambda = 0$ are $f(t) = at$, where $a \neq 0$. Eigenfunctions belonging to $\lambda > 0$ are $f(t) = c \sinh \sqrt{\lambda} t$, where $c \neq 0$. If $\lambda = 1$ we also have $f(t) = c_1 e^t + c_2 e^{-t}$, where not both c_1 and c_2 are zero. Those belonging to $\lambda < 0$ are $f(t) = a \sin \sqrt{-\lambda} t$, where $a \neq 0$.

29. Eigenfunctions belonging to $\lambda = 0$ are $f(t) = a(2 + t)$, where $a \neq 0$. Eigenfunctions belonging to $\lambda > 0$ are $f(t) = a(e^{\sqrt{\lambda} t/2} + \dfrac{\sqrt{\lambda} - 1}{\sqrt{\lambda} + 1} e^{-\sqrt{\lambda} t/2})$, where $a \neq 0$. Eigenfunctions belonging to $\lambda < 0$ are $f(t) = a(\cos \sqrt{-\lambda} t + \dfrac{1}{\sqrt{-\lambda}} \sin \sqrt{-\lambda} t)$, where $a \neq 0$.

30. Eigenfunctions belonging to $\lambda = 0$ are $f(t) = c$, where $c \neq 0$. Eigenfunctions belonging to $\lambda > 0$ are $f(t) = c(e^{-\sqrt{\lambda}} e^{\sqrt{\lambda} t} + e^{-\sqrt{\lambda} t}), c \neq 0$. Eigenfunctions belonging to $\lambda < 0$ fall into two categories. If $\lambda = -4n^2 \pi^2$, where n is an integer $\neq 0$, the eigenfunctions belonging to λ are $f(t) = c_1 \cos 2\pi nt + c_2 \sin 2\pi nt$, where not both c_1 and c_2 are zero. If $\lambda < 0$ but $\lambda \neq -4n^2 \pi^2$, then

$$f(t) = c\left(\frac{\sin \sqrt{-\lambda}}{1 - \cos \sqrt{-\lambda}} \cos \sqrt{-\lambda} t + \sin \sqrt{-\lambda} t\right),$$

where $c \neq 0$.

9.4 Exercises (pages 277–278)

3. (b) $(P_k)' = \sum_{m=0}^{k-1} P^m P' P^{k-1-m}$

9.11 Exercises (pages 287–288)

1. (a) $A^{-1} = 2I - A, A^n = nA - (n - 1)I$

    (b) $e^{tA} = e^t(1 - t)I + te^t A = e^t \begin{bmatrix} 1 & 0 \\ t & 1 \end{bmatrix}$

2. (a) $A^{-1} = \frac{3}{2}I - \frac{1}{2}A, A^n = (2^n - 1)A - (2^n - 2)I$

(b) $e^{tA} = (2e^t - e^{2t})I + (e^{2t} - e^t)A = \begin{bmatrix} e^t & 0 \\ e^{2t} - e^t & e^{2t} \end{bmatrix}$

3. (a) $A^{-1} = A$, $A^n = \dfrac{1 + (-1)^n}{2}I + \dfrac{1 - (-1)^n}{2}A$

(b) $e^{tA} = (\cosh t)I + (\sinh t)A = \begin{bmatrix} \cosh t & \sinh t \\ \sinh t & \cosh t \end{bmatrix}$

4. (a) $A^{-1} = A$, $A^n = \dfrac{1 + (-1)^n}{2}I + \dfrac{1 - (-1)^n}{2}A$

(b) $e^{tA} = (\cosh t)I + (\sinh t)A = \begin{bmatrix} e^{-t} & 0 \\ 0 & e^t \end{bmatrix}$

5. (a) $A^n = O$ if $n \geq 3$

(b) $e^{tA} = I + tA + \frac{1}{2}t^2 A^2 = \begin{bmatrix} 1 & t & t + \frac{1}{2}t^2 \\ 0 & 1 & t \\ 0 & 0 & 1 \end{bmatrix}$

6. (a) $A^n = A$ if $n \geq 1$

(b) $e^{tA} = I + (e^t - 1)A = \begin{bmatrix} 1 & e^t - 1 & e^t - 1 \\ 0 & e^t & e^t - 1 \\ 0 & 0 & 1 \end{bmatrix}$

7. (a) $A^3 = 4A^2 - 5A + 2I$; $A^n = \begin{bmatrix} 2^n & 0 & 0 \\ 0 & 1 & 0 \\ 0 & n & 1 \end{bmatrix}$

(b) $e^{tA} = \begin{bmatrix} e^{2t} & 0 & 0 \\ 0 & e^t & 0 \\ 0 & te^t & e^t \end{bmatrix}$

8. (b) $e^{tA} = e^{at} \begin{bmatrix} \cos bt & \sin bt \\ -\sin bt & \cos bt \end{bmatrix}$

10. $e^{A(t)} = I + (e - 1)A(t)$; $(e^{A(t)})' = (e - 1)A'(t) = \begin{bmatrix} 0 & e - 1 \\ 0 & 0 \end{bmatrix}$;

$e^{A(t)}A'(t) = \begin{bmatrix} 0 & e \\ 0 & 0 \end{bmatrix}$; $A'(t)e^{A(t)} = \begin{bmatrix} 0 & 1 \\ 0 & 0 \end{bmatrix}$

11. $e^{tA} = I + tA + \frac{1}{2}t^2 A^2$

13. $e^A e^B = \begin{bmatrix} e^2 & -(e - 1)^2 \\ 0 & 1 \end{bmatrix}$; $e^B e^A = \begin{bmatrix} e^2 & (e - 1)^2 \\ 0 & 1 \end{bmatrix}$; $e^{A+B} = \begin{bmatrix} e^2 & 0 \\ 0 & 1 \end{bmatrix}$

15. (b) $e^{tA} = (\cosh nt)I + \frac{1}{n}(\sinh nt)A$

16. (b) $f(t) = \frac{1}{5}e^{4t} + \frac{1}{5}e^{-t}$; $g(t) = \frac{1}{5}e^{4t} - \frac{1}{5}e^{-t}$

9.14 Exercises (pages 293–294)

1. $e^{tA} = \frac{1}{2}(3e^t - e^{3t})I + \frac{1}{2}(e^{3t} - e^t)A$

2. $e^{tA} = (\cosh \sqrt{5}t)I + \dfrac{1}{\sqrt{5}}(\sinh \sqrt{5}t)A$

3. $e^{tA} = \frac{1}{2}e^t\{t^2 - 2t + 2)I + (-2t^2 + 2t)A + t^2 A^2\}$

4. $e^{tA} = (3e^{-t} - 3e^{-2t} + e^{-3t})I + (\frac{5}{2}e^{-t} - 4e^{-2t} + \frac{3}{2}e^{-3t})A + (\frac{1}{2}e^{-t} - e^{-2t} + \frac{1}{2}e^{-3t})A^2$

5. $e^{tA} = (4e^t - 3e^{2t} + 2te^{2t})I + (4e^{2t} - 3te^{2t} - 4e^t)A + (e^t - e^{2t} + te^{2t})A^2$

6. $e^{tA} = (4e^t - 6e^{2t} + 4e^{3t} - e^{4t})I + (-\frac{13}{3}e^t + \frac{19}{2}e^{2t} - 7e^{3t} + \frac{11}{6}e^{4t})A$
   $+(\frac{3}{2}e^t - 4e^{2t} + \frac{7}{2}e^{3t} - e^{4t})A^2 + (-\frac{1}{6}e^t + \frac{1}{2}e^{2t} - \frac{1}{2}e^{3t} + \frac{1}{6}e^{4t})A^3$

7. (b) $e^{tA} = \frac{1}{6}e^{\lambda t}\{6 - 6\lambda t + 3\lambda^2 t^2 - \lambda^3 t^3)I + (6t - 6\lambda t^2 + 3\lambda^2 t^3)A$
   $+(3t^2 - 3\lambda t^3)A^2 + t^3 A^3\}$

9. $y_1 = c_1 \cosh \sqrt{5}t + \dfrac{c_1 + 2c_2}{\sqrt{5}} \sinh \sqrt{5}t$, $y_2 = c_2 \cosh \sqrt{5}t + \dfrac{2c_1 - c_2}{\sqrt{5}} \sinh \sqrt{5}t$

10. $y_1 = e^t(\cos 3t - \sin 3t)$, $y_2 = e^t(\cos 3t - 3\sin 3t)$

11. $y_1 = e^{2t} + 4te^{2t}, y_2 = -2e^t + e^{2t} + 4te^{2t}, y_3 = -2e^t + 4e^{2t}$
12. $y_1 = c_1 e^{2t}, y_2 = c_2 e^t, y_3 = (c_2 t + c_3)e^t$
13. $y_1 = 3e^{-t} - 3e^{-2t} + e^{-3t}, y_2 = -3e^{-t} + 6e^{-2t} - 3e^{-3t}, y_3 = 3e^{-t} - 12e^{-2t} + 9e^{-3t}$
14. $y_1 = e^{5t} + 7e^{-3t}, y_2 = 2e^{5t} - 2e^{-3t}, y_3 = e^{-5t} + e^{-3t}$
15. $y_1 = \frac{1}{2}e^t - e^{2t} + \frac{1}{2}e^{3t}, y_2 = -e^{2t} + e^{3t}, y_3 = e^{3t}, y_4 = e^{4t}$
16. $y_1 = 2e^{2t} - 1, y_2 = 2e^{2t} - t - 2, y_3 = 2e^{2t}, y_4 = e^{2t}$

9.16 Exercises (pages 296–298)

2. (c) $y_1 = (b - 1)e^x + 2(c + 1 - b)xe^x + 1, y_2 = ce^x + 2(c + 1 - b)xe^x$
4. $y_1 = -\frac{1}{3}e^t - \frac{1}{6}e^{4t} + \frac{1}{2}e^{2t}, y_2 = \frac{2}{3}e^t - \frac{1}{6}e^{4t} + \frac{1}{2}e^{2t}$
5. (a) $B_0 = B, B_1 = AB, B_2 = \frac{1}{2!}A^2 B, \ldots, B_m = \frac{1}{m!}A^m B$
   (b) $B = -m!(A^{-1})^{m+1}C$

6. (a) $Y(t) = (I + tA + \frac{1}{2}t^2 A^2 + \frac{1}{6}t^3 A^3)B$, where $B = -6A^{-1}C = -\frac{3}{128}\begin{bmatrix} 1 \\ 1 \end{bmatrix}$.

   This gives the particular solution $y_1 = y_2 = -\frac{3}{128} - \frac{3}{32}t - \frac{3}{16}t^2 - \frac{1}{4}t^3$
   (b) $y_1 = y_2 = -\frac{3}{128} - \frac{3}{32}t - \frac{3}{16}t^2 - \frac{1}{4}t^3 + \frac{131}{128}e^{4t}$
7. $E = B, F = \frac{1}{\alpha}(AB + C)$
8. (a) $y_1 = -\cos 2t - \frac{1}{2}\sin 2t, y_2 = -\frac{1}{2}\sin 2t$
   (b) $y_1 = 2\cosh 2t + \frac{5}{2}\sinh 2t - \cos 2t - \frac{1}{2}\sin 2t, y_2 = \cosh 2t + \frac{1}{2}\sinh 2t - \frac{1}{2}\sin 2t$

9.19 Exercises (page 302)

4. (c) $Y(x) = e^x e^{\frac{1}{2}x^2 A}B$
5. If $A(x) = \sum_{k=0}^{\infty} x^k A_k$, then $Y(x) = B + xC + \sum_{k=2}^{\infty} x^k B_k$,
   where $(k + 2)(k + 1)B_{k+2} = \sum_{r=0}^{k} A_r B_{k-r}$ for $k \geq 0$

10.6 Exercises (pages 311–312)

1. (a) $Y(x) = e^x$

   (b) $Y_n(x) = 2e^x - \sum_{k=0}^{n} \frac{x^k}{k!}$ if n is odd; $Y_n(x) = \sum_{k=0}^{n} \frac{x^k}{k!}$ if n is even

2. $Y_3(x) = \frac{x^2}{2} + \frac{x^5}{20} + \frac{x^8}{160} + \frac{x^{11}}{4400}$

3. $Y_3(x) = x + \frac{x^4}{4} + \frac{x^7}{14} + \frac{x^{10}}{160}$

4. $Y_3(x) = \frac{x^2}{3} + \frac{4x^7}{63} + \frac{8x^9}{405} + \frac{184x^{11}}{51975} + \frac{4x^{13}}{12285} + \frac{x^{15}}{59535}$

5. (a) $Y_n(x) = 1 + \sum_{k=1}^{n} \frac{x^k}{k!}$

   (b) $Y(x) = e^x$

6. (a) $Y_2(x) = 1 + x + x^2 + \frac{2x^3}{3} + \frac{x^4}{6} + \frac{2x^5}{15} + \frac{x^7}{63}$

   (b) $M = 2; c = \frac{1}{2}$

   (c) $Y(x) = 1 + x + x^2 + \frac{4x^3}{3} + \frac{7x^4}{6} + \frac{6x^5}{5} + \cdots$

7. (a) $Y_4(x) = x + \frac{x^3}{3} + \frac{2x^5}{15} + \frac{17x^7}{315} + \frac{38x^9}{2835} + \frac{134x^{11}}{51975} + \frac{4x^{13}}{12285} + \frac{x^{15}}{59535}$

   (d) $Y(x) = \tan x = x + \frac{x^3}{3} + \frac{2x^5}{15} + \frac{17x^7}{315} + \frac{62x^9}{2835} + \cdots$ for $|x| < \frac{\pi}{2}$

9. $Y_3(x) = 2 + x^2 + x^3 + \dfrac{3x^5}{20} + \dfrac{x^6}{10}; Z_3(x) = 3x^2 + \dfrac{3x^4}{4} + \dfrac{6x^5}{5} + \dfrac{3x^7}{28} + \dfrac{3x^8}{40}$

10. $Y_3(x) = 5 + x + \dfrac{x^4}{12} + \dfrac{x^6}{6} + \dfrac{2x^7}{63} + \dfrac{x^9}{72};$

$Z_3(x) = 1 + \dfrac{x^3}{3} + x^5 + \dfrac{2x^6}{9} + \dfrac{x^8}{8} + \dfrac{11x^9}{324} + \dfrac{7x^{11}}{264}$

INDEX